"Erudite and appealing, *Cosmic Commons* deserves to enrich many readers' thirst for knowledge and spiritual awakening."
—ELIE WIESEL
Holocaust survivor, 1986 Nobel Peace Laureate, and author of *Night*

"This book is unlike any you have ever read. Whether you believe there is evidence for extraterrestrial intelligence or not, Hart's creative and thoughtful reflections ... will widen your ethical horizons."
—JOHN F. HAUGHT
Distinguished Research Professor of Theology, Georgetown University

"Anyone who takes seriously the possibility that there may be intelligent life beyond the solar system should read John Hart's scholarly and detailed examination of what this might imply."
—WILLIAM R. SHEA
Galileo Professor of History of Science, University of Padua

"Clearly, we desperately need the transformation of consciousness that Hart lays out, not only for the sake of others we might encounter in space, but for our own sakes and those of the remaining others in our world, both human and not."
—CHRISTINE GUDORF
Professor of Religious Studies, Florida International University

"John Hart makes the case for there being life elsewhere in the universe, why humankind should search for it, and what sorts of ecological, social, and ethical interactions would likely develop between humans on earth and ETI, if ever found."
—FRANCISCO J. AYALA
Professor of Biological Sciences, University of California, Irvine

"John Hart invites us to join him in . . . a powerful and moving commentary on human failure and insensitivity where indigenous, terrestrial civilizations have been encountered, suppressed, and even destroyed. Writing as a Christian ethicist deeply concerned about ecological destruction, he resists the shallowness of those scientists who glibly advocate our colonization of other worlds in order to escape a doom-laden earth. You don't have to believe in UFOs to appreciate his compelling argument."

—JOHN HEDLEY BROOKE
Emeritus Professor of Science and Religion, Oxford University

"Hart's driving concern is, given what has been done to indigenous people in the name of 'civilization,' will this past be repeated if new civilizations are encountered. Hart has given us a thorough and thoughtful review of core elements in the social history of colonialism in the hope that past mistakes will not be repeated."

—THOMAS A. SHANNON
Emeritus Professor of Religion and Social Ethics, Worcester Polytechnic Institute

"John Hart cautions us that as humans venture off earth into extraterrestrial exploration, we need to ensure that the tragic history of euro-Christian colonialism and 'discovery' is not repeated in colonizing other terrain."

—TINK TINKER
Professor of American Indian Cultures and Religious Traditions, Illiff School of Theology

COSMIC COMMONS

BOOKS BY JOHN HART

The Spirit of the Earth: A Theology of the Land (1984)

Ethics and Technology: Innovation and Transformation in Community Contexts (1997)

What Are They Saying about Environmental Theology? (2004)

Sacramental Commons: Christian Ecological Ethics (2006)

COSMIC COMMONS
SPIRIT, SCIENCE, AND SPACE

John Hart

CASCADE *Books* • Eugene, Oregon

COSMIC COMMONS
Spirit, Science, and Space

Copyright © 2013 John Hart. All rights reserved. Except for brief quotations in critical publications or reviews, no part of this book may be reproduced in any manner without prior written permission from the publisher. Write: Permissions, Wipf and Stock Publishers, 199 W. 8th Ave., Suite 3, Eugene, OR 97401.

Cascade Books
An Imprint of Wipf and Stock Publishers
199 W. 8th Ave., Suite 3
Eugene, OR 97401

www.wipfandstock.com

ISBN 13: 978-1-61097-318-2

Cataloging-in-Publication data:

Hart, John, 1943–

    Cosmic commons : spirit, science, and space / John Hart.

    xii + 416 p.; 23 cm—Includes bibliographical references and index.

    ISBN 13: 978-1-61097-318-2

    1. Cosmology. 2. Religion and science. 3. Ecotheology. 4. Nature—Religious aspects—Christianity. I. Title.

BT135 .H36 2013

Manufactured in the USA.

The Scripture quotations contained herein are from the New Revised Standard Version Bible, copyright © 1989, Division of Christian Education of the National Council of Churches of Christ in the U.S.A. Used by permission. All rights reserved.

DEDICATION

In appreciation for innovative insights from, and congenial conversation with, friends who have in diverse and complementary ways recognized the dialogic relationship between religion and science, related scientific and spiritual perspectives, and promoted thereby the well-being of the Earth Commons in the Cosmic Commons:

John Opitz, medical geneticist, surgeon, humanist; University of Utah; Humboldt Prize

Arthur Peacocke, physical biochemist, theologian, priest; University of Oxford; Templeton Prize

Edward O. Wilson, biologist, ecologist, humanist; Harvard University; Pulitzer Prize

and

With gratitude for insights, invitations to native peoples' social and spiritual gatherings, decades of friendship and congenial conversation, and collaboration to effect justice and well-being for indigenous populations:

William Means, cofounder and executive director of the International Indian Treaty Council, a nongovernmental organization; community social activist; and diplomat extraordinaire at the United Nations International Human Rights Commission

and

With love, and with hope for the future Commons they are creating and inheriting:

Shanti Morell-Hart, archaeologist, musician, athlete, creative thinker, and beloved daughter

Daniel Morell-Hart, psychologist, artist, athlete, creative thinker, and beloved son

Contents

Acknowledgments xi

Introduction: Cosmic Consciousness 1
Terrestrial Transformation and Extraterrestrial Exploration

PART I: TERRA FIRMA

1. Terrestrial Trauma 29
Economic Roots of Earth's Socioecological Crisis

2. Terrestrial Discovery 59
Political Roots of Earth's Socioecological Crisis

3. Terrestrial Transformation 89
Historical Restoration of Earth's Socioecological Commons

PART II: TERRA CONSCIENTIA

4. Extraterrestrial Discovery 117
Cosmos Community Contexts: In Caelis *and in* Caelo *as in* Terra?

5. Spirit and Science 143
Consciousness and Collaboration

6. Earth Commons and ExoEarth Commons 177
Contextual Considerations

PART III: TERRA INCOGNITA

7. Cosmosocioecological Praxis Ethics 199
Topia, Dystopia, and Utopia in Earth and Cosmos Contexts

8. ExoEarth and Extraterrestrial Transformation 228
Ecosociality in Cosmic Community

9. Cosmic Contact 253
ETI and TI Encounters: Competition and Collaboration

PART IV: TERRA COSMICA

10. Cosmic Coexistence 303
Cosmic Consciousness and Cohesion

11. Cosmic Charter 331
Constructive Consultation and Consociation

12. Cosmic Commons 369
Celestial Cohabitation, Conservation, and Compassion

Conclusion 393
At Home on Earth, Reaching for the Stars

Bibliography 407
Index 413

Acknowledgments

Cosmic Commons has resulted from extended and extensive research because of complex dialogic relationships among ethics, ecology, economics, and ecclesia, and past, present, and future time. The manuscript required almost two years of in-depth research, the latter half of which involved intensive writing. I am grateful for the affirmation and support of family, friends, and faculty colleagues who encouraged me in my exploration of a controversial issue: a literally "far-out" topic, as I said to whoever asked me about my current work.

I particularly thank Boston University School of Theology friends and colleagues Norm Faramelli, Walter Fluker, Andrew Shenton, Nancy Ammerman, Chris Schlauch, Marthinus Daneel, and Dana Robert for their interest in and encouragement for *Cosmic Commons* and my other creative work over the years.

I thank Boston University professors Thomas Kunz, biology and environmental sciences, and George Annas, public health and human rights, and SETI scientist Margaret Race for their insights and support as we worked collaboratively to prepare (and receive funding for) a research grant exploring implications of terrestrial-extraterrestrial intelligent life Contact.

I especially appreciate the support, insights, and spirit-lifting comments of my friend and colleague Elie Wiesel on several occasions; they were needed and helpful in difficult times.

I am grateful for the support, patience, and comments of my Wipf and Stock editor, Charlie Collier, as the book's length increased and submission deadlines passed. I thank Christian Amondson for his editorial and production expertise, and Jacob Martin and Heather Carraher, respectively, for proofreading and typesetting the manuscript.

I received strong support and loving encouragement—as well as insights during our conversations—from my daughter, Shanti, and my son, Daniel, both accomplished professionals in their respective fields.

I am especially grateful for the love and encouragement of my wife, Jane Morell-Hart, and for her patience as I worked on research and writing long hours in the day and late into the night. I am looking forward to a less intensive writing schedule and more enjoyable family time.

Introduction

Cosmic Consciousness

Terrestrial Transformation and Extraterrestrial Exploration

Over millennia, people have speculated that intelligent life exists in places other than on Earth.[1] Throughout human history, oral reports and written narratives about peoples' direct or indirect Contact[2] with extraterrestrial beings have been retained and transmitted transgenerationally. In the past, the speculation was fueled by observation of aerial phenomena, events, or objects whose characteristics did not seem congruent with the familiar workings of nature or biota, as understood at the time. In recent decades, it has been prompted by the preceding and also by the appearance of objects whose maneuvers have defied the possibilities permissible by known laws of physics and current advances in technology—all of which have led to conjecture and even conclusions that these objects have been guided by intelligent beings. Since the mid-twentieth century, reports of such occurrences increased dramatically. Although the vast majority has been explained as weather-related phenomena or other natural events, or as human technology-based, some incidents have remained unexplainable despite rigorous scientific, military, and photographic review.

Many of the reported incidents, and complementary claims purporting direct encounters or interaction of some kind with exoplanetary[3] intelligent

---

1. *Earth* is capitalized because it is one of the planets; to distinguish planet Earth from the soil, earth; to emphasize its importance as the habitat of the biotic community; and by all of this, to promote respect and care for Earth, humans' home planet.

2. *Contact* is capitalized, in current scientific and literary writing, to indicate specifically the encounter of terrestrial and extraterrestrial intelligent life (ETI). Readers might recall Carl Sagan's novel *Contact* and the subsequent movie starring Jodie Foster. Over the years, discussions of contact with ETI have used an uppercase *c*: Contact with ETI.

3. For humankind, the word means external to planet Earth; more specifically, if people are referring to off-planet possibilities generally, exoEarth; and, when people discuss the subject while suggesting that humankind should explore and colonize elsewhere, and establish a base—such as by colonizing Mars (as currently is proposed)—and then

life, were recorded before humans had any thought that someday they themselves would have the capability to venture into space. As that possibility gradually came within reach, speculation increased and has been enhanced as people have reasoned that perhaps what had been deemed impossible for humans to do might have been done already by other beings—especially given that the life of the universe, about 14.82 billion years, is more than three times that of Earth (4.5 billion years) and that more than 17 billion Earth-size planets are estimated to be in the Milky Way Galaxy alone.[4] It is not known how many of these planets may be in the "Goldilocks Zone"—in an orbit in which the planet is the right temperature, neither too hot nor too cold, to have water. Given the cosmic time span and the number of Earth-size planets in the Milky Way galaxy, conditions might be (or have been) right on a planet or planets somewhere in the vast universe, in this and other galaxies, for life to emerge and, in some or many places, for intelligent life to evolve or have evolved. Subsequent science-based inquiries (and science fiction stories) often were reflections on and projections of what the "aliens" or "others" might be like, and how they and we might act and interact on Contact and in its aftermath, here and in the heavens.

Human Place in Cosmic Space

Science's discovery of the immensity of the cosmos can be disconcerting today when humans realize that corporeally they are cosmic minutiae. When the whole "world" was understood to be just Earth and the lights that circled it above—sun by day; moon, stars, comets, and meteors at night—humankind could comfortably consider who they were in this place. In Western religious thought and European cultural settings, people settled into what they believed to be their divinely designated place and role. They were the most intelligent life on Earth. Other life, Earth, and even the visible cosmos itself had been created as background and support for human be-ing and beings; they had been specially created by God in God's "image." In Christian thought, God's solicitude prompted God to become human to save humans from themselves—their universal, endemic, and intergenerational sinfulness.

---

launching further into space, the terminology would be that human "Martians" would be traveling exoMars.

4. Associated Press, "New Estimate Suggests Billions of Planets." Francois Fressin of the Harvard-Smithsonian Center for Astrophysics stated that the Milky Way Galaxy alone is estimated to have some 17 billion Earth-size planets. The large number of planets does not guarantee that any of them have life, let alone intelligent life, but biotic existence seems likely given cosmic dynamics, geodynamics, and biotic evolution as experienced or observed on Earth.

Spiritual, Psychological, and Social Dislocation

Over the centuries, scientific discoveries stimulated a significant reassessment of humans' previous understanding of their place on Earth and in the universe. A profound sense of cosmic displacement occurred twice—through the complementary discoveries of heliocentrism (Copernicus-Kepler-Galileo, sixteenth to seventeenth century) and biological evolution (Charles Darwin, nineteenth century). Humans' new understanding that Earth orbited around the sun effected their conceptual displacement from the center of the universe (where people had thought they were materially and spiritually created on a planet around which the material universe circled), and led to a certain philosophical disengagement with the cosmos (which people had thought was created solely to be the supportive context of human existence). Evolution, in establishing that biota evolve, diversify, and complexify, catalyzed humans' intellectual displacement from viewing themselves as the unique created pinnacle of complex life that is served by all other life. Consequently, people began to search for new meaning in their finite personal and community existence, and a different sense of place.

The speculation about whether or not "we are alone" as intelligent life in the vast cosmos has catalyzed new reflection on the relationship among Spirit, science, and space. The ancient sacred texts of Christians and other peoples of religious faith describe in diverse ways a divine Spirit's presence and creativity, transcendence and immanence; assume that Earth is a fixed "world" about which heavenly bodies orbit in spheres; express their authors' belief that Earth is the center of the cosmos and the home of created beings, humans, who are superior to and above all others; and affirm the uniqueness of human intelligence and its particular personal relationship with the creating Spirit. Science subsequently established that Earth is not a cosmic center nor even a local center; that people and other biota have evolved over billions of years from a common single-celled ancestor; and that space is an immense "frontier" inviting human exploration and eventual settlement. Scientists began, too, to speculate ever more confidently that intelligent life might well have evolved elsewhere in the universe.

When integrated, the three displacements that resulted from scientific data teach humans that they are not preeminent on a planet that is the center of the universe (pre-Copernicus); a specific, unique divine creation at a singular moment in Earth time (pre-Darwin); or the only intelligent life that exists cosmically (pre-ETI). If Contact occurs, they would understand that distinct forms of life evolved independently of humankind during the 14.82-billion-year history of the cosmos and became, because of their longer evolutionary history, more complex, more intelligent, and more technologically advanced than humans, and perhaps also progressed more spiritually—beyond religious institutions and doctrines to a common spiritual consciousness. They might even have evolved to some extent beyond their previous materiality. Contact—the encounter with intelligent extraterrestrial life—could alter forever human self-understanding, religious perceptions, and even ethical practices, to the extent that the latter have been based on belief in human intellectual uniqueness, humans' evolutionary preeminence, and humans' distinct relationship with a creating Spirit.

If people who know now that humankind is the evolutionary result of the natural processes of reproduction, diversification, and complexification of species within a long line of biological ancestors that existed over millions of centuries were to add to that knowledge understanding of a new reality—that humankind is truly not alone as an evolved thinking species in the cosmos, and that they are not the single, extraordinary "image of God" in creation—they would be stimulated to a profound realization and sense that they must not only reject past inaccurate understandings based on their previous ignorance, but explore reflectively new intellectual and spiritual territory as they explore the material universe.

Human Cosmic (Re)Location

The third cosmic displacement and dislocation, were it to occur, would provide humankind with unique opportunities: to come to a greater awareness and knowledge of the extent and complexity of divine creation; to come to a more enhanced spiritual understanding of and appreciation for the modes of divine Presence in the universe; and enlightened by the foregoing, to reflectively consider just where humans "fit" into integral cosmic being—what might be their unique evolutionary niches, their multiple places (spiritual and material) and multiple locations, each of which would have distinctive contextual interrelationships with their abiotic milieu, their biotic community, and the Spirit immanent in all.

Scientific discoveries do not force or lead peoples of religious faith to deny the existence or diminish the creativity of a divine Spirit. Divine Being continues to be understood as the origin of cosmic material being and energies, and of the scientific laws upon which geophysical and biological changes, including evolution, have occurred. Evolution, in that perspective, would be understood as an ongoing unfolding of a divine creativity that is not expressed in a single moment or individual acts of life creation, but as creation-in-process. Put simply: evolution is an autopoietic process of biotic creation as the Creator-initiated Singularity (the explosive origin of the inflating universe) unfolds in dynamic creative freedom within physical, chemical, and biological parameters; the Spirit continues to create, now indirectly, as cosmic and planetary entities and processes interact and integrate. ETI would be, in this understanding, both an example of what has been and is being wrought by divine creativity on a cosmic scale and therefore perhaps a hint of what is to come in future human experience; humankind, too, will continue to evolve into an unknown future in unseen places, adapting to each with wonder and appreciation and a sense of the Spirit to whom all are related who is present in all.

In the twentieth century, as geopolitical, ecological, economic, and population crises developed on Earth, and scientific knowledge and technological accomplishments increased, human extension into the cosmos seemed to be more inviting, and even a pressing need. Displacement from Earth and relocation elsewhere in the solar system and beyond appeared to be welcome in a material and political sense; cultural, psychological, and spiritual fears because of the experience of loss from no longer being a cosmic center, as humans had perceived their status on Earth to be, gave way to practical material considerations necessitated because of pressing survival needs.

Human journeys into space became necessary primarily, it seemed, not to extend research into and appreciation of cosmic mysteries, but to escape the negative consequences of research and technology on Earth. Harm to people and other biota (including extinction of the latter), and pollution of the air, earth, and water of their shared Earth habitat stimulated hope and efforts to find elsewhere a much-needed place, one in which useful abiotic inputs of materials and energy would be available to be used for industrial processes and products to provide for human needs and wants. As has been the case historically on Earth when nations expanded into new territory, including into places where life and even intelligent life with native ethnicities and cultures was already present, little thought was given to human impacts on and responsibilities toward other worlds' environments and their native inhabitants—at whatever stage of evolutionary development and cultural formation they might have achieved prior to when they first discover human beings descending to their planet.

At the present moment in human history, there is no certitude about whether or not Unidentified Aerial Phenomena (UAP)[5] have brought

---

5. Because the acronym UFO (Unidentified Flying Object) has come to mean in the media and popular perception a space vehicle occupied and operated by intelligent beings, numerous scientists, government agencies, and thoughtful individuals have been using UAP. It is also a more accurate term, since many of the reported unusual phenomena—some 90 to 95 percent—have been determined to be, upon thorough investigation, solar illuminated planets, solar reflections, and other natural cosmic phenomena; technologically high-end military aircraft; meteors; or unusual weather events. UAP separates

extraterrestrial intelligent life (ETI) into Earth's skies. Some UAP have been assessed by scientists and military personnel to be intelligently operated, as indicated by scientific analyses of assembled data, and military evidence gathered from radar tracks and military and civilian pilots' experiences—particularly observations of UAP's extremely high velocity and maneuvers, including changes of direction at incredible speeds during direct aerial interactions with military aircraft from Earth nations. Readers might well have their own ideas and opinions on the matter. *Cosmic Commons* takes neither a "pro" nor "con" stance as to whether or not UAP have been observed or ETI has made Contact. It attempts, through a thought experiment, to provide sufficient data as to make UAP and ETI presence (even if assumed from afar) sufficiently credible as to provoke serious and conscientious conversation, and even to stimulate thoughtful and civil debate, when the matter is seriously considered.

What is suggested in these pages is that, in light of current terrestrial ecological crises, political conflicts, and economic hardships, and the possibility that humankind will encounter intelligent life on Earth or in space, a stimulating speculative scenario should be considered. Humanity—intelligent terrestrial life—should reflect on the subject of ETI "as if" some form of Contact has happened already or is on the brink of happening. This would prompt people to ponder the possibility of ETI more objectively, and thereby assist humankind, as individuals and as a species, to develop projected responses to ETI Contact. If people were to regard the likelihood of Contact as a pressing Earth-exoEarth possibility, they would be catalyzed to formulate working guidelines for space exploration, and to provide parameters for human conduct on other worlds when humankind relates to and interacts with the biota (complex and intelligent, and simple and instinctive) and the abiotic environment they encounter.

"As if" considerations might stimulate people's discussion of how, if they think that in exploring and even colonizing newly found worlds humans should and would do better ecologically and socially than they have been doing on their home planet, they propose to explore and develop other worlds' natural environments (including in the way industry will extract and process useful natural goods[6] found there), and to interact with indigenous

---

events from predetermined categories. UAP is an accurate acronym applicable for all of the preceding, since it can refer to one or many unidentified events: Unidentified Aerial Phenomenon; Unidentified Aerial Phenomena.

6. A "good" implies something that can provide some benefit for abiotic nature, flora, or fauna, principally but not exclusively in the place in which it is situated. A "resource" is a planet's, satellite's, or other celestial body's natural good that is viewed as intended for or primarily available to humans or ETI to use in place or to extract, alter, and utilize elsewhere to provide energy, to meet other needs, or to satisfy wants. Since the cosmos and its

people(s) who already inhabit these extraordinarily distant places. Such a thought experiment could prepare at least some segments of humanity, perhaps even a significant group in terms of their numbers or their political and economic power, to advocate exoEarth responsibility.

A consequence of people's considerations and conversations along the preceding lines of thought might be that, simultaneously, they begin to advocate for terrestrial ecological and social responsibility when in a moment of insight they wonder why they are not doing now on Earth what they think they could, should, and would do on another world. Effectively, responsible people in this regard are seeking to foster *socioecological ethics*[7]—an integrated Earth concept that indicates commitment to and action for both social justice for human communities, and ecological justice for the community of all life and for the abiotic common ground of their shared environment.

A thought experiment can prompt humans' creative reflection, insights, innovations in knowledge and beliefs, amendment of previous ideas and beliefs that were strongly held, and adoption of new ideas, ways of thinking, and contextual practices. It can stimulate humankind's commitment to common good conduct to correct existing individual or species deviations from what they should have been doing but have not done. The concept of *utopia* can function complementarily, as a projection of an ideal world in which the well-being of all people, of other biota, and of their Earth home has become a reality: that which is "not yet" could be achieved sooner rather than later if intelligent beings commit themselves to realizing it.

Consideration of an "as if" scenario along the lines suggested would help humankind develop thoughtfully a realistic understanding of the complexity (abiotic and biotic) and extent of the cosmos, and a more informed—and ever-developing—sense of a human place (and, perhaps, of human places) within its integral being.

Stephen Hawking: ExoEarth Worlds and Extraterrestrial Life[8]

Stephen Hawking's ruminations in recent years reveal his preoccupation with exoplanetary exploration and possible Contact with extraterrestrial

---

bodies do not exist exclusively for anthropic appropriation, "good" is a more appropriate term to use than "resource" when discussing ETI use of natural benefits and when considering eco-justice to be an essential component of socioecologically responsible conduct.

7. The concept of *socioecological ethics* originated with me several years ago. I was preparing to present a paper on *praxis ethics*—another of my terms—and wanted to clearly express one of its aspects. My doctoral students have become familiar with the term "socioecological ethics" since I have been using it in class; several have used it in doctoral dissertations.

8. An earlier, summary form of parts of this Introduction and of other chapters in this

intelligent life. In 2006, at the Hong Kong University of Science and Technology, and in 2007 at NASA in Florida, he urged that a moon base and Mars colony be developed in the near future.

In Hong Kong, Hawking stated that humankind's survival requires that science and technology be dedicated to producing transport craft for space travel. This is necessitated by people's terrestrial irresponsibility, which will cause them to destroy themselves and their home planet. Humans must be saved by space, or at least by ships that would shuttle them elsewhere: first within the solar system, then to destinations among distant stars. Earth, he said, is heading for destruction by human-caused or exacerbated catastrophes that might include "sudden global warming, nuclear war, a genetically engineered virus."[9] In the United States, Hawking reiterated his concerns, and declared again that humankind needs some resettlement: "I think that getting a portion of the human race permanently off the planet is imperative for our future as a species."[10] My immediate thought when I read the Hong Kong quote was, "It's the same people!" Hawking's space pioneers would be humans who have already devastated their home planet. How or why would representative members of the human species that has caused terrestrial social and ecological devastation for economic and political reasons act differently in space, on other worlds?[11] My reaction to the Hong Kong quote was reinforced when I read Hawking's Cape Canaveral comment.

Hawking's proposal for and promotion of expeditiously developed human space flight and colonization needs to be carefully analyzed and considerably assessed. He voiced confidence in human consciousness and conscience, technology and uses of technology: humans would begin anew on the moon and Mars. Having escaped human-caused devastation on Earth, people would establish an exoEarth site. He seems to imply that elsewhere in space humans will have a respite from their previous destructive technological and economic practices, and somehow will not reproduce their conduct.

In his statements, Hawking did not discuss how or why humans would suddenly, in space, "get religion," so to speak, about political, ecological, economic, and technological responsibility, after having harmed Earth with their arrogant attitudes and destructive machines. Terrestrial humankind has

---

book appeared in the journal *Theology and Science* as Hart, "Cosmic Commons: Contact and Community."

9. Hui, "Hawking Says," 5A.

10. Boyle, "Hawking Goes Zero-G."

11. Hawking's words stimulated me to focus my next research project on humans' space exploration and, complementarily, on possible extraterrestrial intelligent life explorations on Earth. Hawking's comments were the catalyst for the conception of *Cosmic Commons*, which has come to be my extended reflective response to his remarks.

developed weapons of mass destruction; invented and used mining machinery that rips apart farmland, tears off mountain tops, and poisons the rivers and creeks in which people fish and from which they irrigate agricultural lands to provide food crops—the same aquatic sources that provide rural and urban communities with water needed for their sustenance and livelihood; and generated electricity with nuclear and coal-fired electric power plants that pollute air with their emissions and poison water with their effluents. It is transparently idealistic to think that humankind emigrating from Earth will extraterrestrially, once in space or when living in another planetary place, be converted from its current conduct and consciousness. Somehow, Hawking implies, humans will depart from disruptive and destructive terrestrial practices and act responsibly as part of an extended biotic community.

On arrival elsewhere, apparently people will respect the inherent value and dignity of other intelligent biota, use technology only responsibly (perhaps taking their cues from what they have done or should have done on Earth to benefit humans, biota generally, and their shared planet), and generally care for humanity's new home(s) and habitat(s), the extraterrestrial common ground that they will share with other members of the biotic community. How would such a change of consciousness and conduct emerge in the heavens or on heavenly bodies, in space, or *in situ* on arrival or afterward?

In light of past and present human consciousness and conduct, it is not realistic for Hawking to assume or hope, as he apparently does, that human voyagers and settlers will colonize differently in other places than their ancestors did on Earth in past human history, and behave differently than they are behaving here and now in this time and place. It would be disingenuous to expect that in distant places humankind will not cause or exacerbate global climate change—perhaps even more quickly in other locales than on Earth—or radically remove natural goods from the sites they colonize; or to think that humankind would promote not only the common good of human communities, but the common good of all biota, and in so doing protect ecosystems, and live more simply and responsibly in order to do so. How would humankind come to provide, in new contexts, for common human needs? How would it cease striving principally and without principles to satisfy individual desires, and instead curb already prosperous and economically powerful particular groups' insatiable quest to excessively and exclusively appropriate natural goods rather than use them to ensure the well-being of Earth and all biota?

Hawking did not discuss either, in his cited comments, potential human impacts on and interference in the evolution of life on other worlds, or possible encounters with other intelligent life. He expressed no concern that human explorers and settlers would harm indigenous communities at whatever their level of social and scientific development. (Later, when he declared in a broadcast of his 2010 documentary series *Into the Universe with Stephen Hawking* that humans should avoid Contact with roving ETI because ETI might do to humans as European humans had done to native peoples of the Americas, Hawking did not make a connection to human conduct in the Americas and its likely replication in space.)[12] Superficially, his vision that humans will voyage into space and settle on another world to save our species appears to be a salutary idea and a salvific ideal. But the likelihood of success seems doubtful, given past and present human conduct.

Several aspects of Hawking's proposal are problematic and raise essential questions: What policies and parameters should be established for colonizers on Mars, or anywhere beyond Mars in space, to prevent them from replicating in other places the devastation that humankind has wrought on Earth? Who will go? How will they be selected? Who will provide funding, to what extent, and for what purpose(s) beyond human survival?

Other factors to be considered, should Hawking's space ark be constructed, staffed, and launched, include potential terrestrial impacts such as accelerated ecological devastation of Earth and destruction of human communities on Earth as people who hope for personal survival or species survival elsewhere—which will be an opportunity for very few—stop fulfilling their terrestrial environmental and community responsibilities. To what extent will those left behind become locked in intraspecies struggles for survival? Such a situation would be a sad and ironic twist on humanist-deplored stereotypical assessments of environmental impacts of Christian beliefs: Christians' hopes for a better life in an otherworldly, spiritual heaven have had the outcome that they do not care for this world, the material Earth that is the context of their current life. Whether people think that they are bound to go *beyond* Earth via spacecraft or spiritual transcendence they might equally be unconcerned about what is happening now *on* Earth.

Human planetary consciousness and conduct, then, must be transformed substantially prior to human exoplanetary settlement—so that humans will not, once again, use scientific research and technological development in such a way that the threats of "nuclear war" and a "genetically engineered virus" that Hawking correctly states currently imperil Earth will not recur in distant places. Human settlers on other celestial bodies[13] must

---

12. Cited in several sources, including Wright, "Ethics for Extraterrestrials"; Leake, "Don't Talk to Aliens"; and MSNBC, "Aliens May Pose Risks".

13. "Celestial bodies" is the term used in United Nations documents when referring to any place beyond Earth, including the moon, on which humankind might settle, from which humans might extract natural goods ("resources"), or on which humans might build bases to launch vehicles farther into space.

not do in exoEarth contexts what European settlers in the Americas did on Earth upon encountering existing native populations, and what scientists, engineers, and the transnational corporations by which they are employed now do globally, at times assisted by compliant government officials: disrupt abiotic contexts and ecosystems, and thereby devastate and extinct[14] Earth's populations of flora and fauna.

Although a new *deus ex machina*[15] in the form of space vehicles and artificial habitat colonies envisioned by Hawking might rescue some people from an Earth fraught with peril, the new human consciousness and conduct that Hawking apparently assumes will develop from current human consciousness cannot, on the contrary, be the progeny of existing human thought and practices. If extraterrestrial explorations and economic and military development (which traditionally stimulate scientific explorations on such a scale) follow a "business as usual" mode, they will effect in new worlds conditions similar to what human ideas and inventions have wrought on Earth. Technology, developed by human researchers supported by entrepreneurs' capital or governments' military expenditures, is especially fallible—even in the form of artificial intelligence (AI). Inventive humans, if they retain the same type of dominant and competitive consciousness in extraterrestrial settings that they have had in terrestrial ones, will embed in technology, including into AI, and export into space planetary problems identical with or similar to those that they have caused or exacerbated on Earth.

Care for the cosmos as a whole or in part will not be developed *ad hoc* during extraterrestrial explorations, or when cosmonauts meet other worlds' simple or intelligent biota or discover other worlds' natural goods. If current consciousness is exported to space, the biota or benefits found on other

---

14. Through evolutionary processes such as natural selection, species become extinct; when humans interfere in those processes (through ignorance, or with foreknowledge of the likely or possible deleterious impacts of their actions) to the extent that a species no longer exists that otherwise would still be in existence in its evolutionary niche, it is more accurate to note that in such cases species have not "become extinct" but rather were "extincted."

15. Literally, "god of the machine." In Greek drama, sometimes the protagonist was in an extremely perilous and inescapable situation. Since the drama's script had unfolded to a point where there was no logical, rational course of action or physical means within the story's flow by which the hero could escape, a system of ropes and pulleys was used to lower onto the stage a divine rescuer, who would snatch the endangered person away from what was threatening him. Today, when God and gods are out of fashion, science and technology have become viewed by some as the miraculous means by which humans will be rescued from current and projected ecological catastrophes—global warming being a representative example—which they themselves have brought into being through ignorance, irresponsibility, or a combination of both.

habitable places will be aggressively impacted or commercially extracted because they will be regarded as "resources" to be taken and transformed by human work; there will be little or no consideration of the integral role they play in the "geo"-dynamics[16] and evolutionary processes of distant celestial bodies.

A core consideration before, during, and after Contact is whether or not humankind (and, perhaps, other intelligent life)—given that it has literally "messed up" its home planet ecologically and has not provided for the basic sustenance of all of the human community or of Earth's extended biotic community—can have a change in consciousness and conduct such that it does not replicate elsewhere in the cosmos the socially and ecologically destructive way of life it has imposed in its Earth setting. The complementary danger exists that "alien" (to Earth) intelligent life might well have destroyed its own planetary home, as Hawking fears, roams space in search of other worlds to colonize or to exploit, and is caught in a cycle of existence similar to what humans might have if they do not change their destructive ways prior to departure to explore and settle in distant space.

Anthropocentrism and Anthropomorphism

Stephen Hawking warned that humankind was destroying Earth, and urged space travel to establish extraterrestrial settlements for survival (while not considering that humans causing problems on Earth would likely cause them elsewhere). He also warned that humans should avoid Contact with intelligent aliens roving in spacecraft or settled within range of Contact. He said that space probes, and signals sent into space by, among others, the SETI (Search for ExtraTerrestrial Intelligence) Institute (Mountain View, California) should cease seeking to find and encounter ETI because ETI space travelers likely destroyed their home planet, much as humans are destroying theirs, and voyage through space seeking other worlds to conquer and colonize, or solely to exploit for needed or wanted goods.

In his various statements Hawking presents a very pessimistic view of intelligent life: humans are destroying their Earth home and should desert the neighborhood; aliens have destroyed their home planet and are relocating already. He ponders pessimistically, in fact, whether intelligence itself might lead inevitably to intelligent species' destruction of themselves and their planetary home. In one regard, his pessimism is not universal or absolute: he does urge humans to go elsewhere to do better, with some sort of presumption that better conduct indeed will occur. But other intelligent life

---

16. "Geo" ordinarily refers to something related to planet Earth; it is used here as a transitional term indicating planetary dynamics present exoEarth.

is not granted that same brief note of optimism: they are regarded as likely competitive threats rather than potentially collaborative partners.

Hawking advances substantially in this instance by advocating away from a natural, evolutionary species' self-interest in preservation, toward what might be viewed as an unnatural, intelligent species' selfishness; in doing so, his analysis expresses a strong anthropomorphism and anthropocentrism. His *anthropomorphism* is evident when he projects onto ETI what has been, historically, human conduct on Earth with all its deleterious social and ecological results. His *anthropocentrism* is obvious when he urges humans to avoid Contact so that they, too, might seek and acquire other planets' natural goods before ETI finds and uses them.

In pondering the existence of ETI, unlike many scientists Hawking unreservedly declares that, according to his "mathematical brain," the numbers of planets known, and those theorized to exist, indicate that there is a very strong likelihood of extraterrestrial life, and that intelligent life would have resulted from biotic evolution in some contexts. The diverse forms extraterrestrial life might take and the types of habitat in which it thrives or survives, of course, are not known. Hawking's affirmative assessment and assertion that intelligent life exists elsewhere in space is becoming increasingly more popular among scientists. The US Hubble and Kepler telescopes have discovered, in distant solar systems, planets which meet the requirements needed to support life, as projected from knowledge of Earth biota.

Hawking expresses as a scientist what the film *Avatar* portrays fictionally and what European expansion into the Americas illustrated historically and dramatically. In his BBC television series, cited earlier, Hawking states that impacts of ETI arrival on Earth might be comparable to the outcome of Columbus' arrival: it "didn't turn out very well for the Native Americans." He notes that "we only have to look at ourselves to see how intelligent life might develop into something we wouldn't want to meet." Ironically, even here as he acknowledges the devastation wrought by humankind on Earth, he misses an opportunity to consider the parallel impacts of humans traveling to other planets and replicating what Europeans did on Earth.

Despite this acknowledgment of past colonizing harm on the part of humankind, Hawking continues to give narrow human self-interest a strong emphasis, raising it to a level of anthropocentrism characterized by a dual selfishness: first, only a select number of humans will be able to escape collapsing Earth (he neglects to suggest or at least to speculate about who will decide which people will be saved, from what professions they will be selected, and the extent of their cultural diversity); and second, humans will strive to avoid ETI in their endeavors, but if encounters occur, humans will prioritize their own wants and needs over those of indigenous ET worlds and peoples, and of competitive ETI colonizers. This, of course, is much the same as Europeans' actions upon arriving in the "New World" and US governments' policies and practices ever since, as they unashamedly committed—and continue to commit—physical and cultural genocide against Indian[17] peoples.

European ethnocentrism, and its consequent corollary Euroamerican ethnocentrism, both offshoots of anthropocentrism, might well continue to be exercised by humans in space, as Hawking warns indirectly and unintentionally when he arbitrarily attributes this conduct to ETI in the manner in which he characterizes them. On Earth, such conduct continues to harm indigenous peoples. In *American Holocaust: The Conquest of the New World*, David E. Stannard describes the centuries-old and continuing ethnic cleansing and economic oppression experienced by native peoples. He terms the extensive killing of the original inhabitants of the lands now known as the Americas "the most massive act of genocide in the history of the world."[18] The film *Avatar*, too, suggests and shows that unfortunately and imperialistically just as before, humans will act similarly against other worlds' populations and places in different cosmic regions as they seek to exploit natural goods and indigenous peoples. *Avatar* projects into the future a theme that *Dances With Wolves* elaborated looking into the past. In *Avatar*, the Na'vi people are a new version of (American) Indians as past history on Earth is repeated in the future on other celestial bodies. In doing so, the film presents a stark and realistic contrast to Hawking's romantic projection of what will be human endeavors in space and on other worlds.

Cosmic Commons and Sacred Commons

People's first and primary encounter with the revelatory aspects of the universe occurs at the "local level," in the place(s) of the Earth commons where they have the most immediate and experiential engagement with the Spirit. We humans live, on a cosmic scale, in a sacred "commons of commonses," the space and place where disparate and distant beings "live and move and have their being" as they seek to walk with the Spirit while sharing in the natural goods provided by the Spirit to meet their needs. In contexts

---

17. I have found in my three decades of human rights work with native peoples through the International Indian Treaty Council that most traditional US native peoples prefer "Indian," not "Native American," as a generic reference, and then specify their particular culture to distinguish their own heritage, e.g., Wanapum, Lakota, or Muskogee. Native peoples' human rights organizations include the American Indian Movement, the Indigenous Environmental Organization, and the International Indian Treaty Council; the latter is a nongovernmental organization (NGO) accredited to the United Nations.

18. Stannard, *American Holocaust*, x.

separated by vast distances, intelligent beings share a particular, personal and communal, unique but yet cosmically universal, encounter with the sacred. Spirituality is *personal* because it is an individual's encounter with divine Being; it is *universal* because it is one Spirit, the Spirit of the Cosmos, whom all beings engage. The sacred cosmic commons is a communion of commonses cosmically interrelated and integrated. It is stardust become spirit, it is atoms become life and thought, all in the presence of a transcendent-immanent, Being-Becoming, creating Spirit. The cosmic commons is, for the person of faith, a sacred commons. It is perhaps appreciated most when, on a cloudless night as we contemplate the heavens with their stars, planets, and celestial bodies, the vastness of the universe both draws us outward toward the sparkling distant lights and deep darkness, and focuses us inward as we experience this immensity and wonder about our place in and relation to such immense complexity.

A *commons* is the shared space or place that provides both an abiotic (non-living) home and habitat for the biotic community (the community of all life), and the natural goods ("resources") needed for biotic sustenance and well-being. It is characterized by abiotic dynamics and biotic evolution, which are integrated with Earth and each other through necessary and contingent events. A commons has varied expressions in distinct types of physical space and at diverse levels of human understanding.

The *cosmic commons* is comprised of the interrelated (even when this is not known or acknowledged), distinct, and diverse spatial and local contexts of interactions among members of the integral being[19] of the cosmos. On the local level, a planetary commons is the milieu in which intelligent beings strive to meet their material, spiritual, social, and aesthetic needs, and to satisfy their (often not self-limited) wants—the latter effort at times undertaken with little appreciation of a planet's limits or acknowledgment of people's responsibilities. Acceptance of the idea and ideal of a cosmic commons becomes ever more important as nations and corporations (now transnational; potentially trans-stellar) reach out into the universe through space probes, orbiting space stations, planetary robotic rovers and, in the distant future, satellite colonies on other worlds. The business and government intent of such activities ordinarily is to derive some commercial benefit or military advantage. Acceptance of the cosmos as commons would become even more important should other advanced forms of intelligent life, seeking the same goods or interested in exchanging their human-desired manufactured goods

---

19. "Integral being" is comprised of all the existents of the cosmos. For the theist, integral being includes divine Being; for the humanist, divine Being is not present in the cosmos. Initial presentation of this concept is found in Hart, *Sacramental Commons*, 138, 221, 222 n. 11.

for human-discovered and human-developed goods, be encountered in cosmic explorations.

In the cosmic commons, an intricate and sometimes delicate relationship exists among biota (and between biota and their abiotic context) as they utilize or strive to utilize commons goods and common goods, and as they seek or struggle to preserve the commons good and simultaneously to provide for the common good of their own species or of biota as a whole, all of whom are integrated within an interrelated web of living being.[20] In the cosmic commons, too, intelligent life has particular responsibilities, including respect for forms of life less complex than it is, and regard for and care of common habitat. The cosmic commons includes relationships in home planet environments, in celestial bodies, in space, and among intelligent life's particular communities on its own home planet, on others' home planets, on other celestial bodies it explores or inhabits, and in space itself when it encounters other intelligent life.

Creator Consciousness

Human responsibility in extraterrestrial contexts would be promoted, too, if people of faith were to become and remain conscious of divine immanence throughout the cosmos. Spiritual consciousness will promote a sense of being at home and of having a place not only on Earth but throughout space, and being part not just of the human community, but of a community of all life, particularly sentient life, and especially intelligent life, throughout the cosmos.

In the past, "ET" and "ETI" were discussed primarily in literature, films, and news stories. Today, reflection on cosmic Contact in a cosmic commons, linked to Spirit-consciousness, would help humankind explore relationships among people, between people and other biota, and between humankind and ETI species who encounter each other in space in the presence of the immanent-transcendent Spirit. Anticipated or actual Contact would become less threatening to theology, to Christian and other faith-based ethics, and to human communities than would otherwise be the case. Engagement with life on other worlds, and with intelligent beings who are different "images of God," should stimulate human concern about responsibility to other worlds, and establish dialogic relationships between those worlds and Earth, and between present and future: visions of environmental well-being elsewhere would reinforce Earth caretaking efforts, and visions of social justice on other worlds would reinforce its promotion on Earth. Innovative considerations of

20. The concepts of commons good, commons goods, common good, and common goods are elaborated in depth in Hart, *Sacramental Commons*, chapter 8.

terrestrial-extraterrestrial engagement, as expressed in theological considerations and cosmo-socioecological ethics, would educate in new ways people concerned about ecology and social justice, and fascinated by ETI.

Potentially, religious doctrines and ethics (including, for Christians, a sense of existing in a sacramental commons, and for all religionists, being in a sacred commons) can have a positive impact on terrestrial-extraterrestrial ecological and commercial conduct in theory and in practice; complement scientific thought, research, and development; and promote interplanetary and intercultural ecological and ethical sustainability as a corrective for ecologically and socially harmful, and politically unrestrained, industrial, commercial, and military projects.

Discovery on Earth and in the Cosmos

Since they first set forth on seafaring voyages beyond their continent, Europeans have based their conduct on the Discovery Doctrine, developed in the fifteenth century. Discovery has been a controlling ideology used to justify Europeans' seizure of indigenous peoples' territory and regional natural goods.

Discovery declared that Europe had a civilization superior to those that developed elsewhere, and the only true religion, Christianity. Therefore, the lands Europeans "discovered" were to be considered "empty land"—even if cities, farms, and other marks of urban "civilization" were present. In European eyes, private property, especially when enclosed by fences, was an essential mark of and requirement for "civilization." The Discovery Doctrine rationalized, justified, and guided European expansion into and colonization of the Americas, Africa, and Asia. Although ideologically dominant today, it is usually not cited directly, and remains politically potent but publicly hidden. Exposure of the Discovery Doctrine, and rejection of its past, present, and future use by national and international bodies, could stimulate restitution for past injustices or at least prevent its ongoing use—which will "justify" future injustices.

The egregious and culturally biased Doctrine of Discovery, because it remains operative today in national and international law and policies, has the potential to be activated in and even prior to space exploration. This would have adverse social and ecological consequences in space. It will be explored in these pages because of its prior and current cultural and ecological impacts, and the potential threat it poses for humankind in future times and places.

I became aware of the Discovery Doctrine and its impacts on indigenous peoples in May 2012 when I was invited by the International Indian

Treaty Council (IITC), the first native peoples' NGO recognized by and accredited to the United Nations, to present testimony on the ideology of Discovery at the Consultation of the UN Special Rapporteur on the Declaration on the Rights of Indigenous Peoples document that had been issued by the UN in 2007. The Consultation was hosted by Sinte Gleska (Spotted Eagle) University on the Rosebud Reservation, in South Dakota. Discovery Doctrine became part of my consciousness during my research for and writing of my testimony, and has influenced my thinking ever since.

At the time of the Consultation, *Cosmic Commons* was already in process. It occurred to me a few days afterward that what Discovery had wrought on Earth—injustice toward indigenous populations and economic and ecological harm around the globe—might recur in space. Discovery might be used by segments of humankind not only to devastate intelligent life and cultures, should such be found, but also to seize natural goods present on distant worlds—much as has been done, on both counts, in the United States. It became an important theme woven throughout *Cosmic Commons*, to facilitate understanding of injustices perpetrated in the past upon native peoples and to prevent their continuation. These injustices have continued for more than five centuries to cause economic harm and ecological catastrophe for all people. The invocation and use of the Discovery Doctrine should be rejected during space exploration and in planetary colonization to prevent replication of Earth-like destruction of abiotic creation and devastation of biotic communities.

Government-Generated Controversy and Condescension

Almost since the time reports began to appear about Unidentified Flying Objects (UFOs) the United States government has sought to "debunk" UFO incidents. Eventually, because of ongoing public interest and inquiries, the government resorted to disparagement and ridicule of witnesses, rejected funding for scientific research, and tried to provide alternative explanations for the phenomena. Consequently, ridicule became the standard tool of US government officials, scientists, and military personnel, who rarely, if ever, studied the phenomena themselves seriously and scientifically. Some of the latter, in fact, did not even accept what they had seen personally, choosing instead to self-censor, or kept their experiences to themselves, for fear of losing a job, a promotion, or a higher salary.

Employment consequences became especially true in academia. Professors at institutions of every size found that they dared not mention UFOs if they did not want to suffer ridicule from faculty colleagues. As Leslie Kean observes in her well-researched book *UFOs: Generals, Pilots, and

Government Officials Go on the Record, "The questions raised by the UFO phenomenon were deeply disturbing to our accustomed ways of thinking. The subject carried a terrible stigma and was therefore a professional risk for those publicly engaged with it."[21]

Conversations about UFOs and ETI both in the United States and internationally would be much enhanced and enlivened if people who have had close or distant encounters with UFOs would take the risk or take the stand as credible witnesses, as others have done, and describe their encounters. They might be surprised about the extent to which people they know or professionally respect have had previously undisclosed similar encounters—an experience I had during conversations with Boston University faculty colleagues. It is time for the scientific community, particularly its academic members, to become involved in serious research regarding narratives about Contact with ETI on Earth, and to receive funding for such research from government, foundation, and private sources—without fear of ridicule or negative consequences from academic institutions in which they are presently employed or by which they hope to be employed.

It is my hope that senior scholars, among them especially scientists, particularly astronomers and astrobiologists, will "come out of the closet" and discuss their own experiences or those about which they have become aware because of credible witnesses, perhaps in their own family. In their discussions, they should shift the language of "belief" to the language of intellectual discourse. Rather than saying or denying that they "believe" in ETI or ETI-controlled UFOs, they might state, with genuine scientific (and humanistic) intellectual curiosity, that they "think" the subject is worthy of serious, independent scholarly research and hypotheses because of mounting credible evidence.

Critiques and Constraints

As I considered writing *Cosmic Commons*, because of the preceding I approached the UFO topic somewhat concerned, as a professor in a major research university, about possible professional repercussions. However, my own experiences, along with increasing UFO speculation and even openness and affirmation among noted scientists and religious leaders (including in the Vatican) of the possibility of UFO and extraterrestrial existence, prompted me to continue my research and writing. I have noted since, in discussions with friends and colleagues—including scientists—that several have had UFO experiences of their own and accept the possibility that extraterrestrial

---

21. Kean, *UFOs*, 4.

intelligent life exists and that credible witnesses' accounts of UFO experiences should be taken seriously and investigated scientifically.

I present this book while recognizing that some will reject its ideas, data, and considerations; others will be interested in pursuing the subject further; and yet others will be delighted that they found a teaching scholar who describes what they have experienced themselves, or what they speculated about but feared to discuss even with otherwise congenial colleagues, who might resort to ridicule—or retreat to a position of self-designated, supposed intellectual superiority.

A Reasoned Thought Experiment

Cosmic Commons, as noted earlier, will utilize a thought experiment: neither a pro nor con position will be taken regarding UAPs and ETI but, in order to prompt in-depth consideration, material will be presented under an "as if" scenario—as if Contact has already occurred, or might happen in the near future. In order for this thought experiment to work, as credible a case as is possible must be made, based on reports and narratives by credible witnesses. This will be the operative narrative of *Cosmic Commons*.

Those accustomed to academic works claiming "balance" between opposing views will be disappointed because that is neither the aim nor the presentation mode here, for four principal reasons. *First*, presenting credible reports stimulates the reader to seriously consider the possibility of ETI without being distracted by a "security blanket" of skepticism that says, "Don't worry. It didn't really happen." The focus on considering possible impacts and implications of ET Contact will require taking that possibility seriously, without comforting—psychologically or academically—alternative considerations. *Second*, there has been extensive (and, at times, abusive) US governmental and academic institutional pressure to suppress Contact considerations and to publicly and condescendingly regard and ridicule those who suggest them. Consequently, in US culture, although belief in ETI is high and overt among the general population, it is seemingly low and primarily covert in academia. Culturally, academically, and politically, then, the "imbalance" that needs to be addressed is the overwhelming power of government and academia to dismiss and denigrate UFOs and ETI. *Third*, the heightened focus on Contact implications could stimulate reflective consideration and concern about what *humankind* is doing to its Earth home: at this historical moment, ecologically and culturally destructive actions are wreaking far more harm than could possibly be done by a malevolent ETI. *Fourth*, humankind is gradually becoming extraterrestrial in its space journeys: should humankind arrive at another civilized world, humans would be exoEarth, and thereby "extraterrestrials": they would be "off" or "away from" their Earth home, and should have thought seriously, prior to and during their exploratory voyages, about their responsibilities in exoEarth contexts, and what their consciousness and conduct should be in space and on *terra externa*.

Cosmic Commons relates and integrates a dual narrative: it considers the possible existence of extraterrestrial intelligent beings and corresponding implications for human beings—explored via a thought experiment; and it considers characteristics of responsible human conduct on Earth and in space whether or not Contact is made with ETI in either context, a consideration explored through analysis of and projection from past and present human socioecological conduct.

Cosmic Commons weaves the preceding together in four parts: *Terra Firma* (chapters 1–3) discusses the place and conduct of human communities on Earth, which is humanity's local commons within the vast cosmic commons of creation; *Terra Conscientia* (chapters 4–6) reflects on how human consciousness and conscientiousness influence human conduct on Earth, and might be extended similarly into places in space; *Terra Incognita* (chapters 7–9) considers how humans might reflect on their current sense of cosmic place and engage, interact with, and compete or collaborate with other intelligent life on an exoterrestrial "Earth," that is, other places (planets and "celestial bodies") in space that humankind colonizes; and *Terra Cosmica* (chapters 10–12) explores ways in which human space exploration and settlement, and Contact with extraterrestrial biota, including intelligent life, might be engaged in such a manner as to promote intergalactic or even interstellar cooperation, collaboration, and community. Throughout the book, the ecological, economic, ethical, and ecclesial implications and impacts of Contact will be explored.

Chapter 1, "Terrestrial Trauma," focuses on ecological roots of Earth's socioecological crisis, considering and responding to Lynn White Jr.'s, oft-cited *Science* essay (1967), and in so doing reflecting on how humankind has treated (corresponding to or contradicting religious doctrine) that part of creation that is humanity's home planet. Chapter 2, "Terrestrial Discovery," discusses social roots of Earth's socioecological crisis, focusing on the European Discovery Doctrine that was used to justify seizure of native lands, and its subsequent cultural and ecological impacts on and beyond Indians' territories. Chapter 3, "Terrestrial Transformation," considers a historical restoration of Earth's socioecological commons, as religious and humanist values confront socioecological crises and develop "on the ground" projects to address them. UN Earth-related documents are discussed in depth because of the international collaboration involved in their development of principles for terrestrial socioecological responsibility.

Chapter 4, "Extraterrestrial Discovery," analyzes how Discovery Doctrine might be exported into space explorations, resulting in adverse socioecological consequences that parallel what Discovery has caused on Earth. Chapter 5, "Spirit and Science," describes previous conflicts between the religion and science fields of inquiry, and the contrasting thinking and unifying work of acclaimed priest-scientists Teilhard de Chardin and Georges LeMaître, and priest-"geologian" Thomas Berry; it discusses processes that could inform human consciousness and prompt science-religion collaboration. Chapter 6, "Socioecological Praxis Ethics: Earth Commons and ExoEarth Commons," suggests an integrated contextual-deontological dialogic ethical process to engage socioecological (social justice integrated with ecological well-being) issues in Earth and cosmos contexts.

Chapter 7, "Cosmosocioecological Praxis Ethics: Cosmoethics," focuses on *topia* (place), *dystopia* (dysfunctional, oppressive place), and *utopia* (no place currently; an ideal), and proposes ways to proceed from present contexts toward envisioned ideal contexts, in Earth and cosmic milieus. Chapter 8, "ExoEarth and Extraterrestrial Transformation," analyzes internationally developed and agreed upon UN space documents and principles in order to discern how they might be transformed from their original anthropocentric orientation and amended to be applicable not just to human space ventures, but to human-ETI engagement. Chapter 9, "Cosmic Contact," presents the heart of the "thought experiment." It describes credible witness reports of major sightings of UFOs, particularly at two US sites—Roswell, New Mexico and the Hudson River Valley in New York State—and notes briefly other incidents that occurred around the world, and details the scholarly discussion from diverse philosophical, religious, and theological sources that have speculated for centuries about a "plurality of worlds" and potential philosophical, theological, psychological, and social impacts, collaborative or conflictive, that ETI might have on humankind.

Chapter 10, "Cosmic Coexistence," discusses the potential for terrestrials and extraterrestrials to share a common cosmic consciousness of their interdependence and interrelationship as members of cosmic integral being, and suggests steps that might be taken as a basis for respectful Contact and consequent collaborative and congenial coexistence. Chapter 11, "Cosmic Charter," initiates consideration of an inter-intelligent beings' covenant that would promote constructive consociation to meet common needs through collaborative projects, and provides a preliminary draft of a document that might serve as a foundation for or a contribution to formulation of a dynamic, contextually evolving interspecies and interstellar document, similar to UN documents and the Earth Charter on Earth, to guide TI-ETI relations. Chapter 12, "Cosmic Commons," speculates about and envisions that all intelligent life might, in an advanced stage of evolutionary process, seek celestial cohabitation, conservation, and compassion, or at least view them as ideals for intelligent species and develop dynamic principles and practices to promote mutual well-being.

Cosmic Considerations: Currently Terrestrially Irrelevant?

In light of current multiple ecological and social concerns and catastrophes, it might seem untimely, unseemly, and even irresponsible to be exploring such a "far out" topic as space journeys and terrestrial-extraterrestrial Contact. In the recent past, for example, scenes of ecological and economic devastation wrought by British Petroleum's Deepwater Horizon oil fiasco filled the airwaves. Locally and globally, millions of people have experienced employment and economic trauma, many of whom have been added to the ranks of the suffering poor as a result of the Great Recession that impacted global economies; many still lack jobs and are denied access to adequate housing, health care, and basic goods to fulfill their subsistence needs. Ecologically across the United States, in January 2013 the National Climatic Data Center released scientific data that indicated that 2012 was the hottest year in US history, a full degree hotter than 2011, and 3.2 degrees warmer than the average for the entire twentieth century.[22] Consequences have included massive droughts and the related loss of agricultural products throughout the world, and farms and forests ablaze from fires upon which little substantial rain has fallen to assist firefighters. Extreme weather events, such as fierce hurricanes, are increasing in number and intensity, impacting human lives and livelihoods: recently, Isaac on the Gulf Coast and Sandy in the Northeast were especially devastating in the United States.

Consideration of the prospect of Contact could help to eliminate, over time, such socioecological irresponsibility. Reflection on Contact's potential impacts might prompt humankind to envision what it hopes to do in the heavens to prevent recurrence of what it has done on and to Earth. The envisioning process might catalyze people not only to formulate plans for living according to their new consciousness on distant planets in the future but also to evidence, in their concrete conduct on Earth in the present, what they hope their actions will be on Mars, other planets, and celestial bodies decades hence. If this does occur, a dialogic relationship between the present and the anticipated future might well be established, and spur conscientious human actions to socioecologically renew Earth even prior to space voyages and Contact. This will, in turn, generate an ongoing, positive cycle of interactive socioecological responsibility—between people living on Earth

---

22. Gillis, "Hottest Year Ever"; Borenstein, "Heat Record."

and people living in distant places, between present place and anticipated future place.

The prospect of engaging ETI, and concern about conserving the environments and socioecological relationships that humans encounter on other worlds, should have, then, immediate and long-term benefits on Earth and be well worth human consideration for now as well as for the future. As people consider how they should act elsewhere, they might realize that their ideas about respecting celestial bodies and other living beings are not being put into practice here. The new consciousness that would provide a base for action in space–a needed corrective, as seen earlier, for Hawking's idealistic escapism–could serve humankind well now, by stimulating people to care for their Earth home and habitat. Considerations of extraterrestrial places and life, in that case, will not be just speculations about the future, but practical proposals for projects in the present. Humankind will make Earth a better place when pondering its responsibilities in space. In all places, people will understand that God "is not far from each one of us," because in God we terrestrial humans and extraterrestrial intelligent beings "live and move and have our being" (Acts 17:27–28).

Cosmic Commons does not provide a "blueprint" or a "road map" for "getting from here to there," a means for conversion from current deleterious human consciousness and conduct to a people-friendly, Earth-friendly, biota-friendly, and cosmos-friendly mode of future thinking and acting. It presents a summary of historical developments (including in progressive thinking and ethics) that brought us "from there to here," going from the past to the present, and provides thereby a foundation for present and future historical projects to go "from here to there." It presents an idealistic vision, a future *utopia* understood not as an impossible dream but as a realistic possibility that can be realized over time—if needed but difficult changes are made in Earth economic structures, ecological practices, individualistic ethical ideologies, and exclusivist ecclesial doctrines.

Science, Spirit, and Space

In recent years, beginning toward the end of the twentieth century, ecology has come to serve as a bridge between religion and science. The term "natural environment" used in scientific circles came to be understood as a complementary expression of "God's creation" in Christian (and other religions') thought. Similarly, one might hope, collaborative science-religion ideas, insights, and interaction can result in cooperative or at least complementary development of considered approaches to journeys in the cosmic commons, and to modes of terrestrial-extraterrestrial intelligent life Contact.

During the research for and writing of *Cosmic Commons* I understood the complexity and interrelationship of diverse issues that should be considered and knew I could not explore the breadth and depth of discussion that they required—in themselves and in relation to others. I recognize that readers might think, "Why did he not consider this, or incorporate the ideas of these scholars?" I leave it to those with considerably more expertise in scientific, philosophical, theological, biblical, economic, and socioecological fields of inquiry than I to develop more fully what I have presented in seminal form.

I developed *Cosmic Commons* in context, in my present time and historical place as a socioecological ethics teaching scholar and a human rights activist advocating for those whom Jesus called "the least of these," and for all biota, whom Indian people name "all my relations." I am a spiritual man who has experienced the Presence of Spirit Immanent-Transcendent, a scholar who explores intellectually and reflectively the deeper meaning of what I see, read, and experience, and a social justice thinker who relates theory and practice in praxis: in context, on the ground, with others who envision and are dedicated to realizing communities' utopias embodying justice, harmony, and peace.

I hope that others will find the original ideas I have presented to be stimulating and worthy of consideration; perhaps the concepts will prompt creative thought and concrete action on their part. The intent of *Cosmic Commons* is to provoke discussion, not to "prove" the case. I continue exploration of these issues in a complementary manuscript: *Encountering ETI: Aliens in Avatar and the Americas*.

The cosmic commons awaits humanity's material, ethical, and spiritual journey, as humankind extends its contexts and expands its consciousness. When people explore the integral being of the cosmos in all its complexity and all its perceivable (to the mind or to the physical senses) dimensions, they might well find common ground with other intelligent life on material, social, intellectual, and spiritual levels. Should that happen, in the cosmic commons Contact could catalyze cosmic community.

PART I
TERRA FIRMA

1

Terrestrial Trauma

Economic Roots of Earth's Socioecological Crisis

In landscape paintings through the ages and in diverse global cultures, Earth is portrayed in idyllic scenes. Flora of exceptional and colorful beauty, lushly abundant, grace bucolic settings. Fauna browse peacefully or predate powerfully and gracefully on prairie and mountain and among forest trees. Earth's air is clear, except for sufficient clouds to provide flowing depth, contrast, and texture. A beaming sun is strategically placed to highlight skies above, streams below, and rich soil on which flora, fauna, and forests flourish. Serene communities are nestled in and integrated with their local environment, on mountainsides or lowlands. Earth's waters flow pure and free, whether cascading among cliffs or freely flowing through forest glens. All is serene. Usually, as Wendell Berry, author of *The Unsettling of America: Culture and Agriculture*, observes in a perceptive insight, "landscapes are viewed as more beautiful when no people are present in them; there's no sense that people are part of nature, and [that] a farm well-integrated with Earth's rhythms and natural goods is a worthy part of a 'landscape' painting."[1] Even when the painting's title does not include the word, such art evocatively suggests Eden, a "paradise lost."

Edenic "memories" persist, to a certain extent, even in the contemporary era, where knowledge of stellar dynamics and biotic evolution is well

---

1. Wendell Berry is a creative social thinker, community and ecology visionary, and best-selling author of numerous books, essays, novels, and poetry. He has been a University of Kentucky professor, a writer-in-residence at Stanford University, and an international lecturer; he is a fifth-generation Kentucky farmer. The quote is in my notes from years ago, with no source cited. I called Wendell, an old friend, seeking the source. He replied that it is probably from an environmental organization newsletter or other publication.

accepted not only by secular humanists, but also by members of mainstream faith traditions. In the present era (and in times to come) biblical criticism (analysis) teaches that the "Story of the Garden" or "Story of the Fall" is more a statement about human failings, and a projection of hope for a better human (and planetary) future, than a historical narrative recalling actual past events. In this perspective, "Adam" and "Eve" are viewed as "rising apes" rather than "fallen angels": humankind has evolved from primate ancestors, and did not originate as a couple created specially to live serenely in paradise. (The Bible is neither a history book nor a science text; it is a religious teachings text. Some teachings are related through poetry, others through stories bearing a deeper meaning than might at first seem apparent.)

In the past two centuries scenes such as those described above have continue to be portrayed in distinct places—perhaps particularly, in some cases, not just to celebrate nature's beauty and bounty but, as rural and urban landscapes changed, to present a "what things were like before humans messed them up" view. This would have an implied intention to promote care for the rare parts of the Earth garden where such scenes actually could still be viewed, or to restore other places to a more "pristine" condition (which can never be entirely "pristine," given how emissions-laden air and effluents-laden waters circle Earth constantly, and will do so for centuries or millennia to come, a current reality portrayed in subsequent landscapes, described below). In both instances and intents, a specific hope and vision permeates preservationist and conservationist views and socioecological[2] plans and projects: that we humans as a species can get beyond current socioecological crises and continue to "progress" ethically, economically, ecologically, and, for Christian believers in particular, ecclesially.

Over the past century, in order to present and catalyze consideration of humans' devastation of land, air, and water, art has more frequently expressed concern over negative human impacts on nature—to portray the current state of affairs on Earth, to counter romanticism with realism, and

---

2. The concept and term *socioecology* (and its cognate *socioecological*) integrates culture and context: human social milieus in particular places and times and historically through time and through and across cultures; Earth locales identified as bioregions or ecosystems evolving co-relationally: in them, humans relate to each other, to other biota, and to Earth (as themselves and as members of the biotic community, the community of all life, as a whole), independently of and interdependently with each other in specific environments or across environments. Ecology studies relationships; environment is a local place, or a planetary place as an integrated whole comprised of all local places, and thereby the context of all ecological relationships. "Socioecological ethics" advocates responsibility for, commitment to, and efforts toward achieving social justice interrelated with environmental well-being. I coined the concept and initiated its discussion in essays, public presentations, and class lectures and conversations a decade or so ago. It is elaborated more fully in chapter 6.

thereby to balance the bucolic nineteenth-century paintings with sobering art in which scenes are not as lovely. Such art (including photography), in contrast to earlier work, does have humanity very much "in the picture." However, rather than portray the kind of human integration with nature that Berry envisioned, characterized in part by family agriculture, the art displays human attempts to conquer nature, to forcibly adapt nature (Earth and biota) to human wants rather than adapt humans to nature in interrelated interdependence. Berry himself offers contrasts in *The Unsettling of America*. He describes "nurturer" and "exploiter" attitudes and acts, a characterization that aptly applies today not only in agriculture, but in other fields of human endeavor. He contrasts *agriculture*—owner-operated farms whose families' agricultural operations and local, community-oriented efforts are integrated well with the particularity of Earth's places and rhythms—with *agribusiness*—corporate controlled, owned from afar, and manager operated industrial enterprises that coerce and exploit Earth for company profits, no matter the impacts on local places and local communities.[3]

An ever-increasing awareness has emerged that humankind has dramatically been altering not only the biotic evolution and even the very existence of other species, but also the abiotic context, the nonliving Earth environment, in which we and all biota "live and move and have our being" (Acts 17:28). Earth—as an experiential, material setting and dimension of what is expressed by Paul's words as he cites divine immanence—is undergoing a dramatic transformation, catalyzed not least by human conduct exhibited presently most evidently in global climate change. As will be discussed later, the change in ecology is intimately interconnected with historically developed and developing economic systems and ecclesial doctrines, and the ethics that emerged from, with, or against them.

Earth Environments Endangered

News media are replete with stories about ways in which Earth has been experiencing dramatic geologic and climatic events in the twenty-first century.

---

3. W. Berry, *Unsettling*, 7–8. I noted and wrote about the agriculture versus agribusiness contrast in an article titled "Crisis on the Land" for *Christianity and Crisis*; it was reprinted in the *Utne Reader*. I was then director of the Heartland Project, a twelve-state effort of the Heartland Regional Catholic Bishops that was begun to save owner-operated agriculture. It was expanded to include, in the bishops' 1980 pastoral letter *Strangers and Guests: Toward Community in the Heartland*, mining and forestry practices, land ownership issues, and Indian spirituality and treaty rights. See Heartland Regional Catholic Bishops Conference, *Strangers and Guests*; an analysis and elaboration of the statement is found in Hart, *Environmental Theology*, 40–43. I was editor and principal writer of the bishops' document.

In recent years, sporadic regional earthquakes toppled communities in China and elsewhere; an entire nation was impacted by the Haiti earthquake. Hurricanes and tornadoes are increasing in number and force, drought and floods alternately plague agricultural areas, forests are infested by voracious insects and fierce fires, and tsunamis spawned by earthquakes ravage coastal areas and have devastating impacts on human life, especially when destroying or damaging human structures and causing far-reaching impacts that can ruin vast areas. This was the case in Japan in 2010 when the Fukushima nuclear plant failed: radiation irradiated local soil, spread by air and in water from its locale, and impacted negatively not only the health and welfare of Japanese people, but farms and food supplies, energy generation, and industrial production; irradiated debris and nonnative organisms eventually made landfall on the West Coast of the United States, posing potential health risks to local citizens and endangering native biota.

There is a majority and still growing scientific and political consensus that some of these major irruptions, enhancements, and extensions of natural forces are being caused or at least exacerbated by human conduct. Scientists from the Intergovernmental Panel on Climate Change (IPCC), the Union of Concerned Scientists (UCS), and the US National Oceanographic and Atmospheric Agency (NOAA), as well as in research universities around the world, agree that climate change is occurring now. Global warming on an alarming scale is its most obvious impact currently, and will continue to be so in the foreseeable future. Planet Earth is imperiled by ecological catastrophes brought on by past and current human consciousness and conduct, which includes a lack of conscientious commitment to creation care, and a lack of conscience-catalyzed compassion for community, particularly for its most economically, nutritionally, and medically vulnerable members.

Earth in Socioecological Crisis

Earth's present and ongoing "ecological crisis" is simultaneously Earth's "economic crisis," and more: Earth's people are experiencing a *socioecological* crisis. It is a crisis experienced unequally by members of diverse social classes. Scientific studies, UN documents, and religious faith traditions' statements all assert that those must vulnerable to and affected by global climate change are the poor, who experience a disproportionate share of harmful ecological impacts, including from droughts which scorch family farms, and floods which inundate family homes, communities, and even entire populated islands. The poor already have a disproportionate, inequitable share of Earth's natural goods and manufactured goods, and often of monetary resources needed to have even minimal sustenance; their socioecological injustices are being further exacerbated by global heating. The historical roots of Earth's socioecological crisis must be analyzed in order to begin taking steps to mitigate and eliminate it.

"Celestial" Causes of Climate and Community Crises

Representative thinkers from diverse historical periods and intellectual traditions have provided insights regarding the historical roots of Earth's socioecological crisis. In distinct but complementary ways, and sometimes indirectly and unintentionally, they provide warnings about where humanity is—ecologically, economically, ethically and, for some, ecclesially—and where humanity should aspire to be—or suffer drastic consequences. Their ideas focus sometimes on community, and sometimes on community's context: creation.

The work of thoughtful and perceptive scholars suggests that humankind, in order to begin progressing toward the desired future in which Earth and humans' well-being will be restored, should work as individuals and communities, and communities of nations, to undo or alleviate ecological and social devastation resulting from or exacerbated by human ideologies and actions. Expressed another way, people should not depend on divinities such as a *deus ex machina* ("god of the machine": trust in technological innovations), a *deus ex caelum* ("God in heaven": trust in divine intervention), or a *deus ex Wall Street* ("god of the market": trust in receiving profits from wise financial investment in the capitalized labor of others). The latter false faiths in particular types of nonsacral "divine" being provide bases upon which scientists and technologists, corporations, religious leaders, and politicians can build belief systems associated with their particular deity. In so doing, they attempt to justify existing socioecologically harmful technologies and economic systems and practices, and are assisted by assertive political parties and their policies and posturing. The religion of the market complements and often supersedes traditional religions as a factor that contributes to socioecological degradation, and imperils societal well-being. Trust in any of these divinities (the last of which is especially worshipped by believers who are already or aspire to be wealthy) has had consistently adverse social, economic, and ecological impacts.

Socioecological devastation since the early twentieth century has resulted particularly from worship of the *deus ex Wall Street*. The God of the Market was uncovered, described, and analyzed by Harvard theologian Harvey Cox in his 1999 essay "The Market as God—Living in the New

Dispensation."[4] Cox observes that in reading the business pages he discovered, embedded within them,

> a grand narrative about the inner meaning of human history, why things had gone wrong, and how to put them right. Theologians call these myths of origin, myths of the fall, and doctrines of sin and redemption. But here they were again, and in only thin disguise: chronicles about the creation of wealth, the seductive temptations of statism, captivity to faceless economic cycles, and, ultimately, salvation through the advent of free markets.[5]

Cox goes on to describe how The Market has become the Supreme Deity, the "only true God," who is manifested in bear and bull theophanies, tolerates no rival gods, and must be universally worshipped. In phrasing particularly apt for socioecological considerations, he notes that under The Market's rule, "creation becomes a commodity." While Psalm 24 teaches that "the earth is the Lord's and the fullness thereof, the world and all that dwell therein," in the Market religion's dogma "human beings, more particularly those with money, own anything they buy." In an era when different religious conflicts occur throughout the world, Cox notes a clash of religions and civilizations that is not recognized, in which The Market is a formidable foe for older religions.

The *deus ex Wall Street* continues to extend its power and influence today, despite market crashes or "adjustments" and even during the "Great Recession." Its devotees are instructed to have faith that it will "correct itself" and bring prosperity in the future. It resembles the *deus ex machina* in that it depends on human constructs—economic theory and practices—that parallel scientific theories and the technologies derived from them. It resembles the *deus ex caelum* in that great faith is required to believe in these divinities' existence and power, since they are not visible to human sight.

Given the current state of world economies, and current stresses on world ecologies, people throughout the world might grasp at and hope in deities that serve as "opiates of the people," distracting humanity from focusing on the causes and effects of present material problems and diverting people to hope instead for some savior to change Earth and human conduct. Or, people might hope for a better life in a heavenly realm, or for a divine rescue from what they have done on Earth—in the form of a *deus ex caelum* who will miraculously restore Earth or rapture them up into heaven—because

---

4. Cox, "Market as God." Cox presents a sardonic and insightful critical analysis of the "religion" of the stock market, including its faith, divinities (bear and bull), priesthood, doctrines, and eschatological expectations.

5. Ibid., par. 2.

they are in denial about what their responsibility might be for what has happened to Earth, or are in despair about their powerlessness to change things. Concurrently, Stephen Hawking's advice to build a moon base and a Mars colony promotes belief in establishment of a refuge from Earth's ills by escape via a *deus ex machina*—a space vehicle—to a "heavenly" realm in the sky. He envisions not a future life in an otherworldly spiritual place, but a near-future present life in a distant material home where socioecological conditions will be better than on endangered Earth (without elaborating how or why humanity's new creation will be effected, given that the same humankind that has caused devastation on Earth will journey on his proposed salvific arks to the moon and Mars).

The most critical injustices of the socioecological crisis—ecological and economic—merit specific, in-depth analysis of their roots and present status, in order to strive to resolve them in the future.

Historical Roots and Resolution of the Ecological Crisis

In a much-quoted article in the journal *Science* in 1967, "The Historical Roots of Our Ecological Crisis," religious historian Lynn White Jr. declared that Christianity utilized texts from the biblical book of Genesis to justify and promote environmental degradation and New Testament passages that advocated caring about getting into heaven to avoid responsibility for caring about and for Earth. Church and Bible were, effectively, the "roots" of environmental degradation. Christians were so focused on attaining a spiritual life in heaven in the future that they did not act responsibly on Earth, the context of their material life in the present. White's analysis rightly criticized Christians for not taking better care of Earth, and wrongly blamed Christianity as the major cause of the global ecological crises. He did stimulate, however, significant Christian assessment—by churches and individuals—of their attitudes and actions on Earth, ecologically speaking.

Consequently, strong ecological statements and environmental caretaking projects emanated from churches, nationally and globally—something White advocated. Complementarily, Pope John Paul II named St. Francis of Assisi the "patron saint of ecology," a suggestion White made in his essay. In "The Ecological Crisis: A Common Responsibility," his 1990 World Day of Peace message, Pope John Paul II declared that "Christians, in particular, realize that their responsibility within creation and their duty towards nature and the Creator are an *essential* part of their faith."[6] "Essential" means something integral to, necessary and required for, and expected of the practice of faith, not an optional practice that might or might not be part of Christian

---

6. John Paul II, "Ecological Crisis," pars. 15, 13 (my italics).

life. Duties toward creation and Creator must be observed by all those who call themselves "Christians," by definition and without exception.

Yet, despite the flurry of church pronouncements about and involvement in efforts to promote Earth's well-being, the ecological crisis worsened.

What White did not discuss or even note in his assessment was that politics, economics, and science have had a far greater impact on promoting an ecological crisis and on whether or not ecological issues are even addressed when the crisis is rampant. *Contra* White, might Christianity have been co-opted by science and political ideologies as the latter caused human estrangement from Earth and ecological destruction of Earth? In that case, Christianity would have been a conduit and even an affirmation, rather than the cause, of scientific and political conduct.

In the United States currently, transnational corporation–funded denials of climate change (financed especially by the energy corporations and organizations they support), dutifully parroted by pandering politicians, have deftly shifted public attention away from considering, let alone condemning, human-caused contributions to global climate change. This shift has been effected despite record droughts, fires, frequent major storms, and diminishing water supplies globally; record ice melts in Greenland, the North Pole and the Arctic region generally; melting of ice in, and generation of massive icebergs from, Antarctica; and the continually melting, and soon to disappear, remaining glaciers in Glacier National Park, Montana, in the United States. Concern about global warming became endangered and near-extincted by extraordinary efforts to deny the existence of climate change. While churches have been deploring and fighting against creation crises and catastrophes, often science, technology, economics, and politics have pressed on with "business as usual."

White's original thesis about "historical roots" needs to be supplemented in depth, if not supplanted, by consideration of the economic, financial, scientific, technological, and political factors that preceded, prompted, and partnered with biblical and Christian doctrines—those that White cited and those he omitted.

Enlightenment?

The "Age of Enlightenment" began in the eighteenth century. It was an historical era in which European philosophers and scientists expanded Western peoples' belief in their superiority to other peoples and cultures, and dug an intellectual chasm between humankind and Earth, and between humankind and other biota. Cultural theorists and ruling European monarchs continued to develop and aggressively assert that European culture was superior:

because of its intellectual creativity in the fields just cited, its social arrangements centered on property ownership and control, and its religious ideology—it had the only "true" religion—and practices (these will be considered later in discussion of the Europe-generated and -disseminated Discovery Doctrine).

During the Enlightenment, in Western thought Earth came to be viewed primarily in mechanistic terms. Humans were regarded as the central, superior beings on a planet that was still regarded by some, despite Galileo, as the center of the universe—at least in terms of importance and of divine selection and approbation. Earth was to be altered and exploited at will to meet human needs and satisfy human wants. While scientists of the period declared that they were "thinking God's thoughts after Him," they did not include in this thought God's affirmation of the goodness of Earth and all creatures, not just humankind, a perspective expressed in the Christian (the dominant religion in Europe) Bible: from Genesis, the first book, through Revelation, the last book. Consequently, the sacral sense of nature that had been understood biblically and expressed in religious thought and practice was excised from European consciousness.

Contemporary religious scholars have documented this shift and called for a return to traditional understandings of the sacredness of nature and of the divine Presence permeating nature. Seyyed Hossein Nasr, a religion scholar and science thinker with a considerable scientific background, and Alister McGrath, a molecular biophysicist, theologian, and ordained priest in the Church of England, provide complementary perspectives in their respective Muslim and Christian analyses of and disagreement with the ways that the Enlightenment separated Creator from creation and promoted human domination of creation. Each suggests that both the human community and the Earth environment need religion to reverse humans' relatively recent sacrality-denying consciousness. They consider the historical roots for, and suggest a historical resolution of, the ecological crisis.

Seyyed Hossein Nasr

Nasr declares in his now-classic *Man and Nature* that "the blight brought upon the environment is in reality an externalization of the destitution of the inner state of the soul of that humanity whose actions are responsible for the ecological crisis."[7] People are acting in ways that correspond to scientific theories and historical philosophies that objectify creation and remove humankind from its relationship to the rest of creation, rather than promote human be-ing in accord with their spiritual nature and longing, interrelated

7. Nasr, *Man and Nature*, 3.

with creation. He laments a "lack of acceptance of the spiritual dimension of the ecological crisis," caused by the "survival of a scientism which continues to present modern science not as a particular way of knowing nature, but as a complete and totalitarian philosophy which reduces all reality to the physical domain and does not wish under any condition to accept the possibility of the existence of nonscientistic world-views. . . . [A]lternative world-views drawn from traditional doctrines remain constantly aware of the inner nexus which binds physical nature to the realm of the Spirit."[8] Nasr asserts that "the pure monotheism of Islam . . . never lost sight of the sacred quality of nature as asserted by the Quran" while Western Christianity remained aloof from "the theological significance of nature and the need for its 'resacralization'" after the "gradual de-sacralization of the cosmos which took place in the West" linked to the "rationalism and humanism of the Renaissance which made possible the Scientific Revolution."[9] Nasr hopes in and calls for a "spiritual rebirth" by which humankind will "attain a new harmony with the world of nature."[10] This is especially needed, he says, because "to destroy the natural environment is . . . to fail in one's humanity. It is to commit a veritable crime against creation."[11] Nasr observes that in medieval cosmology, "man had been placed at the centre of the Universe, not as a purely terrestrial and earth-bound man but as the 'image of God'. His centrality was due not to anthropomorphic qualities but to theomorphic ones. By removing him from the centre of things, the new astronomy did not bestow upon man the transcendent dimension of his nature; rather it affirmed the loss of the theomorphic nature by virtue of which he had been placed [by God] at the centre."[12] Knowing the age and vastness of space caused not an extension of appreciation of creation, but a significant human displacement from a creation and cosmos which became overpowering, and from the hierarchical position that God had given to humankind when God placed people on Earth with a special role as God's vicegerents. (Despite his own impressive scientific background, Nasr accepts neither what science states are the age and dynamics of the cosmos, nor evolution on Earth: everything that exists originates in divine creativity rather than by cosmic dynamics or terrestrial biotic evolution, and is present in God before and until God places it into God's creation.)

Nasr notes that "with the Renaissance, European man lost the paradise of the age of faith to gain in compensation the new earth of nature and natural

8. Ibid., 4.
9. Ibid., 5–6.
10. Ibid., 9.
11. Ibid., 9–10.
12. Ibid., 68.

forms to which he now turned his attention. Yet it was a nature which came to be less and less a reflection of a celestial reality. Renaissance man . . . became wholly man, but now a totally earth-bound creature. He gained his liberty at the expense of losing the freedom to transcend his terrestrial limitations."[13] Nasr laments the "transformation of cosmology into cosmography, a movement from content to form. . . . All that is left is the body without its inner spirit and meaning."[14] It is essential to understand symbolism in nature—all nature points beyond itself to God: "[T]o understand fully the meaning of symbolism, of the symbolic meaning of forms, colours and shapes, of all that surrounds us, is a way to see God everywhere. It is thus a way of making all things sacred."[15] Nasr does not think it is possible to see and accept Earth and cosmos as sacred while holding on to contemporary scientific beliefs.[16]

Nasr returns to these themes in *Religion and the Order of Nature*. In the book's opening sentences he declares: "The Earth is bleeding from wounds inflicted upon it by a humanity no longer in harmony with Heaven and therefore in constant strife with the terrestrial environment. The world of nature is being desecrated and destroyed in an unprecedented manner. . . . The environmental crisis now encompasses the entire Earth."[17] Earth is threatened by a chaos unlike any before: all human life faces destruction. In this crisis situation: "To preserve the sanctity of life requires remembering once again the sacred quality of nature. It means the resacralization of nature, not in the sense of bestowing sacredness upon nature, which is beyond the power of man, but of lifting aside the veils of ignorance and pride that have hidden the sacredness of nature from the view of a whole segment of humanity."[18] He discusses how new understandings of the solar system and the universe caused loss of a sense of place for humans:

---

13. Ibid., 64.

14. Ibid., 66.

15. Ibid., 131.

16. In this regard, Nasr's thought ironically complements the idea expressed by some atheist scientists—that no one can be both a scientist and have a religious faith: a scientist who worships must "leave his brain outside the church doors." Nasr's rendition would be, conversely: no one can have a religious faith and be a scientist: a believer who does scientific research on evolution must "leave his faith outside the laboratory doors." The absoluteness of these assertions—beliefs?—impedes efforts to bridge the distinctions between scientific and religious approaches in order to find common ground on resolving shared concerns; and both perspectives contradict the reality in which some scientists are participants in the life, teaching, rituals, and ministry of their specific religious traditions—including those scientists who are ordained clergy in their faith tradition.

17. Nasr, *Religion*, 3.

18. Ibid., 7.

The advent of Copernicus marks the beginning of the destruction of the traditional idea of cosmic order, which was to culminate in the Newtonian vision of the world as a machine. With the destruction of the Ptolemaic model the aspect of order as hierarchy was destroyed, leading to the loss of man's "home" and sense of "place" in the vast Universe that surrounded him. The Copernican Revolution implied not only changing the center of the cosmos from Earth to the Sun or destroying the significance of the role of man at the center of the world, but it denied finally that the cosmos had any center at all.[19]

Human dis-placement became so complete, in Nasr's view, that the cosmic center itself became nonexistent. The cosmos now has no center—at least in human consciousness when it accepts the new cosmology.[20]

Nasr declares that "if evolutionism is taken seriously any species is simply an element in the flowing river of process with no ultimate value whatsoever. It is difficult in fact to defend the rights of creatures to life, if one accepts the prevailing evolutionist view, save by appealing either to sentimentality or biological expediency, neither of which are theologically pertinent." Nasr continues this assertion by declaring that "many current eco-theologians, as well as philosophers of ecology, remain *rabid* followers of Teilhard de Chardin, who sought to create a religion out of evolution."[21]

---

19. Ibid., 133–34.

20. In contrast to Nasr on the last point, it might be stated that the "center" of the cosmos need not be a cosmogeographical location: it can be the Presence of the Spirit that enables the center to be everywhere and nowhere, lacking local specificity. The Presence in the cosmos of the Being-Becoming Creating Spirit means the Spirit is its center from every local place, as the center of a sphere is connected to every point within it, to form together a holistic and whole form. Similarly, from cosmic beginnings: the Singularity from which all began—priest-scientist Georges Lemaîtres' "Day Without Yesterday" and "Primeval Atom"—was initiated by and occurred in divine Presence. The cosmos is presenced by the Spirit, who is its creative, ordering Center, and who continues to be that Center in an inflationary universe (or multiverses); there is no need for intelligent species in distinct regions of space to localize the cosmic center in the spatial place they inhabit, although this might be the philosophical perspective of numerous intelligent species in diverse and distinct areas of the cosmos. Many different suns in different areas of space might be viewed, as each species determines its own temporal and spatial center, as the center of the universe. Many suns might have emerged through the creative processes of celestial dynamics to be local centers of and central to solar systems and even galaxies, but there can be only one cosmic center. Spirit is the center of the universe(s), from whatever part of the cosmos this is pondered, and in the integral being of all interrelated cosmic spaces.

21. Nasr, *Religion*, 216 (my italics). While the priest-scientist-mystic Teilhard de Chardin did in fact propose the integration of religion and evolution (for which the Catholic Church hierarchy and the Jesuits exiled him from his native France for most of his life), he did not "create a religion" out of his thoughts and theories. Rather, on the

In the final chapter, "Religion and the Resacralization of Nature," Nasr elaborates consciousness-altering steps that must be taken in order for humanity to dramatically diverge from, and form a diametrically different position from which to dismiss, the operative "scientism" of the West. He has brought the reader "to this point of affirmation of the sacred quality of nature, now forgotten and in need of reassertion. Nature needs to be resacralized."[22] Touching on the economic causes of current environmental degradation and destruction, Nasr notes and rejects the "illusory satisfaction of a never-ending greed without which consumer society would not exist."[23] Nasr observes that the Quran teaches that "all creatures in fact share in man's prayers, and praise God, for it states, 'The seven Heavens and Earth and all beings therein celebrate His praise, and there is not a thing but hymneth His praise'" (XVII: 44). People's prayers "form parts of the chorus of the praise of God by the whole of Creation and a melody in the harmony of 'voices' celebrating the Divine, a celebration which on the deepest level is the very substance of all beings.... [T]o destroy nature and cause the extinction of plants and animals as a result of human ignorance is to murder God's worshippers and silence the voice of the prayer of creatures to the Divine Throne."[24]

Seyyed Hossein Nasr, then, describes historical roots in Western thought that have led to the ecological crisis, and suggests that nature, through the assistance of religion, needs to be resacralized in human thought. This would help to conserve nature as a whole and in its component parts, each and all of which—not humanity alone—praise God.

Nasr's concluding thoughts provide an excellent segue to the complementary thinking, on several points, of scientist-theologian-Anglican priest Alister McGrath.

Alister McGrath

In *The Reenchantment of Nature*, Alister McGrath analyzes the current ecological crisis and its roots in Western history from the perspective of a scientist and Christian, and discusses the change in consciousness needed

---

contrary, he sought to find extended scientific and cosmic significance—in time and in space—for belief in God and for consideration that the universe was destined for and evolving toward God at an Omega Point. At the Omega Point, evolution would culminate in integration with a cosmic Christ. Teilhard hoped to strengthen *Christian* faith (specifically, Catholic faith), not provide an alternative faith.

22. Ibid., 270–71.

23. Ibid., 271.

24. Ibid., 281. Quranic texts and Nasr's thought complement very well, in the Bible of Jews and Christians, the verses of Psalm 148, which sings about all creatures praising God: animate—in all biota's complexity and diversity—and inanimate.

to effect a change in conduct. In a chapter titled "A Manifesto to Exploit: The Enlightenment and the Master Race," McGrath, *contra* Lynn White Jr.'s assertion that Christianity is "the most anthropocentric religion" in human history, states that "the most self-centered religion in history is the secular creed of twentieth-century Western culture, whose roots lie in the Enlightenment of the eighteenth century and whose foundation belief is that humanity is the arbiter of all ideas and values."[25] The Enlightenment has contributed substantially and negatively, in the views of both Nasr and McGrath, to twenty-first-century Western ecological consciousness and conduct. It has aided and abetted the human irresponsibility which has led to the current socioecological crisis on Earth. McGrath comments further on the Enlightenment: "The roots of our ecological crisis lie in the rise of a self-centered view of reality that has come into possession of the hardware it needs to achieve its goals."[26] Further, "The 'disenchantment of nature' is one of the most worrying cultural developments of our time. What once evoked a sense of awe from appreciative and respectful human beings has now been explained away, deconstructed and desacralized."[27] McGrath suggests that "a right attitude to nature rests on the revival of our capacity for wonder, resting on our appreciation of the nature of reality itself. If nature has become *disenchanted*, the remedy lies in its *reenchantment*."[28]

McGrath declares that "to study nature is to catch glimpses of the divine in the ordinary."[29] If humankind disrespected nature, and sought solely to dominate it to satisfy human wants, it would not be possible to have these "glimpses" in the consequent degraded nature. Technology came to mean that "Humanity could conquer the nature that had once held it captive. Humanity had triumphed over both God and nature, proclaiming its mastery of the former by declining to believe in him, and over the latter by forcing it to serve humanity's ends. The Enlightenment was wedded to the idea of human supremacy over all its rivals and set itself the goal of defeating nature and proclaiming the death of God."[30] However, the Enlightenment backfired: "The Enlightenment aimed to achieve human liberation by domination of nature. It has ended up by enslaving people to a dying earth and offering them no alternative home."[31]

25. McGrath, *Reenchantment*, 54.
26. Ibid.
27. Ibid., xi.
28. Ibid., xii.
29. Ibid., 20.
30. Ibid., 84.
31. Ibid., 93–94.

In his concluding thoughts, McGrath laments that "We have lost touch with the world of nature and have constructed our own worlds in its place."[32] The theme of his book, he says, is that "we reclaim the idea of nature as God's creation and act accordingly, bringing attitudes and actions in line with beliefs." In addition, "we must see nature as a continual reminder and symbol of a future renewed creation."[33] In terms of intergenerational responsibility, he states that "we must remember that others who are yet to come also need to glimpse these flashes of glory in everyday things, and preserve [the 'present beauty of nature'] for their sakes."[34] McGrath concludes: "To reenchant nature is not merely to gain a new respect for its integrity and well-being; it is to throw open the doors to a deeper level of existence."[35]

Seyyed Hossein Nasr and Alister McGrath fault the Enlightenment as the principal cause for the current ecological crisis. They celebrate nature's beauty, the diversity of creatures to whom we are related as God's handiwork, and the Presence of the Spirit who permeates all. They call for a religious recovery of traditional attitudes toward and actions upon Earth, so that nature might be "resacralized" (Nasr) and "reenchanted" (McGrath). Then, creation would be respected and renewed.

Historical Roots and Resolution of the Economic Crisis

A community crisis parallels the ecological crisis. Globally, the gap between the rich and the poor increases. Over the centuries, gradually developing economic assumptions and even "doctrines" have been developed that justified the hegemony of a particular class of people over all other peoples.

Key thinkers in the Christian tradition such as Thomas Aquinas (1225–1274) stated that humans were "social beings" who joined together in communities for mutual benefit and support. Their earthly home was guided by "natural law" originating from God, which stated that all property was common, not individually owned, and ultimately "owned" only by God. Property might become private in civil law, but because of the priority of "natural law" over human legal constructs people might, in times of great deprivation, take from this common property the goods they need to provide for their basic sustenance—even if the goods had become private property through civil laws. In his discussion of the seventh commandment, "You shall not steal," Aquinas states in Question 66, titled "Whether It Is Lawful to Steal Through Stress of Need": "In cases of need all things are common property, so that

32. Ibid., 183.
33. Ibid., 184.
34. Ibid., 185.
35. Ibid., 186.

there would seem to be no sin in taking another's property, for need has made it common. . . . [I]t is not theft, properly speaking, to take secretly and use another's property in a case of extreme need: because that which he takes for the support of his life becomes his own property by reason of that need. . . . In a case of a like need a man may also take secretly another's property in order to succor his neighbor in need."[36]

People as social beings in community tried to develop laws that were supposed to protect their common and individual interests, particularly in areas of political power and economics, through the responsible conduct of conscientious political rulers. The latter leaders were supposed to uphold just laws. But over time even "Christian" nations, because of the emergence of particular economic, political, and religious interests that came to exercise power, departed from and even came to contradict Aquinas's core social teachings.

In the sixteenth century, the theology of Protestant Reformer John Calvin (1509–64) emphasized divine predestination of humans' location (heaven or hell) in the afterlife, and proposed that individual salvation was possibly indicated by a person's faith and their adherence to Scripture. His ideas were fostered even more by the teachings of his followers who emphasized that "signs" (not proofs) of possibly being predestined by God for heaven included hard work, thrift, and material prosperity—God was blessing them in this life and the next. These theological ideas that focused on individual salvation and "signs" thereof led to the disparagement of the poor, who were obviously not being blessed by God.[37] This dramatic shift in Christian attitudes occurred, ironically, despite extensive biblical teachings, including from the Hebrew prophets and Jesus, which exhorted people of faith to have compassion for and take solicitous care of the poor; and, the lifestyle of Jesus and his apostles, and of the Christian community described in Acts who "shared all things in common."

Other historical factors that disrupted human community after the time of Aquinas included the imperialistic political development and global implementation of the Discovery Doctrine, and the ideological development and social implementation of capitalism. Both were born in Europe and extended abroad forcibly because European nations had more advanced technologies that enabled them to dominate other peoples, and an extreme sense of ethnocultural and religious superiority. (Discovery will be described and discussed extensively in chapter 2. Discussion of capitalism follows below.)

36. *Summa Theologica*, II-II, Q. 66, art. 7, 552.

37. British economist and social thinker R. H. Tawney provides an in-depth study and extensive analysis of the impacts of Calvin and his followers in *Religion and the Rise of Capitalism*.

The Europe-based economic perspectives and ideologies of the sixteenth through eighteenth centuries that complemented or supplemented Discovery ideas have had a continuing negative influence on Western ideologies and cultures. Although they merit extensive consideration, this has been done by others in numerous books and articles; a brief summary of altered economic views in Britain, which extended into the British colonies, will suffice here to indicate ideas that have reinforced Discovery ideology and its implementation, and contrasting perspectives.

In the eighteenth century, in particular, significant new economic perspectives were developed, and old ones amended. These intellectual and ideological changes were expressed in particular by era representatives Adam Smith (1723–90), John Wesley (1703–91), and Thomas Paine (1737–1809), whose lives overlapped. They offered contrasting but at times complementary approaches to economic issues: Smith as a university professor in Scotland, Wesley as an Oxford-educated priest of the Church of England who founded the Methodist Church, and Thomas Paine as a fiery rhetorician, key thinker, and social activist for British colonies seeking independence from England.

Adam Smith: Capitalism

In the eighteenth century, Adam Smith began to teach and write about a new approach to economics. He became the theoretical founder, forerunner, and apologist for the future of the capitalist system and its social impacts. He claimed that despite social and economic inequalities that would be produced by this economic construct, the poor would not lack necessities of life. Smith proposed, in effect, and without naming it, the original "trickle down economics"—the excess wealth accumulated by the rich will be provided to the poor by an "invisible hand" that would ensure their socioeconomic survival, and not by a redistribution of wealth from the affluent few to the impoverished many. The rich resist losing any of this excess, and want to *increase* it by *decreasing* availability of needed goods for the middle and lower classes.[38]

Adam Smith first presented the principal ideas in *The Wealth of Nations* as a series of public lectures. In the book, in his discussions of international commerce, he describes an "invisible hand," mobilized unknowingly by private entrepreneurs, that will effect public good by promoting, unintentionally, social well-being. The passage in which he posits an invisible hand, much (ab)used, bears citing at length:

---

38. A twenty-first-century variation on Smith and "trickle-down" theory is that taxes for the wealthy should be cut because when the rich become richer they use their wealth in such a way that jobs are created and unemployment declines.

> As every individual, therefore, endeavours as much as he can both to employ his capital in the support of domestic industry, and to direct that industry that its produce may be of the greatest value, every individual necessarily labours to render the annual revenue of the society as great as he can. He generally, indeed, neither intends to promote the public interest, nor knows how much he is promoting it. By preferring the support of domestic to that of foreign industry, he intends only his own security; and by directing that industry in such a manner as its produce may be of the greatest value, he intends only his own gain, and he is in this, as in many other cases, led by an *invisible hand* to promote an end which was no part of his intention. . . . By pursuing his own interest he frequently promotes that of the society more effectually than when he really intends to promote it.[39]

Smith here advocates that businesses maximize profits; in so doing they will benefit themselves, their industry, and their nation. The business owner who is pursuing profit will unknowingly provide a benefit to society at large. Business will promote "the public interest." Smith has a strong focus on individual labor and accomplishment in his work, which is understandable given his social position as an educated professor and not one of the laborers, who were generally poor and illiterate, in Scotland and the rest of Europe. He does state that the "invisible hand" *frequently* becomes operative; it is not always at work. Some economists have used Smith's phrase to argue against any government assistance to people in need, or any government efforts to establish a living wage for workers, or, today, to provide universal health care for all citizens. Some have tried to argue that the "invisible hand" is divine guidance, ignoring the fact that in Christian traditions God is solicitous of the poor, and that the Bible prompts its readers to aid the poor directly (in the Last Judgment story in Matthew 25:31–46, for example). When the marketplace is controlled by those with disposable income, it cannot be influenced by the poor who are expending their subsistence income for whatever basic nourishment, clothing, and shelter are available.[40]

Adam Smith had proposed the workings of an "invisible hand" in an earlier work, *The Theory of Moral Sentiments*, published originally in 1759, a book which brought him extensive public recognition. At the time, he was

---

39. Smith, *Wealth of Nations*, 484–85 (my italics).

40. The fact that poverty is rampant in the United States and other industrialized nations indicates well that there has not been in the past, and still is not in the present, an unconscious or divine "invisible hand" rectifying the excesses of the affluent. It indicates, too, why the Catholic Church and other religious institutions advocate compassion, and even a preferential option, for the poor.

professor of moral philosophy at Glasgow University. Smith discusses benefits accruing to the poor through the unwitting assistance of the rich:

> The produce of the soil maintains at all times nearly that number of inhabitants which it is capable of maintaining. The rich only select from the heap what is most precious and agreeable. They consume little more than the poor; and in spite of their natural selfishness and rapacity, though they mean only their own conveniency, though the sole end which they propose from the labours of all the thousands whom they employ be the gratification of their own vain and insatiable desires, they divide with the poor the produce of all their improvements. They are led by an *invisible hand* to make nearly the same distribution of the necessaries of life which would have been made had the earth been divided into equal portions among all its inhabitants. . . . When providence divided the earth among a few lordly masters, it neither forgot nor abandoned those who seemed to have been left out in the partition. These last, too, enjoy their share of all that it produces. In what constitutes the real happiness of human life, they are in no respect inferior to those who would seem so much above them. In ease of body and peace of mind, all the different ranks of life are nearly upon a level.[41]

The shortcomings of this statement are readily apparent. (It should be remembered that Smith is writing centuries ago, in a different era and locale; but it should be noted, too, that Smith's ideas are quoted today as if they were written today and apply universally.) The rich "select from the heap" as much as they can, even if it means the poor must lack necessities thereby. The rich certainly consume *much* more than the poor, not *little* more. He does speak frankly about the "natural" selfishness of the rich, and their attempt to gratify their "vain and insatiable desires." However, he draws erroneous conclusions from this, that during such activity the rich "divide with the poor" what results from their greed, doing so to the extent that the "invisible hand" provides "nearly the same distribution of the necessaries of life which would have been made had the earth been divided into equal portions." Smith claims, too, a divine establishment of class status: "providence divided the earth among a few lordly masters." His assertion is in direct violation of biblical texts, in which there is not only no division of Earth among "lordly masters," but the contrary: Yahweh gives the land to the community as a whole when they enter the place they consider to be their "Promised Land"; private property is an Israelite addition but even then, the Jubilee Year

---

41. Smith, *Theory of Moral Sentiments*, 182 (my italics).

required a periodic redistribution of land back to the poor if they had lost their inheritance in land in the years intervening since the previous Jubilee.

The result of all this, for Smith, is that the poor "enjoy their share of all that [Earth] produces" and are happy because their "ease of body and peace of mind" is about "level" with that of the wealthy. Adam Smith obviously had no real contact with the poor, otherwise he would not have made such sweeping assertions. Neither Smith nor his contemporary admirers wish for any change to the status quo that might benefit the poor and jeopardize the position, power, and purse of the affluent. Consequently, they resist government and Church efforts to enable the poor to have a more equitable share of Earth's land and natural goods.

Smith Contradicted: John Wesley and Thomas Paine

In the same century that Adam Smith provided the bases for the subsequent development of the economic system of capitalism, two contemporaries, one the founder of a new Christian movement in England and the other a fiery writer and orator of the revolution to separate thirteen American colonies from England, offered diverse counterpoints to Smith's theories and the social ideologies and practices that would follow from and embody elements of his thought: John Wesley, compassionate founder of Methodism, and Thomas Paine, passionate revolutionary rhetorician. Each envisioned a far different society than that proposed by Smith.

John Wesley: Methodism and Christian Social Responsibility

John Wesley had a religious upbringing in the Church of England; his father was an Anglican priest. He studied at Oxford and, after he graduated from Oxford (1728), was ordained an Anglican priest. Eventually, Wesley and his followers separated from the Anglican Church, and the Methodist Church came into being.

Wesley's most significant teaching on economics, social well-being, and a Christian way of life was presented in his sermon "On the Use of Money" (1744). In it, he suggested "three plain rules" to guide Christian employment, savings, and compassion: first, "Gain all you can"; second, "Save all you can"; and third, "Give all you can."

The first rule, "Gain all you can," was intended to promote among his followers gainful employment that promoted individual and social well-being. One was not, however, to earn a salary "at the expense of life, nor at the expense of our health."[42] Work must not result in harm to the

---

42. Beach and Niebuhr, *Christian Ethics*, 373.

worker's mind or body. Gaining all we can was to be done, too, "without hurting our neighbor," which follows, Wesley says, if we "love our neighbor as ourselves."[43] It was to be "honest industry"[44]: time was not to be wasted, and the worker was to be faithful to their particular calling. Workers must have a "common sense" approach to work, carefully measuring its impacts on self and neighbor.

The second rule, "Save all you can," prompted the common people who were Wesley's principal audience to spend the money they earned on necessities, not on superfluities that he called "idle expenses,"[45] nor on "pleasures of life" such as gluttony, drunkenness, and "elegant epicurism," nor for expensive apparel, furniture, books, or pictures.

The third rule, "Give all you can," declared that God placed people on Earth "not as a proprietor, but a steward": no matter the extent of what they acquired, God retained ownership of Earth, and people were to use Earth's places and goods to meet their own and others' needs.

For Wesley (as was the case with Thomas Aquinas seven centuries earlier), everything is God's, but God has permitted people to have and to use natural goods—judiciously, responsibly, and thriftily—to meet their needs; beyond that, they have a responsibility to help others, rather than acquire ever more things for themselves while others suffer from lack of life's necessities (waiting in vain for an "invisible hand" to provide relief).

The contrast between Wesley and Smith is evident in even the most superficial analysis. These British contemporaries had radically different understandings—and writings that flowed from these understandings—regarding the acquisition and distribution of money and property. Smith was concerned first and foremost about the moneyed class: not about how they had become part of that class but about how they might retain their wealth and enhance their status through ideological reinforcement and material accumulation, particularly business enterprises. Wesley was concerned first and foremost about the moneyless class: also not so much about how they became part of that class, but, by contrast, of how they might emerge from poverty as individuals, families, and communities. Smith taught that God had ordained a classed society. Wesley taught that poverty was not a divine intention but the result of social injustice perpetrated against the poor, and of the poor's lack of striving to overcome, through their industriousness and community compassion, this externally imposed social condition. Smith stated in *The Wealth of Nations* that the entrepreneur or business owner does not intend "to promote the public interest"; directs his industry "in such

43. Ibid., 374.
44. Ibid., 375.
45. Ibid., 376.

a manner as its produce may be of the greatest value"; and "intends only his own gain"—all very individualistic and self-centered business practices through which, somehow, an unintended "invisible hand" would provide for the needs of the poorer segments of society, including needed benefits which working people could not procure with the remuneration received from their employer. Wesley believes not in some *demiurge* that will succor the poor, but in a *Deity* who commands community concern and an active compassion for the "least of these," as expressed by providing the poor with money and other goods they need.

Thomas Paine: Revolution and Social Transformation

Another counterpoint to and contemporary of Smith was the American revolutionary Thomas Paine, a pillar of the US Constitution, whose statements about the requisite role of government to provide for the needs of the poor, and advocacy of property ownership as a natural right, stand in stark contrast to the views of Smith—and in supportive congruence with the teachings of Aquinas.

Thomas Paine was born and raised in England, in a working class family, and was primarily self-educated. He served in the navy and then was employed in low-paying government positions. He came to know firsthand the plight of the poor in England. With his upbringing, experiences, and participation in the American Revolution as a background, Paine published *The Rights of Man, Part Second* in 1792. In it, he declares:

> When it shall be said in any country in the world, "My poor are happy; neither ignorance nor distress is to be found among them; my jails are empty of prisoners, my streets of beggars; the aged are not in want, the taxes are not oppressive; the rational world is my friend, because I am a friend of its happiness"–when these things can be said, then may that country boast of its constitution and its government.[46]

While one may argue with Paine's apparent acceptance of a condition of poverty, as expressed in the statement that the poor are "happy," one may note also that at least if the poor *are* "happy," it is presumably because they have finally received some basic education and needed subsistence goods in order to be without "ignorance" and "distress," and without the need to be beggars on the streets.

Paine wrote the preceding ideas just sixteen years after the 1776 publication of his pamphlet *Common Sense* (a major stimulus to the American

---

46. Paine, *Rights of Man*, 446.

Revolution), the US Declaration of Independence, and Adam Smith's *The Wealth of Nations*.

In contrast to Smith, Thomas Paine did not pronounce a providential establishment of a classed society, but declared in *The Rights of Man* that "there ought to be a limit to property" and the wealth of vast estates is "a prohibitable luxury" if it exceeds what is "necessary or sufficient for the support of a family."[47] In *Agrarian Justice* (1796) Paine declared that "the earth, in its natural, uncultivated state, was, and ever would have continued to be, *the common property of the human race*; in that state, every man would have been born to property."[48] In obvious disagreement with Smith, he states that property in land should not exist in perpetuity, since "Man did not make the earth . . . neither did the Creator of the earth open a land office, from whence the first title-deeds should issue."[49] Paine advocated government intervention to eliminate inequitable property arrangements, since (and here is an obvious contrast to Smith's idea of an "invisible hand") if left with a choice regarding whether or not to provide for the needs of the poor, the rich would be unwilling to act justly: ". . . with respect to justice, it ought not to be left to the choice of detached individuals whether they will do justice or not."[50] In the meantime "the great mass of the poor in all countries are becoming an hereditary race, and it is next to impossible for them to get out of that state of themselves."[51] The rich increase their wealth, but "all accumulation" of "personal property" that does not result from "what a man's own hands produce" is derived from living in society, and the "man" owes "a part of that accumulation back again to society from whence the whole came.[52] . . . [T]he accumulation of personal property is, in many instances, the effect of paying too little for the labor that produced it; the consequence of which is that the working hand perishes in old age, and the employer abounds in affluence."[53]

Thomas Paine's thought complements the Christian teaching of John Wesley regarding compassion for the poor, the "hereditary race," despite

47. Ibid., 434.
48. Paine, *Agrarian Justice*, 613 (Paine's italics).
49. Ibid., 611.
50. Ibid., 618.
51. Ibid., 619. Here the ideas of Thomas Paine the fiery revolutionary anticipate and complement, to some extent, those of the Catholic Church today: the poor (the "hereditary race") will rise from their oppressed state if people of means exercise a "preferential option for the poor," and if laws are legislated that provide for their needs.
52. This should be a reminder to those who think that they did everything themselves to acquire vast wealth or to succeed in business, and therefore owe nothing to government, not even a proportionate, just share of tax payments that would benefit society as a whole, as society has benefitted them.
53. Paine, *Agrarian Justice*, 620.

their philosophical differences on Christianity and Christian doctrine; Paine the deist and Wesley the ordained Anglican priest both express compassion for the poor, and suggest complementary ways for the poor to rise from their immoral placement and retention in a financially subordinate socioeconomic position in class-stratified societies.

Contemporary Implications and Impacts

Over the centuries since Smith's book was published and his theories were accepted—while Paine's vision and his ideas about poverty and property were largely ignored, and Wesley's social justice advocacy had minimal societal impact on the thought and actions of most Christians, including many Methodists—the "invisible hand" often has been a clenched fist for the poor in the United States and around the world. From exploited working people to dispossessed farmers, from young single mothers to elderly men on street corners who "will work for food," the poor suffer from the impacts of "the marketplace." A major reason is that they do not have the wealth to guide the marketplace, let alone to buy into it or to influence politicians via campaign contributions. Economics, rightly called the "dismal science," is scarcely scientific, and certainly promotes the dismal plight of the poor.

What Paine wrote about principles of commerce is relevant here for economics: "It is one thing in the counting house, in the world it is another."[54] It would do economists well to emulate, in this regard, one of Paine's most famous sayings: "My country is the world, and my religion is to do good."[55] Instead, they act as if the "market" were divinely inspired or worthy of worship, an attitude described by Harvey Cox in "The Market as God," elaborated earlier. The market religion declares that people will attain "salvation through the advent of free markets." Writes Cox: "At the apex of any theological system, of course, is its doctrine of God. In the new theology this celestial pinnacle is occupied by The Market, which I capitalize to signify both the mystery that enshrouds it and the reverence it inspires in business folk."[56]

Unfortunately, devotees of this new religion confuse selfishness with self-love and self-interest. Smith, one of their prophets for profits, theorizes that when the rich act out of "self-interest" they will help the poor indirectly. It is obvious in the world around us that this is not so.

---

54. *Rights of Man*, 401.
55. Ibid., 414.
56. Cox, "Market as God," 19.

Karl Marx: Poverty and Social Progress

In the nineteenth century, Karl Marx (1818–83) provided perspectives that were counterpoints to Adam Smith on economic issues (though not entirely), and to John Wesley on religious beliefs and practices (again, not entirely). Effectively, Marx probably saw Smith as accurately presenting the capitalist perspective, while thereby inadvertently making the case for Marx's proposal that working people could not depend on capitalists' sympathy or largesse in terms of their low wages, and must effect a change in economic practices, through revolution if necessary. Simultaneously, he viewed religion, particularly Christianity which was the dominant religious faith perspective in Europe, as an ideology which distracted workers from their need to change society in their life on Earth by diverting them with belief in a future life in a different world, to which they would have access; for him, it was more important to alter this than to hope for compassion from Christian employers.

On the issue of religion, social injustice, and social change, Marx's most well-known statement (in religious circles) appears in the Introduction to his "Contribution to the Critique of Hegel's Philosophy of Right" (1843): "*Religious* suffering is at the same time an *expression* of real suffering and a *protest* against real suffering. Religion is the sigh of the oppressed creature, the sentiment of a heartless world, and the soul of soulless conditions. It is the *opium* of the people. The abolition of religion as the *illusory* happiness of men, is a demand for their *real* happiness. The call to abandon their illusions about their condition is a *call to abandon a condition which requires illusions*."[57] In this statement, Marx asserts that because people experience extreme hardship and oppression, and lack any hope of overcoming their social conditions, they seek comfort in religious beliefs and rituals, hoping that in a future, heavenly realm they will have peace and well-being.

Karl Marx's most noted statements on economic equality are expressed in the "Contribution" regarding private property, and in the "Critique of the Gotha Program" (1875) in regard to the social responsibilities and expectations of the working people.

Regarding private property, an issue prominent in the *Communist Manifesto*, Marx declares, "When the proletariat demands the *negation of private property* it only lays down as a *principle for society* what society has already made a principle for *the proletariat*, and what the *latter* already involuntarily embodies as the negative result of society."[58] Marx here states that when the workers are successful in abolishing private property, they are abolishing something which they do not have currently, since they have not the means

---

57. Marx, "Contribution," 12 (Marx's italics).
58. Ibid., 23

to acquire property because of the injustice of their social situation. Thus, after the social revolution all people will be on the same level playing field, with no one having the advantage of private property. With private property abolished, the people as a whole can develop new, communal relationships to property.

In regard to mutual socioeconomic expectations, Marx writes in "Critique of the Gotha Program":

> In a higher phase of communist society, after the enslaving subordination of the individual to the division of labour, and therewith also the antithesis between mental and physical labour, has vanished ... after the productive forces have also increased with the all-round development of the individual, and all the springs of cooperative wealth flow more abundantly—only then can the narrow horizon of bourgeois right be crossed in its entirety and society inscribe on its banner: *From each according to his ability, to each according to his needs!*[59]

In all likelihood, the vast majority of the general population throughout the world, many of whom do not have the resources to provide for themselves through their labor—particularly when they have no work, or are paid exploitative wages—would have welcomed in Marx's time and would welcome today implementation of the "banner" statement made by Marx. They would be happy to contribute to society and to the economy "according to their ability," and would hope to be compensated sufficiently (or, to be provided with what they need out of the store of natural goods that are common goods in times of need, in Aquinas' thought), "according to their needs."

Economic Crisis: Twentieth-Century Steps toward Historical Resolution

A crisis whose roots are in history ought to be resolved in history. Economic systems were born out of particular historical contexts, under the guidance of those whom they would most benefit—indeed, who were determined, and had the power to ensure, that they were going to be the principal beneficiaries. As the gap between rich and poor widened, the wealthy took often brutal steps to retain their social control. Movements for social change became in part, in response to the unfolding situation, movements for revolution or at least major structural transformation.

Theologies of liberation emerged amid this turmoil. One idea and practice that emerged from their consciousness of the state of the poor, was accompanied by compassion for the poor, and was expressed in conscientious

---

59. Marx, "Critique," 388 (my italics).

efforts to assist the poor, was the "preferential option for the poor," which originated in Latin America.

The "Preferential Option for the Poor"

The phrase "preferential option for the poor" was coined by Gustavo Gutiérrez, Peruvian priest and theologian of liberation. Gutiérrez suggested that instead of exhorting the rich and comforting the poor the Catholic Church in Latin America should continue its spiritual mission for all social classes but simultaneously have a special, not exclusive, focus on the basic needs of the poor, including by pressing for concrete systemic changes that would help the poor meet their basic needs. Gutiérrez proposed that the Church should advocate a "preferential option for the poor," and use its substantial political and economic power to support efforts by the poor to acquire justice, particularly through a change in operative and oppressive social systems—political, economic, and ecclesial—that worked together to maintain the socially poor in their conditions of poverty. Oppressive political structures coerced those seeking representation in political bodies, or even aspiring to public office; oppressive economic structures impoverished, kept impoverishing, and kept impoverished the majority population, especially those disparaged and discriminated against because of race, ethnicity, culture, economic class, and gender, particularly members of native cultures.

Gutiérrez's concept was subsequently accepted and adopted by the Latin American Catholic hierarchy at their CELAM III international bishops' conference, also called the Puebla Conference (1979). In their Final Document (later called the Puebla Document), in the chapter "A Preferential Option for the Poor," the bishops stated: "We affirm the need for conversion on the part of the whole Church to a preferential option for the poor, an option aimed at their integral liberation."[60]

At Puebla, Pope John Paul II suggested a practical and provocative process to address and ameliorate the plight of the poor. In his journey around Mexico, while in the rural village of Cuilapán, the pope addressed a multitude comprised primarily of landless and impoverished *indios* and *campesinos* (agricultural workers). After noting that "there is always a social mortgage on all private property, so that goods may serve the general assignment that God has given them," John Paul declared that "if the common good demands it, there is no need to hesitate at expropriation itself, done in the right way."[61]

60. Final Document, §1134, in Eagleson and Scharper, *Puebla and Beyond*, 264.
61. *Puebla and Beyond*, 82. Interestingly, the idea of a "social mortgage" on private property, which had been expressed previously by Pope Paul VI, has a complement in a proposal by Thomas Paine in *Agrarian Justice*: that "every proprietor, therefore, of

Subsequently commitment to the poor expressed in the phrase "preferential option for the poor" was accepted by the Vatican. Other bishops throughout the world took up the concept and cause, and it attracted significant international ecclesial and secular interest. Theologians, ethicists, parish members, and community activists appropriated the concept as their own, and sought to concretize it in their respective (and overlapping) educational and pastoral endeavors. It found expression in bishops' statements from around the world, including in the United States.

Ecology, Economics, and Racism

Ethnic minorities and all peoples of color are especially harmed by integrated economic-ecological impacts on financial well-being and environmental context. The residue of Discovery impedes racial and ecological well-being.

African American theologian James Cone states very clearly the link between race, economics, and ecology: "People who fight against white racism but fail to connect it to the degradation of the earth are anti-ecological—whether they know it or not. People who struggle against environmental degradation but do not incorporate in it a disciplined and sustained fight against white supremacy are racists—whether they acknowledge it or not. The fight for justice cannot be segregated but must be integrated with the fight for life in all its forms."[62] Cone cites a 1987 United Church of Christ Commission of Racial Justice report that notes that "40 percent of the nation's commercial hazardous-waste landfill capacity was in three predominantly African American and Hispanic communities."[63] In the same vein, womanist theologian Emilie Townes declares that "toxic waste landfills in African American communities" are "contemporary versions of lynching a whole people."[64]

The most thorough studies of environmental racism in the United States have been done by sociologist Robert D. Bullard, in his landmark work *Dumping in Dixie*. Bullard states from the outset his "assumption that all Americans have a basic right to live, work, play, go to school, and worship in a clean and healthy environment."[65] He then laments that "People of color

---

cultivated lands, owes to the community a *ground rent* (I know of no better term) for the land which he holds" (611). In the same essay, Paine proposed using a 10 percent inheritance tax to redistribute land and wealth from the affluent to the poor.

62. Cone, "Whose Earth Is It, Anyway?," in Hessel and Rasmussen, *Earth Habitat*, 23–32. Cone deepens the meaning of "white supremacy" beyond media usage, to mean white cultural, economic, political, theological, and social hegemony and domination.

63. Ibid., 27.

64. Cited in ibid., 26.

65. Bullard, *Dumping*, xiii.

in all regions of the country bear a disproportionate share of the nation's environmental problems. Racism knows no geographic bounds."[66] In an observation that complements the ideas of Cone and Townes cited above, he states that "mainstream environmental organizations were late in broadening their base of support to include blacks and other minorities, the poor, and working-class persons."[67] Although environmentalists and the public at large in the dominant culture are unaware of or do not care about the link between environmental degradation and racism or other forms of discrimination, such as classism, for those suffering from it, "[e]nvironmental discrimination is a fact of life. Here, environmental discrimination is defined as disparate treatment of a group or community based on race, class, or some other distinguishing characteristic."[68]

Racism is part of national structures: "Institutional racism continues to affect policy decisions related to the enforcement of environmental regulations. Slowly, blacks, lower-income groups, and working-class persons are awakening to the dangers of living in a polluted environment. . . . [B]lacks and other minority groups must become more involved in environmental issues if they want to live healthier lives."[69] In the past, "Environmental risks were offered as unavoidable trade-offs for jobs and a broadened tax base in economically depressed communities. Jobs were real; environmental risks were unknown. This scenario proved to be the de facto industrial policy in 'poverty pockets' and job-hungry communities around the world."[70]

In addition, "Racism influences the likelihood of exposure to environmental and health risks as well as of less access to health care."[71] These findings are supported by the 1987 national study by the Commission for Racial Justice, which found that "Race was by far the most prominent factor in the location of commercial hazardous-waste landfills, more prominent than household income and home values."[72]

Bullard concludes that "There can be no environmental justice without social justice," and that "The richest nation on earth can no longer afford to sacrifice any of its people and communities to environmental pollution. The solution is environmental justice for all."[73]

---

66. Ibid., xiv.
67. Ibid., 1.
68. Ibid., 7.
69. Ibid., 15.
70. Ibid., 27.
71. Ibid., 98–99.
72. Ibid., 35.
73. Ibid., 159.

Terrestrial Ethics, Ecology, Economics, and Ecclesia

The historical roots of Earth's socioecological crisis are ethical, ecological, economic, and ecclesial. It has become evident that by the time humankind ventures from the Earth commons into the greater cosmic commons those roots and the contemporary fruits of their maturation should have been removed. If humankind is not to do in the heavens as it has done on Earth, people should have developed a new consciousness, conscientiousness, conscience, and conduct.

In *ethics*, people should have accepted and implemented principles of compassion for humankind in community (and rejected individualism as a creed, and greed as a supreme virtue, as capitalism teaches in theory and embodies in practice). In *ecology*, people should have come to understand that they are a part of and related to creation and creatures, and are not distinct from and superior to Earth and biota; they should recognize the inherent value of the extended biotic community and the common Earth home shared by all life (and should have eliminated Earth-destructive exploitive industrial practices and extravagant consumption of Earth's natural goods). In *economics*, people should develop and implement financial arrangements that promote community well-being, which ensures that individuals as individuals and as members of a community and a common human family will benefit (and will not permit Earth's goods and human-crafted goods to be appropriated by a few to the detriment of many). In the *ecclesia* understood more as a community of believers committed to following the life and example of Jesus than as a rigidly organized institution, Christians should recover the compassion and sense of sharing exhorted by the prophets and Jesus, and exemplified in the Acts community.

Humans' compassion and caretaking informed by an active conscience expressed in conscientious conduct would prevent negative social and ecological impacts not only on Earth from which they will depart into space, but on celestial bodies on which they disembark. Humankind must not merely look to an anticipated future in which people will do better in other places than they have historically done on Earth. Considerations generated now would prompt continuing dialogic relationships among and in *places* (Earth-exoEarth-Earth . . .) and in and through *time* (present-future-present . . .) and, well before interstellar or eventual intergalactic journeys begin, would enable people realistically to hope for and strive to effect what humans aspire to do in the heavens.

2

Terrestrial Discovery

Political Roots of Earth's Socioecological Crisis

The historical roots of Earth's *ecological* crisis are economic, ecological, ethical, and ecclesial. The historical roots of Earth's *social* crisis are political, economic, Eurocentric, and theological. The integration and influence of these roots continues today. It causes denigration of peoples and cultures, social classes and religions, and it imperils Earth as a whole and in its bioregional locales. Individuals and groups assert that their own particular, self-designated, and prioritized characteristics—as well as those of their nation, religion, and economic class—are superior to every other's. If aggressive individuals or special interest groups wield power, directly or indirectly, it will be exercised to continue hegemony. If they have insufficient power to control their nation, state, or other political entity in order to solidify their own interests above all others, they will make efforts to secure it. They are not concerned about potential harmful consequences for others who are not "one of us," however "we" define "them." Racism, ethnocentrism, classism, cultural elitism, and religious affiliation have been especially egregious, particularly but not exclusively in the United States.

An intellectual journey back to the European Age of Discovery (early fifteenth to seventeenth century) will unearth political aspects of these historical roots, which were embedded deep in European cultures and nations. They were exported overseas by the nations who developed long distance seafaring capabilities, beginning with monarchic rivals Portugal and Spain. Territories and peoples were encountered whose existence had not been known previously.

During that period European nations came to be more externally defined in relation to each other, and internally cohesive, if not entirely unified. Their shared cultural characteristics, particularly their monarchical political systems, private property-based economic systems, and dominant Christian religion (which was understood to be the only "true faith"), came to instill in them an arrogant and assertive sense of civilizational superiority over all other nations and peoples—those already known through travel and trade, and those to be found elsewhere in a world ever-expanding in European consciousness.

Europeans justified to and for themselves their colonial expansion to other places, and their attempted co-optation or conquest (political, military, economic, cultural, or any combination thereof) of other peoples. They claimed that their expansion to and seizure of new (to them, not to the existing inhabitants) worlds and territories was done for other peoples' material and spiritual well-being—as judged and decided by the new arrivals. Their pretext and rationalization was that Europeans were politically and economically "civilizing" all other peoples in this life (for their cultural benefit and material well-being in the visible world), and religiously "saving" them for eternal life (for their spiritual benefit, everlasting, in a world unseen).

In the Americas, this European attitude toward and attempts to control newly engaged cultures was characteristic not only of the initially imposed *European* hegemony that ended in the Americas with colonial revolutions that achieved independence. It was carried forward by its successors, the dominant political and cultural ideologies and powers in North, South, and Central America, in the then-emerging and now established era of *Euroamerican* hegemony.

In order to have a self-satisfying religious and supposed legal basis for imperialism, European rulers (including popes) and their advisors devised a Discovery Doctrine[1] that, developed and strengthened over time (from the fifteenth century to the twenty-first), has been instrumental in depriving other peoples of their lands, cultures, religions, and rights. The Doctrine continues to be used today by Western nations, usually without being referenced. Its utility has prompted nations on other continents to appropriate its "principles" for their own nationalistic purposes. In years to come, its impacts might well, barring ideological transformation, color activities and claims during space exploration (even through telescopes) and colonization. In that possible scenario intelligent beings with organized social arrangements, residential and commercial structures, and established agricultural operations, seen from afar or close at hand, might be judged outright as "inferior" by human explorers expressing the hopes and ideology of humankind, or at least those of the industrial, military, and political leaders who fund their space voyages in a hunt for new territory to colonize and natural goods to exploit.

---

1. Sometimes in written or oral usage the term is shortened simply to Discovery.

Discovery Doctrines developed by European nations and Christianity gave a new twist to a phrase in Christians' shared Lord's Prayer. Instead of petitioning God that the divine will be done "on Earth as in heaven," Christian nations effectively claimed, in Discovery, that God's will was embodied already by European culture, philosophy, religion, and theology; therefore, God's will was to be done "on Earth as in Europe." Christians thereby added a new understanding of God as divine Being and as Creator to speculation on divine action in Genesis 1 about which biblical scholars and theologians have argued for millennia. In the Bible it is unclear whether Yahweh creates from nothing, because no being exists except for God—*creatio ex nihilo*—or if the "waters" of Genesis 1:1 suggest some pre-cosmic creation matter which God reshapes—*creatio ex materia*. Discovery Doctrine "theology" remakes God as *deus ex Discovery* while simultaneously divinizing Discovery as *Discovery ex Dei*. Eurocentrism blinded popes, philosophers, and theologians to the provocative (and heretical) theological doctrine they formulated, and its actual and potential spiritual and social implications.

Historical Origins and Development of Terrestrial Discovery Doctrine

In the late fifteenth century, Portuguese and Spanish explorers ranging far in sailing ships declared, on their return, that they had "discovered" a "New World" across the Atlantic Ocean. From that time forward European rulers, and the explorers and colonizers that they sent to find and settle "discovered" lands (which were also "empty lands" by European definition—empty, that is, of a Europe-like civilization and the "true faith"), engaged in a twofold assault upon indigenous peoples and the territories or nations they inhabit around the world.

First, they perpetrated an "American holocaust," in the words of David E. Stannard in *American Holocaust* and Ward Churchill in *A Little Matter of Genocide*, against those they came to call generically "Indians" without honoring any distinctions among peoples of resident native nations. Stannard states that "the destruction of the Indians of the Americas was, far and away, the most massive act of genocide in the history of the world."[2] Churchill concurs, noting that

> The American holocaust was and remains unparalleled, both in terms of its magnitude and the degree to which its goals were met, and in terms of the extent to which its ferocity was sustained over time by not one but several participating groups.

---

2. Stannard, *American Holocaust*, x.

The ideological matrix of its denial is also among the most well developed of any genocide.[3]

After citing pertinent statistics, Churchill declares, "All told, it is probable that more than one hundred million native people were 'eliminated' in the course of Europe's ongoing 'civilization' of the western hemisphere."[4]

Second, they promoted devastation of Earth's water, air, and earth, spreading around the globe environment-destructive agricultural, forestry, property relationships, and commercial practices that they had developed in their homeland.[5]

The twofold assault continues today, in terms of both ongoing genocidal attitudes toward and oppressive actions against native peoples,[6] and ecologically harmful industrial practices, particularly irresponsible mining, forestry, agribusiness operations, and transfers to corporations and individuals, as private property, regional places that once were shared, used, and cared for by indigenous peoples.[7]

Voyages of Discovery

When explorers on sailing ships sent forth by rival European nations came upon the extensive lands of the "New World" (which natives called Turtle Island), although this world was already replete with native peoples, some of whom had villages, agricultural fields, and community-based fishing

---

3. Churchill, *Matter of Genocide*, 4.

4. Ibid., 86. Throughout his well-researched scholarly text, Churchill documents genocide committed against Indians in the Americas, North through South. His discussion is based on the definition of *genocide*, a term first coined and described by Raphael Lemkin; he cites Lemkin and his definition (70).

5. Portions of this chapter were presented as "Discovering Doctrines, Supporting Sovereignty" at the United Nations Consultation held by James Anaya, UN Special Rapporteur on the Rights of Indigenous Peoples, at Sinte Gleska University, Rosebud Reservation, Mission, South Dakota, May 1, 2012. I am honored to have been invited, by Lakota elder William Means of the International Indian Treaty Council, to present to the Special Rapporteur a statement on the Doctrine of Discovery and ways in which Christian churches, in particular the Catholic Church, should both reject past and present invocation of this doctrine, and return to Indian peoples native lands that have been unoccupied or that have been abandoned over the years, as required by treaties signed with them—or, alternatively and by mutual agreement, to make comparable restitution.

6. Documentation of continuing violations of the human rights of native peoples might be found in the research and United Nations testimony of the International Indian Treaty Council (IITC), the first indigenous peoples' nongovernmental organization (NGO) recognized by and accredited to the UN almost forty years ago; see http://treatycouncil.info/home.htm.

7. See news articles and essays published online by the Indigenous Environmental Network (IEN): http://www.ienearth.org.

operations, the Europeans claimed to have discovered them. Although native peoples simultaneously discovered Europeans arriving on their territorial shores, there was not a balance of discovering, each of the other, because Europeans were culturally and assertively acquisitive; more aggressive ideologically; and had developed a superior military technology and a new ideology related to the word *discovery*: the Discovery Doctrine. This ideology-infused concept meant far more than merely coming upon something hitherto unknown. Its tenets included European assertions that their monarchs had rights to ownership and use of, and exclusive commercial rights over, the territories and natural goods in lands newly found by explorers whom they financed, and who flew their particular sovereign flag.

When they became settled immigrants on Indian[8] lands, Europeans began to carve up the territory into parcels of property "owned" by their respective monarchs or by colonies chartered by them. The newcomers waged war upon indigenous populations when it became the only means by which Europeans could seize lands from natives who did not want to part with their homelands and traditional hunting, agricultural, and spiritually significant places. European governments declared these actions to be rights exercised under their "Doctrine of Discovery," and claimed that they were legal—but only for Europeans.

The "legitimation" of Discovery was based originally on papal authority. Steven Newcomb, in *Pagans in the Promised Land: Decoding the Doctrine of Christian Discovery*,[9] ascribes the Discovery Doctrine's origins to the Catholic Church and calls it *Christian* Discovery.

Robert Miller, in research complementary to that of Newcomb, documents the long history of the development of the "Doctrine of Discovery." He traces the earliest form of the doctrine back to the 5th century,[10] when

---

8. In the United States native peoples prefer to use the term "Indian" as a generic term for all indigenous peoples. Even though the currently "politically correct" term is "Native Americans," elders and activists note that they are neither "American" nor "Indian," but that they will continue to use the latter term to affirm their ongoing historical identity, because of which they have been politically, culturally, religiously, and economically oppressed. The principal native peoples' advocacy groups in the United States for several decades have been the American Indian Movement (AIM) and the International Indian Treaty Council (IITC). The Treaty Council's creative formulations and politically astute involvement in the UN led to the development and passage by the UN General Assembly of the Declaration of the Rights of Indigenous Peoples in 2007.

9. Newcomb utilizes the categories of cognitive theory to analyze the distinct cultural values and thought processes operative respectively among Europeans and Indians during the "Age of Discovery" and in its aftermath. He illustrates ways in which Discovery tenets have continued to permeate political systems and judicial processes, particularly in Western cultures and especially in the United States.

10. This doctrine was carried forward by, among others, Pope Innocent IV, whose

popes sought to promote global papal jurisdiction in order to have a single Christian civilization. This ideology was expressed in and enhanced by the Crusades in the eleventh through thirteenth centuries.[11]

In the fifteenth century, Portugal became the first major seafaring and exploring European nation, having been first to develop sailing ships capable of traversing long distances, and began to establish colonies, particularly in Africa. Spain followed toward the end of the fifteenth century; it received a significant initial impetus to its Discovery voyages with the travels of Christopher Columbus, who arrived in 1492 at what was called the "New World," and later the "Americas."

Spain and Portugal became intense and hostile rivals seeking new lands to Discover. Each could use papal documents to support their respective claims to territory in the New World. In 1493, Alexander VI had stated in the papal bulls *Inter caetera divinai* and *Inter caetera divinai II* that Spain had a right, by Discovery, to lands discovered by Christopher Columbus and to all lands west of a specified dividing meridian: "a line from the Arctic pole, namely the north, and to the Antarctic pole, namely the south . . ."[12] Portugal cited papal bulls prior to *Inter caetera II* (often referred to in contemporary scholarship as simply *Inter caetera*) which specifically named Portugal's Discovery rights. The vague wording of the bulls, taken together, seemed to indicate that Spain would have title to all lands "discovered" west of the meridian specified by Alexander VI, and Portugal would receive title to all lands "discovered" east of the line. In order to resolve their differences, Portugal and Spain signed the Treaty of Tordesillas in June 1494. It moved the dividing and binding meridian 370 leagues west of the Cape Verde Islands, off the west coast of Africa. As a consequence, Portugal's Discovery claims to Africa and Brazil would be honored by Spain, and Spain's substantial Discovery claims in the Americas would be reciprocally respected by Portugal. The Treaty enabled both nations to continue colonization without interference by the other in their respective Discovered territories.

A new "international law" began to be developed to provide justification for seizure of foreign lands not ruled by "Christian princes." Initially, it integrated existing European laws that had been developed to avoid conflicts among European nations competing for "discovered lands" with desirable

---

1240 teachings influenced Alexander VI in the fifteenth century (as noted here); formal legal statements of the Discovery Doctrine by Franciscus de Victoria (a priest and advisor to the king of Spain, and a pioneer in international law development) in the sixteenth century; and Hugo Grotius in the seventeenth century. See Miller et al., *Discovering Indigenous Lands*, 9, 11–13.

11. Miller, *Native America*, 15.

12. The document, in English translation, may be found online; see http://www.nativeweb.org/pages/legal/indig-inter-caetera.html.

territory or desired natural goods. Discovery "doctrine" was, of course, a European-developed pretext to unjustly take territory from peoples already in place.

Indigenous peoples did not have and did not need, in their eyes, a papal stamp of approval to occupy and use their traditional territories. In some areas, people whose ancestors had arrived a dozen millennia or more earlier than the Europeans were enslaved by the new arrivals, forced from their territories, or killed outright in order to take by force their lands and whatever Earth natural goods were located on them, which included fertile soil, virgin forests, clean water, and beneficial minerals (especially, in the colonial era, gold and silver; in later times, coal, oil, and uranium).

Under the Discovery Doctrine the invasion and seizure of indigenous peoples' lands by a European power was made legal under European international law in two ways which evolved gradually as a result of Discovery experiences, claims, and tactics. First, by being the first nation on site, demonstrated initially by seeing (established by a declaration of a visual discovery *per se*), then by seeing and symbols (such as by landing and planting a flag, and burying a carved stone or engraved plate), and finally by seeing, symbols, and settling (as represented by a token group, usually soldiers in a fort). Second, by whether or not the peoples who were encountered had each of two cultural requirements to be accepted as "civilized" in European eyes: a European-like political-economic system, especially in terms of possession of private property whose ownership was delineated by human owner-placed boundaries such as fences; and, the Christian religion.

Where Europe-mirroring culture or the Christian Religion was not present, the lands found were declared *terra nullius* ("vacant land"), or *vacuum domicilium* ("vacant home")—despite existing populations of native inhabitants, or the extent of their agricultural and urban development.[13] Europeans invoked these Discovery provisions to declare that lands were empty and their supposed ownership had not been validly established by residents (not even by virtue of occupancy for millennia and by constructed cities). In a rationalization and justification not hitherto expressed or employed in human history, popes and kings and their religiously, ethnically, and culturally biased advisors stated that "empty" or "vacant" did not mean that no one was dwelling on the land, but that indigenous peoples were not using or working their territory in the manner of Europeans, and so were not in compliance with European customs and laws (no matter that this was not possible, since cultures around the world developed over time their own laws and understanding of property relations, if any, and responsible use) or faithful to Christianity (even if Christians, let alone Christian missionaries, had

13. Miller et al., *Discovering Indigenous Lands*, 7–8.

never before come into contact with natives, which would have provided at least an opportunity for them to accept or reject Christianity). Absent either or both of these requirements—a Europe-mimicking culture and Christianity—peoples encountered were subjects of and subjected to European rulers, and their lands were open to possession and use by the first European nation to come upon them.

Miller discusses representative European *charters*, granted by English monarchs in the colonial era, whose language evidences the Discovery ideology regarding ownership and use. Using the rationale of having a superior European civilization and the true faith, the charters enabled colonists to take possession of, settle on, and use Discovered lands:

> [I]n the 1606 First Charter to Virginia and the 1620 Charter to the Council of New England, James 1 granted the colonies property rights in America because the lands were "not now actually possessed by any Christian Prince or People" and "there is noe other Subjects of any Christian King or State . . . actually in Possession . . . whereby any Right, Claim, Interest, or Title, may . . . by that Meanes accrue." English monarchs also invoked other elements of Discovery when they granted colonial charters because they ordered their colonists to take Christianity and civilization to American Indians for the purpose of "propagating Christian Religion to those [who] as yet live in Darkness and miserable Ignorance of the true Knowledge and Worship of God and [to] bring the Infidels and Savages, living in those Parts, to human civility, and to a settled and quiet Government."[14]

Note how native peoples need both "true civilization" and "true religion" to retain their respective indigenous cultures and lands, and are to change from "Savages" to a "settled and quiet Government"—to resigned, cooperating subjection to immigrant European culture and rule.

Tenets of Discovery were abandoned by English monarchs, colonies, and colonists as their population expanded and European-originated diseases decimated Indian populations; eventually, they resorted to taking lands they claimed without recognizing or acknowledging Indians' inherent sovereignty status, even when this was specifically stated in treaties, and without recognizing Indians' supposed Discovery right to decide when and where they would agree to part with territorial lands and natural goods.

---

14. Miller, *Native America*, 19.

England in the Americas

The gradual transition in wording between the earliest treaties Indians signed with European immigrants in New England, and their subsequent iterations, reveals two sharply divergent understandings of "property" as embodied in concrete human relationships with land itself when occupied and used, and with the general nature of the human community-creation relationship as a whole, as represented in attitudes toward and actions on Earth. The treaties subtly shifted in wording from "use" of land that continued to be in Indian hands, to "purchase" and "ownership" of land: these terms and practices were not in Indians' lectionaries or worldviews.

In Indian thought and practice, land, water, and air could not be owned—the latter had intrinsic value, an inherent part of their being that should be respected and could be negated only when people had to provide for their needs: necessity transformed intrinsic value (value in itself) into instrumental value (value for others' uses, either in its natural form or as transformed by human labor) in that particular place and time. Integrated relationally with Earth, native peoples farmed agricultural fields, hunted in forests, and fished waters communally. The Earth and other natural goods were regarded as a sacred provider, Mother Earth, whose holistic integrity was to be respected. Earth was never private "property" that could be parceled, and bought and sold. Natives' territorial sovereignty was mutually respected; it was usually defined by a village's or extended family's uses of an area to provide for material needs. Natural goods converted by human labor to individual goods were regarded as the personal property of the individual—but these too were shared with others at particular moments to meet others' specific needs, because cultural attitudes of mutuality and reciprocity prevailed. The native peoples' ideal and practices were to "share all things in common," much like the early Christian community in Jerusalem founded by the Apostles most closely associated with Jesus, as described in Acts 2 and 4.

George Tinker, Osage scholar, activist, and spiritual elder, elaborates another key area of distinct and contrasting thought processes and cultural perceptions between Turtle Island-based Indian perspectives and Europe-based (and its successor, Euroamerican-based) Western ideologies: their respective basic worldviews. In *American Indian Liberation: A Theology of Sovereignty*, Tinker states that in Indian thought, the primary consciousness is *spatial*: community life in Earth places is viewed principally *in situ*, with an insightful awareness of humans in context in specific places, as in community they relate holistically with Mother Earth, each other, and the other living beings among whom they are situated. They do not lack temporal awareness, but that is secondary to spatial consciousness. In Western thinking, the primary consciousness is *temporal*: a linear approach to human present and future presence guided by a sense of historical time, renders present context and relatedness to Earth and creatures secondary. Tinker describes "four fundamental, deep structure cultural differences between Indian people and the cultures that derive from european traditions that separate American Indian cultures (including religions, traditions, social structures, politics and so forth) distinctly apart from amer-european cultures and religion." The differences include "spatiality as opposed to temporality; attachment to particular lands or territory; the priority of community over the individual; and a consistent notion of the interrelatedness of humans and the rest of creation."[15] Tinker's summary discussion of the first and fourth distinctions is particularly pertinent here:

> First, these indigenous traditions are spatially based, rather than temporally based. The euro-western world has a two-millennia history of a trajectory shifting decidedly away from any rootedness in spatiality toward an ever-increasing awareness of temporality. Whether in its capitalist or socialist (marxist?) guise, history and temporality reign supreme in the euro-west, where time is money and "development," or progress, is the goal. On the other hand, Native American spirituality, values, social and political structures, even ethics, are fundamentally rooted not in some temporal notion of history but in spatiality. This is perhaps the most dramatic (and largely unnoticed) cultural difference between Native American intellectual traditions and euro-western thought processes. The euro-western intellectual tradition is firmly rooted in the priority of temporal metaphors and modes of being. Native Americans think inherently spatially and not temporally. The question is not whether time or space is missing in one culture or the other, but which of these metaphoric bases functions as the ordinary and which is the subordinate. Of course, Native Americans have a temporal awareness, but it is subordinate to our sense of spatiality. Likewise, the euro-western tradition has a spatial awareness, but spatiality lacks the priority of the temporal. . . .
>
> Finally, spatiality, a community-centered sense, and the notion of interrelatedness lend themselves to yet a fourth categorical difference between these indigenous cultures and the west. In native North America, indigenous people find their primary

---

15. Tinker, *American Indian Liberation*, 7. Tinker is Professor of American Indian Cultures and Religious Traditions, Iliff School of Theology, Denver, Colorado, and an enrolled member of the Osage Nation.

attachments in terms of particular lands and territory. Individual ownership, even group ownership, of land is a concept foreign to Indian peoples. Rather, there is a firm sense of a group filial attachment to particular places that comes with a responsibility to relate to the land in those places with responsibility.[16]

In *Changes in the Land: Indians, Colonists, and the Ecology of New England*,[17] William Cronon describes conflicting Indian and European attitudes, ideologies, and practices regarding land occupancy and use. In words complementary to Tinker's, Cronon states that

> When the Europeans first came to New England, they found a world which had been home to Indian peoples for over 10,000 years. But the way Indians had chosen to inhabit that world posed a paradox. . . . Many European visitors were struck by what seemed to them the poverty of Indians who lived in the midst of a landscape endowed so astonishingly with abundance. As colonist Thomas Morton wrote, "If this Land be not rich, then is the whole world poor."[18] Here was a riddle: how could a land be so rich and its people so poor? . . . Thomas Morton had posed his riddle knowing full well that his readers would recognize its corollary: if Indians lived richly by wanting little, then might it not be possible that Europeans lived poorly by wanting much? The difference between Indians and Europeans was not that one had property and the other had none; rather, it was that they loved property differently.[19]

For the colonists, land that was not enclosed by fences and altered for a particular use was not owned, and was therefore available to be claimed and taken as private property. The boundary-forming and property-designating and dividing fence

> represented perhaps the most visible symbol of an "improved" landscape: when John Winthrop had denied that Indians possessed anything more than a "natural" [as distinct from "legal"] right to property in New England, he had done so by arguing that "they inclose noe Land" and had no "tame Cattle to improve

---

16. Ibid., 7–9. An extended elaboration is provided in chapter 3 of this volume, "Ecojustice and American Indian Sovereignty," under the heading "Spatiality: Place versus Time."

17. Cronon, *Changes in the Land*, especially 56–81, 130.

18. Morton is not posing a question with "then": his intended meaning is, "then the whole world is poor."

19. Cronon, *Changes in the Land*, 33, 80.

the Land by." Fences and livestock were thus pivotal elements in the English rationale for taking Indian lands.[20]

By contrast the Indians' lives and practices were intimately related to place because of their primarily spatial worldview. Some lived in seasonal mobility between agricultural and hunting areas; others stayed in a place and used practices such as periodically burning forests to create new habitats for diverse free (not domesticated by humans) creatures (such as deer, elk, and bear) that were especially needed periodically, and then hunted for their instrumental value (e.g., food and clothing). The colonists did not view this as Indians' exercise of control over land, since such practices were done by and for the community as a whole, not for some individual commercial benefit that would have been indicated by demarcation, particularly with fences, into private property. The Rhode Island colonist Roger Williams, who favored just treatment of native peoples, and gave them money for the use of their lands (not viewing this payment as a purchase of Indian property) declared that the Indians did in fact have title to their territory. Williams stated that the King of England had perpetrated "injustice, in giving the Countrey to his *English* Subjects, which belonged to the Native *Indians*." Regarding signs of sovereignty over particular places, such as occurred when forests were burned to enhance hunting grounds to provide food, Cronon notes that

> Burning the woods, according to Williams, was an improvement that gave the Indians as much right to the soil as the King of England could claim to the royal forests. If the English could invade Indian hunting grounds and claim right of ownership over them because they were unimproved, then the Indians could do likewise in the royal game parks.[21]

Williams's comments were not well received by colonial governors. Acquiring land for their king and for themselves (as individuals or as members of a trading company or other commercial enterprise) was their primary objective in the colonial period, and using a European ideology of property and the Discovery Doctrine enabled them to rationalize what would not otherwise be acceptable by European law: taking another's property without compensation.

20. Ibid., 130.
21. Ibid., 57.

Inclosure Policies and Practices: In the Americas as in Europe

There were important precedents in Europe for the English colonists' ideas on private property vis-à-vis common holdings. In their homeland, *inclosures* of previously commonly used land were undertaken arbitrarily and unilaterally in the early sixteenth century.[22] Initially, over the protests of the common people, European rulers—monarchs and members of the lesser nobility—imposed laws of "inclosure" to enforce land reallocation in order that wealthy individuals would have exclusive use of estate land that had been commonly used, and land redistribution so that community held or controlled land would be owned by the nobility. The "unenclosed" land had been used by generations of peasant farmers to cultivate crops or maintain orchards, graze livestock, fish in bodies of water to provide food, or gather wood to heat their homes and cook their food. The social and material results of inclosures were that peasants, who did not have the resources to buy private land, were reduced to serfs on the manors. A series of inclosure riots erupted across the English countryside from 1520–49. Thomas More, in *Utopia*, describes inclosure as the cause of social problems, unrest, and destruction of communities (both physically and socially) to satisfy the greed of the nobility:

> Your sheep . . . have become so greedy and fierce that they devour human beings themselves. . . . For in whatever parts of the land the sheep yield the softest and most expensive wool, there the nobility and gentry, yes, and even some abbots—holy men—are not content with the old rents that the land yielded to their predecessors. Living in idleness and luxury, without doing any good to society, no longer satisfies them; they have to do positive harm. For they leave no land free for the plow: they enclose every acre for pasture; they destroy houses and abolish towns, keeping only the churches, and those for sheep-barns. . . . Thus one greedy, insatiable glutton, a frightful plague to his native country, may enclose many thousand acres of land within a single hedge. The tenants are dismissed; some are stripped of their belongings by trickery or brute force, or, wearied by constant harassment, are driven to sell them. . . . By hook or by crook these miserable people—men, women, husbands, wives, orphans, widows, parents with little children, whole families

---

22. Summary information about *inclosures* (later spelled *en*closures), upon which this analysis primarily is based, may be found in three Wikipedia articles from the United Kingdom version of the online encyclopedia: for "Inclosure Acts," see http://en.wikipedia.org/wiki/Inclosure_Acts; for "Enclosure," see http://en.wikipedia.org/wiki/Enclosure; and for "Common Land," see http://en.wikipedia.org/wiki/Common_land.

(poor but numerous, since farming requires many hands)—are forced to move out. They leave the only homes familiar to them, and they can find no place to go.[23]

Ironically, in their "new world" English colonists did to others—native peoples—what had been done to them in their "old world." For centuries in England the nobility and peasants had a land use arrangement beneficial to both social classes: the farmers paid an equitable rent for the use of part of the nobles' estates, upon which they earned their livelihood; the nobles added the rent to their accumulated wealth, acquired without labor. But then, in order that the nobles and gentry (including abbots) could raise more profitable sheep, the poor farmers were suddenly and forcibly compelled to leave lands on which their families, through generations, had produced agricultural goods; indeed, as in More's words, the sheep "devour human beings" by displacing them: sheep now roamed where humans had resided. The new arrangement generated greater wealth for the idle nobility, and great suffering for the once hardworking poor—because of their sudden joblessness, homelessness, and consequent great poverty and inability to provide for themselves, by their labor, life's necessities. During this era, thousands of financially desperate urban thieves, who had previously been thriving farmers and agricultural laborers, were hung as punishment for their crimes; however, as *Utopia* pointedly states, "no punishment however severe can withhold those from robbery who have no other way to eat."[24]

More published *Utopia* in 1516, just a quarter-century after Christopher Columbus landed in the "New World." His book contains several oblique references to the "discovered" world and its native inhabitants, including speculation about their communal way of life as described by several explorers, including Amerigo Vespucci. In *Utopia*, common ownership of property—communism—is advocated as necessary for justice in society. The narrator Raphael who describes the island of Utopia, having lived there for five years, declares, "What if I told them [people who disagree with his suggestions for just and stable European societies] the kind of thing that Plato advocates in his republic, or that the Utopians actually practice in theirs? . . . [P]rivate property is the rule here, and there all things are held in common."[25]

As is noted by Raphael, communism as an ideal ideology and economic arrangement was advocated by Plato: in his consciousness, as expressed philosophically in his words, and in his conduct, as evidenced practically in his public work. In his *consciousness*, Plato teaches in the *Republic* that the

23. More, *Utopia*, 49.
24. Ibid., 16.
25. Ibid, 34.

Guardians, entrusted with maintaining an ideal society (who "devote their lives to doing what they judge to be in the interest of the community"[26]), "should have no houses or land or any other possessions of their own, but get their daily bread from others in payment for their services, and consume it together in common"[27]; and, "since they have no private property except their own persons (everything else being common)" litigation would "disappear."[28] Plato's ideas were refined in *The Laws*, in which he extends common possession of goods to all citizens in his proposed "ideal society": "You'll find the ideal society and state, and the best code of laws, where the old saying 'friends' property is genuinely shared' is put into practice as widely as possible throughout the entire state.... [I]n such a state the notion of 'private property' will have been by hook or by crook completely eliminated from life. Everything possible will have been done to throw into a sort of common pool even what is by nature 'my own.'"[29] In his *conduct*, Plato remained adamantly principled in this regard, according to Diogenes Laertius (3rd century CE). When Arcadia and Thebes constructed a city together, they wanted Plato to be its legislator; he set the precondition that they accept communism. They refused his requirement; he rejected their invitation.[30]

Despite ideas on an ideal society as expressed in the works of More and Plato, and the historical precedent of peasants working land in common as exercised in pre-Inclosure England, English colonists whose ancestors had for generations used lands communally and had revolted against enclosures of common land, imposed inclosures of their own, replete with fences, on Indian lands—and excluded Indians from using the land as they had used it for generations, over thousands of years. The colonists effectively became in practice, though not in title or status, the "nobility" of the New World: they had royalty-granted Charters in the newly named "Americas" that enabled

26. Plato, *Republic* 412e, 113.

27. Ibid., 464c, 179.

28. Ibid., 464e. A negative element in Plato's ideal society is that he accepts slavery as a social reality; a positive element is that he proposes sexual equality between women and men, including among the Guardians: "[W]omen should in fact, so far as possible, take part in all the same occupations as men, both in peace within the city and on campaign in war, acting as Guardians and hunting with the men.... [T]here is nothing unwomanly in this natural partnership of the sexes" (466d, 181).

29. Plato, *Laws* 739c, 161. Plato did not advocate farming in common, but each man [*sic*] had an equal allotment of land that he would pass on to his "favorite son"; the land could not be sold by the farmer or his descendants because "each man who receives a portion of land should regard it as the common possession of the entire state" (740a, 162). In these stipulations, his society differs from the English common owned by the nobility, who may enclose it or otherwise dispose of it as they choose.

30. The incident is described by Laertius in *Lives of Eminent Philosophers* 3.23, as cited by George Logan, in More, *Utopia*, 36 n. 4.

them to enclose lands and evict people who had more claim on New World land than the peasants had had in England. Indigenous Indians were forced to move elsewhere so that the exploding settler population could be accommodated. It was not "sheep" but colonists who "devoured" indigenous peoples. Eventually, the situation developed where Indians were forced to buy from the king or colonies (and, later, from US states and private owners) lands that had been unjustly seized from them on the Discovery pretext.[31]

Spain in the Americas

Spain's discovery and Discovery of the New World began in earnest following the accomplishments of, and the information provided by, the voyages of Christopher Columbus in 1492. The Spaniards initiated small settlements on "their" lands. The colonizers hoped to establish new trading routes and partners and to acquire natural goods (gold being a principal one) from the lands they claimed by Discovery.

Prior to widespread colonization of "New Spain," the Spanish monarchy had outlawed slavery in their homeland. The Laws of Burgos in 1512 did not prohibit slavery per se, but did require that Indians receive necessary food, clothing, and beds. Since the Laws of Burgos were insufficiently worded to provide for conquest by force (as Major was teaching, citing Aristotle as his justification), in 1513 the Spanish monarchy issued the infamous *Requerimiento* ("Requirement") to be read to peoples encountered in the New World. The proclamation was very specific. When the Spaniards approached indigenous peoples, however friendly they might be, the militarized explorers were to read to the natives in Latin (which, of course, they did not understand) a document which embodied a Discovery rationale. The Indians were told that they must submit to Spanish rule, with the Spanish monarchs as their sovereigns, and adopt Christianity as their religion. As the natives observed and listened politely, patiently, and curiously, not comprehending what was being said, they were warned that if they did not accept Spanish rule and religion, "We shall take you and your wives and your children, and shall make slaves of them, and as such shall sell and dispose of them, as their Highnesses may command; and we shall take away your goods, and shall do all the harm and damage that we can, as to vassals that do not obey."[32] Sub-

---

31. This practice continues in the twenty-first century: Indian nations in the United States that try to acquire back lands once theirs for thousands of years, and unjustly taken from them through a form of legal theft, must buy them back—at the purchase price set by state or federal governments, or by private owners and realtors, often at current market value.

32. Hanke, *Aristotle and the American Indians*, 16.

sequent genocide and plunder were then, in Spanish eyes, permissible since Indian peoples did not immediately or even subsequently submit and obey.

In the Spanish colonies, then and thereafter, when early Spanish military personnel, governors, merchants, and settlers who followed in the wake of Christopher Columbus and other explorers sought to free themselves from manual labor and to acquire needed free labor to work agricultural plantations and mines, they tried to enslave native peoples by claiming that the *indios* were "beasts who talked."[33] This would classify natives as subhuman and thereby subject to enslavement. Pope Paul III intervened. On June 9, 1537, he issued the papal bull *Sublimus Deus*, which declared that the *indios* were in fact "truly men [who were] capable of understanding the Catholic faith." He ordered that "Indians and all other people who may later be discovered by Christians, are by no means to be deprived of their liberty or the possession of their property . . . nor should they be in any way enslaved."[34] Paul III's papal bull (so called because the document was written on leather made from the hide of a bull) confronted injustices perpetrated against native peoples; it was a papal political act that advocated, for the *indios*, limited liberty and retention of their property. In the colonial era, "property" meant property in land as well as personal property.[35]

In the colonial period several Spanish bishops confronted the colonists and advocated for the rights of the *indios*. Bartolomé de Las Casas (1474–1556), the first Catholic priest ordained in the "New World," was named "Protector of the Indians" in 1516 (coincidentally, the year that More's *Utopia* was published) to fight Indians' oppression. Subsequently, he was appointed bishop of Chiapas, Mexico. However, the Spanish landholders in his diocese withheld the keys to the cathedral and the bishop's residence to deny him entry into his church and home—all because of his advocacy of the rights of indigenous peoples. His policies included refusal of absolution for sins—a major punishment for Catholics of the time—for anyone who kept Indians as property. He returned to Spain as a consequence of the colonists' control over church property, and provided to the Spanish monarchy and to Catholic Church officials written documentation of the oppression of Indians. The words and actions of Las Casas so moved King Charles V in Spain that he issued "New Laws," among which was *Law 35* that abolished slavery in the

---

33. Casas, *Selection of His Writing*, 13.

34. Hanke, *Aristotle and the American Indians*, 19.

35. Today, the equivalent of Paul III's statement would be a declaration stating that the integrity of Indians' territorial sovereignty should be restored and conserved, and that Indians' ownership of and control over the development and use of natural goods of creation present in their territory should be implemented.

Spanish colonies; subsequently, the king ordered cessation of military conquests in the New World.

Miller discusses the debate that arose in Spain, even after Discovery conflicts between Portugal and Spain were supposedly resolved. He describes the extent to which papal documents provided a firm foundation for Discovery claims. During the debate Franciscus de Victoria, Catholic priest, royal advisor, and the first theology professor chair at the University of Salamanca, considered the issue. (Victoria's legal opinions regarding Discovery led him to be considered afterward as a seminal thinker in international law.) Victoria issued three conclusions based on his analysis. They have been incorporated, virtually unchanged, in the "European Law of Nations" regarding American Indians' rights and standing: "First, the natives of the Americas possessed natural legal rights as free and rational people. Second, the pope's grant of title to lands in America was invalid and could not affect the inherent rights of the Indians. Third, violations by the Indians of the natural law principles of the Law of Nations (as determined by European Christian nations) might justify a Christian nation's conquest and empire in the New World." Victoria shifted the basis for Discovery claims from papal authorization to a more universally accepted and potentially legally enforceable document, based on "natural law." Discovery continued to be, in this formulation, Eurocentrically defined and elaborated. Victoria concluded that "if infidels prevented the Spanish from carrying out any of their natural law rights, then Spain could 'protect its rights' and 'defend the faith' by waging lawful and 'just wars' against the natives."[36] In Miller's words, "The reasoning that natives were bound by the European definition of the natural-law rights of the Spanish was an ample excuse to dominate, defraud, and then engage in 'just wars' against native nations that dared to stop the Spanish from doing whatever they wished. Consequently, Victoria limited the freedom and rights of the natives of the Americas by allowing Spain's natural law rights to trump native rights."[37]

In Spain, England and other European countries, the Eurocentric imperialistic ideology expressed and legitimated in the Discovery Doctrine was invoked time and again, and applied against people and places in the "new" worlds they discovered. A "law" that was really a provincial law, for all its claims to be "international" law, was declared to be a global law so as to privilege Europeans and European customs and culture over wherever, whenever, and whatever native peoples' cultures were encountered.

---

36. Miller, *Native America*, 16.
37. Ibid., 17.

Dissecting Discovery Doctrine

When Europeans set sail first as explorers and then as colonizers, they carried to the shores of the Americas instructions on how to claim lands they found—even if other peoples were there already. Rituals to make such claims included planting a flag, inscribing stones on site, and burying lead plates with the sovereign's name.[38] The justification for their attitude and action, the "Doctrine of Discovery," stated that if a civilization, no matter how advanced in other matters, did not have European political and economic structures and values and the Christian religion, it had to be "civilized," with deadly force if necessary, into a European model. European culture was, by self-description, "superior" to other cultures, and the newly encountered citizens had to be "Christianized" because the only "true" religion was Christianity.

Newcomb states that

> On the basis of a biblical viewpoint that the chosen people are providentially assigned the task of subduing the earth and exercising dominion over all living things, the Christians considered themselves to be chosen people divinely obligated to "save" the heathen nations by subjugating them, euphemistically referred to as "civilizing" them. . . . [T]he heathens are destined by God to be saved and *reduced* to Christian European "civilization."[39]

Miller et al. describe the elements of the Doctrine of Discovery once operative for European explorers and that continue today, however disguised or unreflectively, in Euroamerican-based and -dominated cultures:

> 1. *First discovery*. The first European nation to find land about which other European nations are unaware.
>
> 2. *Actual occupancy and current possession*. After first discovery, a European nation had to become contextually established, as

---

38. The most extensive description and analysis of the origins, development, dissemination, and implementation through centuries of the Doctrine of Discovery is found in the works of Robert J. Miller and Steven T. Newcomb. Miller is Professor of Law at Lewis & Clark Law School, Portland, Oregon; Newcomb is the indigenous law research coordinator for the Sycuan Band of the Kumeyaay Nation, San Diego, California, and cofounder and codirector of the Indigenous Law Institute. Their respective significant works in the field include: Miller, *Native America*; Miller et al., *Discovering Indigenous Lands*; and Newcomb, *Pagans in the Promised Land*. The present discussion of Discovery is indebted to these works, which are scholarly but very accessible to a general readership. Miller is an Eastern Shawnee (Oklahoma) enrolled citizen. Newcomb is Shawnee/Lenape. I am grateful to William Means, International Indian Treaty Council, for referring me to Miller, and to George "Tink" Tinker for referring me to Newcomb.

39. Newcomb, *Pagans in the Promised Land*, 113.

evidenced by a fort or settlement, to solidify a first discovery claim.

3. *Preemption/European title.* European nations in possession as per the preceding acquire the property right of preemption; only the established European nation, and no other European nation, is allowed to buy land from native peoples.

4. *Indian title or Native title.* Indigenous peoples no longer have full property rights, only rights to occupy and use once native lands. (Supposedly, for as long as they chose to live on their lands; they were not to be forced off or coerced from these lands.)

5. *Indigenous nations limited sovereign and commercial rights.* Native peoples lost international rights to diplomacy or trade with any European nation other than the 'discovering' nation.

6. *Contiguity.* Lands contiguous to settlements or even to Discovered lands not yet settled belonged to the discovering nation. A corollary was that a European nation that 'discovered' the mouth of a river was entitled to claim as its own all lands through which the river ran; this might include thousands of square miles, as was the case with France's Louisiana Territory, through which ran the Mississippi River whose origin was Discovered by French explorers.

7. *Terra nullius.* This term (vacant land) and *vacuum domicilium* (vacant home) were used to declare that lands were empty, and any supposed ownership null and void. The lands in question were vacant not because they had no people dwelling in them, but because they were not being used in a way that would have been the case if the lands had been in Europe, utilized in ways compliant with European laws and customs.

8. *Christianity.* Native peoples lost numerous ownership and use rights when "Christian" nations' representatives "discovered" them.

9. *Civilization.* Europeans believed they had a superior culture, and had a divine mandate to civilize, educate, and convert Indigenous peoples, sometimes via paternal and guardian roles over them.

10. *Conquest.* This term came to have two meanings because of historical developments. At first, it meant military subjugation of Indigenous peoples, including by applying a doctrine of "just

war." In Europe, conquest was solely a military and political operation; once an invasion succeeded, the conquered people would transfer their allegiance and submission to the new rulers from the conquering country, but would retain their individual and family property rights. A second meaning was added by the U.S. Supreme Court in the nineteenth century: "conquest" was a "term of art" with special significance: property that was sovereign indigenous territory before conquest, was located politically now in the United States of America, and was *not* exempt from seizure by the conquerors.[40]

In the United States after independence, a key legal factor in fostering an internal, US application of the Discovery Doctrine, specifically against Indians, was the case *Johnson v. M'Intosh*, settled in 1823 by the US Supreme Court. The Court said that Discovery was part of US law because it had been operative previously as English law for the English colonies. The Court stated that "the United States . . . [and] its *civilized* inhabitants now hold this country. . . . [D]iscovery gave an exclusive right to extinguish the Indian title of occupancy, either by purchase or by conquest. . . ." (my italics). The Court asserted that native peoples were not "civilized"; therefore, taking away their lands and sovereignty was justified by the "character and religion of its inhabitants . . . the superior genius of Europe . . . [and] ample compensation . . . [which included] bestowing on them civilization and Christianity, in exchange for unlimited independence."

The Court had arbitrarily and unilaterally determined that the benefit of Discovery to native inhabitants was that they were forced to change their own cultures, and religious beliefs and rituals, for those of the invaders who had murdered their people and stolen their lands. The Court also employed the concept and practice of *terra nullius*—"empty land"—to reach its decision: empty, that is, of a European-like culture and the Christian religion. Chief Justice John Marshall, who wrote the Court's opinion, and his court colleague Justice Joseph Story acknowledged that Discovery rights could only be "maintained and established . . . by the sword"; this asserted the "right of the strongest"—in other words, "might makes right." *Johnson v M'Intosh* introduced also a new twist to European "Discovery Doctrine." As noted previously, historically in Europe when one nation conquered another, ordinarily the conquered peoples' private property in land was not taken: they kept their individual or family titles, but were expected to obey their new government. However, in the United States, the Court stated, as it added a second meaning to "conquest," this did not apply because of the conflict of American

40. Miller et al., *Discovering Indigenous Lands*, 6–8. Here I have substantially paraphrased rather than quoted Miller et al.

and Indian cultures and religions, and the Indians' "savagery."[41] The Court thereby broadened the concept, terms, and reality of "conquest" centuries after "discovery" and "conquest" originally had been devised and developed; the justices sought thereby to justify the expansion of the newly emerged US nation onto additional, extensive native lands, and in the process to provide new territory and additional terrigenous goods for a new "superior" but still philosophically and ideologically Europe-based US "civilization."

Using cognitive theory categories, Newcomb demonstrates how in the United States "people of European ancestry have historically succeeded in projecting their own imaginative categories and concepts onto the indigenous nations and peoples of this hemisphere, now known as the Americas. The ideas known as federal Indian law are a product or result of this multigenerational cognitive process. Categories and metaphorical concepts such as *Indians, tribes, primitive, heathen, pagan, infidel, backward, savages,* and *uncivilized* are no more descriptive of objective qualities or inherent characteristics of the indigenous peoples of the Americas than the terms *front* or *back* are descriptive of objective qualities or characteristics of trees."[42]

A demonstration of how arrogant and self-serving the Doctrine of Discovery was and is may be discerned through consideration of a thought experiment in which the "discoverers" are reversed. Imagine that before Europeans had left home on voyages of discovery, Indians or other peoples indigenous to Turtle Island had developed sailing ships and weapons superior to those in Europe, and crossed the Atlantic west to east. If indigenous peoples had had a "Doctrine of Discovery" it would have asserted that a superior civilization was one in which, first, the well-being of the community as a whole—which includes all individual members of any community—takes precedence over individuals (who share in community well-being); second, that equitable sharing of Earth's natural goods ("resources") was a defining characteristic of civilization; and third, that no one could claim private ownership of Earth's air, waters, or earth—these were the Creator's creation, to be shared by all biota, interrelated and interdependent; and that these understandings originated in the principles and natural laws given by the Creator to native peoples—to which they were faithful, as evident in their spiritual consciousness and contextual conduct. Consequently, indigenous peoples would conclude, European civilization was inferior, and had to be subjected to the superior culture and religion that had been born on Turtle Island—for the Europeans' own good, for a global common good, and to embody transculturally the Creator's intent and instructions to all humanity.

41. Ibid., 8.
42. Newcomb, *Pagans in the Promised Land*, 18.

An objective thinker today would judge that the Turtle Island characteristics just elaborated on behalf of Indians did indeed describe a superior civilization (in fact the communal civilization, that had common property and no private property, was the ideal presented in philosophies that originated in Europe: by Thomas More in *Utopia* and by Plato in *The Republic* and *Laws*). This superior civilization was more likely to have internal peace and, if its way of life were internationally practiced, external peace as well.

The Discovery Doctrine's past history and present use raise important questions regarding the continued use of Discovery today, in the original nations in which it was applied and beyond. These questions should prompt concern about possible extension of Discovery into space exploration and settlement, and the socioecological implications and impacts of its consequent trans-contextual, transnational, transcultural, and trans-species application vis-à-vis extraterrestrial intelligent life, and on commonly sought celestial bodies that have desired terrigenous goods.

Racism in Discovery

Racism has been largely unexplored thus far in most discussions of Discovery. It permeates the exercise of Discovery in the Americas not only as an undercurrent in the treatment of and injustices toward Indian peoples, but as an overt justification of treatment of another important racial-ethnic population in the United States: African Americans. Discovery enabled and justified Europeans' voyages to conquer new lands and goods, including on the continent of Africa.

Portugal's forays along Africa's west coast began during the life of King Henry the Navigator (1394–1460). In 1443, almost fifty years before Columbus landed on Turtle Island, Nuno Tristão arrived on the island of Arguim, west of Mauritania. In 1444, the Portuguese slave trade began from Lagos. The following year later, Prince Henry established a trading post on Arguim to acquire gum Arabic and slaves. Henry justified slavery on the grounds that he was converting slaves to Christianity. He was brutal toward the indigenous peoples of Africa, setting a devastating precedent for other Discovery practices toward native peoples, in Africa and the Americas.[43]

In 1482 Portuguese explorer Diogo Cão while sailing along the western coast of Africa discovered a river of immense size—which came to be called the Congo River—flowing out into the ocean, retaining its fresh water some twenty leagues into the vast salt water expanse.[44] Using Discovery Doctrine

---

43. Entries for "Arguin" and "Henry the Navigator" may be found on Wikipedia: www.en.wikipedia.org/wiki/Arguin and www.en.wikipedia.org/wiki/Henry_the_Navigator.

44. Hochschild, *King Leopold's Ghost*, 7.

phrasing, Cão made a European Discovery claim on the river (and thereby on the lands contiguous to it on both sides) for the Portuguese monarch by erecting a limestone pillar, on top of which was an iron cross, and on one side the monarch's coat of arms and an inscription that stated:

> In the year 6681 of the World and in that of 1482 since the birth of our Lord Jesus Christ, the most serene, the most excellent and potent prince, King João II of Portugal did order this land to be *discovered* and this pillar to be erected by Diogo Cão, an esquire in his household.[45]

Elements of Discovery are evident: the area was observed and explorers had set foot upon it; a pillar to establish their presence and Discovery claims was erected; the pillar had the coat of arms of the claiming monarchy; the Portuguese explorer had inscribed his name and the monarch that he served.

In 1491, the Portuguese sent an expedition of emissaries and clergy on a ten-day hike to Mbanza Kongo, the capital of the Kingdom of the Kongo. The ruler of the three hundred square mile kingdom greeted them cordially to his capital and his well-run country. The kingdom, as with other regional entities, had slavery as a social institution. The Portuguese integrated European and African cultural practices when they encountered slaveholding kingdoms: they began a slave trade that was lucrative for both cultures. Subsequently, Portugal launched its substantial slave trade from its African colonies. Africans' slavery came to be widespread globally.

Newcomb states that during the Era of Discovery, a "clear example of how Christian Europeans identified themselves with the Hebrews of the Old Testament is found in Portugal's crusading efforts to conquer areas along the western African coast during the Age of Discovery. Prince Henry of Portugal dedicated his entire adult life to fighting infidels in his bid to make Christendom victorious throughout the world." He cites a scholarly article by C. Raymond Beazley, who states that "in the fervor of the Sacred War" the Portuguese "take into their mouths the very language of the Chosen People" by asking God to assist them in battle by making the flood tide rise before its time to do so. "If God, they cried, had once made clear the way for the children of Israel through the Red Sea . . . could he not show as great a favor to his Chosen People [the Portuguese], and make the waters of Arguim Bay to rise before their time?"[46] (The Supreme Court of Tennessee, in *Cornet v. Winton* in 1826, used Portugal's Discovery invasions of Africa as support for England's and the US government's claims over Indian lands in what became the United States. Said the court: "Under this law of nations, they sent for

---

45. Ibid., 8 (my italics).
46. Newcomb, *Pagans in the Promised Land*, 44, citing Beazely, "Prince Henry," 21.

slaves to Africa, and consigned the captives and their descendents to perpetual bondage. *Under these auspices, was European dominion over the soil and over the bodies of men interwoven into the codes of American jurisprudence. It was deemed a title of the highest authenticity throughout the whole christian [sic] world.*"[47])

In 1500, a Portuguese ship blown by winds to what became South America landed in Brazil. Since the Treaty of Tordesillas had granted Portugal any discoveries in the region in which the ship landed, Portugal could claim and colonize Brazil. Complemented by Portuguese claims to Africa, Portugal's soon-burgeoning and lucrative slave trade was strengthened by the Africa-Brazil link. In a few years, slaves by the millions were sent from Africa—including from a village near Cão's Discovery pillar—to Brazil to toil in mines and on coffee plantations. Several priests who had been sent as missionaries abandoned their ministry, acquired concubines, and became slave traders. Previously free Africans were captured, baptized, given simple clothing, and shipped as slaves, chained in ships' cargo holds. Many perished en route.

Africans were soon enslaved en masse by European colonizers. Portuguese explorations of Africa, in particular, enabled Portuguese and other slave traders to enlist the aid of African middlemen to secure slaves from the African interior for export to the Americas via European ships sailing initially from Portuguese ports.

In South America, after Spain abolished slavery in its colonies, slaves remained only in the Portuguese colony of Brazil: enslaved Brazilian indigenous peoples, and Africans who had been transported in Portuguese ships.

In the early seventeenth century, African slaves were exported to the southern British colonies in North America; approximately 25 percent of slaves in these colonies came from territories in Africa that were either claimed under Discovery by the Portuguese, or were contiguous to Portuguese territories and provided slaves to them.[48]

Discovery thus played a role in the African slave trade, and through the centuries resulted in the oppression not only of Africans in their homelands and as slaves in the Americas, but eventually of African Americans into the twenty-first century. Discovery, then, as a root of racism, has historically contributed, since the colonial era, to the social crisis of racism in the United States. African Americans have suffered economic distress and political

---

47. Newcomb, *Pagans in the Promised Land*, 78, citing *Cornet v. Winton*, 2 Yerg. (1826), at 152–53 (Judge Haywood's opinion; Haywood's italics).

48. Hochschild, *King Leopold's Ghost*, 8–11. Linguists have found that the Kikongo language, from the area near Cão's pillar, is discernible today in Georgia and South Carolina in African Americans' Gullah dialect.

exclusion, even after slavery was abolished and equal opportunity laws were enacted. These injustices had Discovery roots, in the use of Discovery tenets in the colonization of Africa and in the carving of Africa into European colonies, which had led to social crises for Africans. By the twentieth century, apartheid was the most visible form of the crisis in human dignity and well-being that came to characterize much of Africa because of Discovery.

In the United States, African Americans conscious of the continuing discrimination and injustices that they endure today might consider, along with Africans, calling not only for reparations to slaves' descendants, but for international rejection of the Discovery Doctrine and mitigation of its impacts. Understanding the role of Discovery in fomenting their past and present injustices, rejection of its ongoing use in the present, and establishing a Discovery-free future would enable African Americans to uproot this ongoing cause of the social crisis that they experience.

Discovery Past and Present

The original, unjust bases for the Discovery Doctrine—religious imperialism, political imperialism, economic imperialism, cultural imperialism, and racism—have remained unexamined and unquestioned for more than half a millennium. Discovery has been reaffirmed time and again in US Supreme Court decisions issued over decades, and by US laws, policies, and practices toward Indians in particular. Similarly, in Canada, Australia, New Zealand, and Europe legal decisions and governmental laws, policies and practices against First Nations, aborigines, and native peoples continue to reinforce existing injustices and perpetuate new injustices.

Over the centuries since Discovery was first formulated, then, tenets of the doctrine have been invoked periodically, particularly in judicial proceedings. Cases continue to be adjudicated, on issues such as land and natural goods ownership and use, solely on the basis of prior courts' decisions regarding the issue at hand. The reasoning behind prior decisions, the Discovery elements that were cited pro or con to settle a case in the past, has not been critically examined to discern whether the original court decisions were just or unjust in themselves. As a consequence, past injustices are perpetuated through time, and perpetrated on new persons or social groups—especially Indian peoples. Indians as individuals and nations continue to bear the brunt of Discovery applications. Legal ingenuity has developed ways to put old wine into new wineskins, and thereby not only exacerbate past injustices but execute new and original forms of material and cultural harm.[49]

49. In his writings, Miller provides a lengthy list and in-depth discussion of such ongoing miscarriages of justice, including especially by the US Supreme Court.

In books he authored or coauthored, Miller provides and analyzes historical examples of five centuries of Discovery applications throughout the English-speaking world of former British colonies: in the United States, Canada, Australia, and New Zealand. When the UN Declaration on the Rights of Indigenous Peoples was approved by the UN General Assembly in 2007, the vote was 143–4 in favor, with seventeen abstentions; predictably, the four nations named, each of which has a significant population of indigenous peoples, voted against the Declaration. Subsequently, in part because of political changes in some governments' leadership, these same countries changed their original vote from negation to affirmation—but with "reservations," retaining the right not to rigorously implement all of its requirements because of concerns about implications of what they were pledging to do to fulfill the principles and policies of the Declaration.

The native peoples and native places of Turtle Island[50] on planet Earth have been engaged in a sustained social struggle for "life, liberty, and the pursuit of happiness" since the time of European "voyages of discovery." Indians have fought being oppressed by doctrines, both religious and political, that have been used to justify the injustices to which their respective peoples and places have been subjected.

Discerning and Dissenting from Discovery Deception

In US history, elements of Discovery remain not only in attitudes toward and treatment of Indians and African Americans, but also in certain forms of residual Social Darwinism regarding the economically poor. Currently, the rich resist and resent attorneys' and church leaders' efforts to enable the poor to benefit from equitable treatment under the amended US Constitution, and from the ideals of democracy that counteract Discovery uses today. The wealthy advocate, instead, upper economic class "rights" vis-à-vis the middle and lower economic classes. The upper class has embraced the European monarchs' claims to priority rights to land and economic benefits from the nations in which they live; they have adapted and hidden Discovery in US political and economic systems, which they largely control or at least manipulate to their benefit.

Vestiges of continuing Discovery elements are evident in US invasions of countries, particularly in Latin America, as described by historians throughout the twentieth century and into the twenty-first. Although the real rationale for foreign invasions was to acquire territory (at times for military bases) and natural goods (particularly oil), and maintain regional political

---

50. The Indigenous Environmental Network Web site logo portrays Turtle Island, with the Americas silhouette on the turtle's shell.

hegemony, the publicly expressed rationale has been to safeguard "national interests" and "national security."

The Discovery Doctrine in its ongoing iterations, directly or indirectly, has been the foundation for efforts by nations to range far from their borders to seek natural goods in others' territories or, on the pretext of "national security," to take military (sometimes through mercenaries), political (sometimes through local political elites as surrogates), and economic (sometimes through powerful local wealthy elites) control of other countries. It might well become operative in space exploration and colonization, UN treaties and other international agreements notwithstanding.

Occasionally a strong voice of reason has emerged from within the presumed solid front of the military-industrial elite that enforces ideals of Adam Smith and his contemporary US admirers. During the Great Depression, in his words and actions, retired Marine Brigadier General Smedley Darlington Butler (1881–1940) became such a voice. At the time of his retirement in 1930, Butler was the most decorated veteran in US history. In his postservice work, which included nationwide tours on behalf of the rights and needs of World War I veterans, Butler ran for the US Senate unsuccessfully in 1932 as a candidate in the Republican Party primary in Pennsylvania.

After his military career, Butler reflected back on his accomplishments. He realized that he had not really been fighting for "democracy" in his military expeditions to Latin America, but for US economic interests. He delivered a speech titled "War Is a Racket" in 1931, two years into the Great Depression, to the American Legion Convention in New Britain, Connecticut; it was elaborated subsequently in other venues and in his writings. In General Butler's words:

> War is just a racket. . . . It is conducted for the benefit of the very few at the expense of the masses. . . . It may seem odd for me, a military man, to adopt such a comparison. Truthfulness compels me to. I spent thirty-three years and four months in active military service. . . . And during that time I spent most of my time being a high-class muscle man for Big Business, for Wall Street and for the Bankers. In short, I was a racketeer, a gangster for capitalism. . . . I helped make Mexico, especially Tampico, safe for American oil interests in 1914. I helped make Haiti and Cuba a decent place for the National City Bank boys to collect revenues in. I helped in the raping of half a dozen Central American republics for the benefits of Wall Street. . . . I helped purify Nicaragua for the international banking house of Brown Brothers in 1909–12. . . . I brought light to the Dominican Republic for American sugar interests in 1916. In China I

helped to see to it that Standard Oil went its way unmolested.... The best [Al Capone] could do was to operate his racket in three districts. I operated on three continents.[51]

Butler's words serve as a continuing caution today regarding wars (in the past and present and, potentially, in the future): determine if their primary intent is to protect people or pursue profits, and who profits most from them. As robots explore planetary or lunar surfaces, and eventually are followed by astronauts, responsible citizens in all countries must seek transparency from their political leaders to ascertain the real purposes of such expeditions—perhaps especially when a government declares that ventures are needed to "protect national security." A key question to them should be: Earth's nations have signed space exploration and colonization treaties which state that no nation may own territory or goods exclusively for itself, because everything is part of the "common heritage of [hu]mankind" and natural goods ("resources") found should benefit first the least economically developed nations and the most disadvantaged groups within all nations: Is this particular voyage intended to fulfill those treaty obligations or will it benefit primarily transnational and, eventually, transgalactic corporations that will seek taxpayer-financed military accompaniment? The question will help to discover if Discovery is still undermining peoples' well-being. One might well wonder what the correlation might be between Butler's actual responsibilities in the field and the US publicly expressed rationale for them, and the responsibilities and public rationale for his military counterparts on space voyages, lunar bases, and planetary colonies, in the twenty-first century and beyond.

Discerning Developing Dangers: Discovery in Space?

Europeans' and Euroamericans' genocide against Indians in the Americas and near-terracide against Earth should be kept in mind during discussions of potential impacts by Earth explorers and settlers in cosmic space. If current ideologies and policies on Earth are sustained, and human consciousness and conduct remain unchanged prior to space voyages to "discover" habitable or natural goods-laden worlds, terrestrial ecological and cultural devastation will be exported to extraterrestrial places and peoples. Merely shifting segments of humanity from being Earth inhabitants to becoming

51. Butler's speech, his booklet "War Is a Racket," and a video produced by James G. Butler in which actor Graham Frye portrays Smedley Butler delivering the speech, are available on the Web: excerpt (with USMC photo): www.resistersbook.org/newsite/pdfs/butler.pdf; booklet: www.informationclearinghouse.info/article4377.htm; video: www.youtube.com/watch?V=F3_EXqJ8f-0.

occupants of new planets or celestial bodies might bring them temporary respite from human attitudes and actions on Earth, but these would soon surface on habitable places in space, and harm them and other places as they devastated Earth, humans' home planet.

In space, the operative rationales just noted, in which Discovery is embedded, could be used to justify invasion and seizure of celestial bodies, in whole or in part. UN documents' goals include preventing that possibility, and promoting use of newly found places and goods to benefit humankind as a whole. It remains to be seen whether or not this will be the case. Private entrepreneurs who hope to find riches in space currently are developing plans and projects to do so. They have released press statements to generate publicity for their private enterprises for corporate profit, engaged in lobbying efforts to secure political allies, and recruited long-range speculative investors.[52] Discovery could become operative because of a key provision of UN space treaties: any nation might choose to withdraw from a treaty at any time, after giving notice, and would no longer be bound by it.

The extended discussion of Discovery discloses what it meant in past and present human history to foment the current socioecological crises imperiling the well-being of the human community and the Earth environment. Repudiation of Discovery will eliminate one barrier to intercultural, international, and inter-intelligent species collaboration, conservation, and community. Reflecting on the past and present and re-visioning the future will help to provide a foundation for conceptualization, development, and establishment of an interdependent, integrated, and intergenerational relational cosmic community, and stimulate collaborative projects to effect its realization.

---

52. *New York Times* reporter Kirk Johnson has written about one company, Planetary Resources, that hopes to launch a rocket within the next decade to mine platinum and other precious metals; see Johnson, "Gold Rush Among the Stars."

3

Terrestrial Transformation

Historical Restoration of Earth's Socioecological Commons

In the twenty-first century ethical principles and guidelines related to care for the creation commons have been discussed in greater depth and breadth than had been the case previously. While the plight of the Earth commons became particularly pressing, socioecological considerations increasingly began to include references to terrestrial-extraterrestrial Contact. Reflection focused on how Contact might be mutually constructive—rather than destructive—for humankind and extraterrestrrial intelligent beings. These considerations came to be influenced by credible reports of ETI encounters; in light of scientific data that the cosmos was more than three times the age of Earth; and, because potentially life-friendly planets had been viewed through telescopes. It was speculated that ETI would likely be far more advanced culturally, socially, spiritually, and technologically than humanity.

Insights and speculation about Contact, and principles for conduct at and after Contact (as developed by humans), reflect terrestrial understandings current only at the time of their formulation. They are informed and limited by the extent and availability of up-to-date information provided by Earth's natural and social sciences and cultural histories; religious and humanist reflection; and initial human space explorations through telescopes, space probes, moon and Mars rovers, astronauts, and analysis of the information these have provided. Still, reflection on and from this current socially, spiritually, and scientifically dynamic context leads to a preliminary formulation of foundational terrestrial-extraterrestrial ecological and social principles. Some principles might become eventually intergenerationally, inter-specially, interculturally, and intercontextually normative, on Earth and in space; other principles might be rejected or revised in, for, or from new situations, or as a result of new data, new ethical understandings, or new working theories.

Dialogic Contextual Historical Projects

The development and implementation of dynamic practical projects, which will be *historical* in two senses—developed in history; gradually implemented through history—can both stimulate consideration of important ethical and contextual constraints and constructs, and accelerate efforts to contextualize ethical theory in specific places on Earth and in space. In fact, a dialogic relationship might result: consideration of both how humankind can do better on Earth, and how humans envision doing better in space, can result in thought and action in which present consciousness and conduct influence and even control, to some extent, what humanity will do in the future in distant places; and, the idealized future influences in the present what humanity does or must do on their home planet in order to realize what is now solely its vision of its ideal better place, as projected from its utopian thinking.

All of the preceding proposes to promote the socioecological well-being of present and future places in which humans live, work, and interact—among themselves and with other intelligent beings as they are encountered. In this ongoing dialogue, theory and practice, present and future, consciousness and conduct, and existing Earth place and projected celestial bodies' places, influence and alter each other—even before actual Contact or colonization. Consequentially, a transformation of human consciousness and conduct can at least begin, if not become fully actualized, when humankind envisions *Contact* elsewhere in the cosmos (whether it is initiated by humankind or by other intelligent beings), and when humanity considers possible *colonization* elsewhere in the cosmos, even on worlds yet unknown except from afar through optical and radio telescopes, and communication media.

In order for such a transformation to take place at least on a philosophical or intellectual level, let alone a material level, humankind must develop spiritual and scientific concepts and perspectives that build upon materiality and spirituality, and are not constrained by materialism or dogmatism (religious or scientific). *Materiality* is corporeal or material existence; *materialism* is (a) ideologically, belief that solely material objects, including matter-energy pheonmena, exist; or (b) economically, a focus on wealth acquisition and accumulation, or on attaining objects that money can purchase: consumerism as ideology. Human reflection should be influenced by humility, curiosity, and relatedness, not fundamentalism or anthropocentrism, when considering cosmic creativity and reality, and be open to considering potentially highly beneficial and original (from a human standpoint) and initially surprising and even provocative concepts suggested by either new space contexts or new space Contacts. In all of this, humans' regard for Earth as home, habitat, and provider; humankind-originating visions to renew and restore Earth places; humanity's contextual conduct in its efforts to realize (make real and present: *real*-ize) what is envisioned; and humans' striving to achieve ever better *topias* by realizing ever more advanced *utopias*, can carry over into initiating collaborative relationships in new places, on planets or other habitable celestial bodies, or on mutual discovery with other intelligent beings of each other's spacecraft.

A new Earth consciousness that transcends anthropocentrism and religious ideologies, and stimulates cosmocentrism (creatiocentrism, in a religious perspective) and an imaginative vision for improved future prospects for Earth could, then, be transferred to, and transformed as needed in, particular space contexts. Proposals for apprehending and concretizing the new consciousness and vision in space—even prior to departure from Earth—will provide a foundation for establishing Contact and living in new contexts in ways that benefit humankind, intelligent extraterrestrials who may be encountered, and new cosmic milieus in themselves, with their intrinsic value.

Conceptualization of what the future of Earth might be and striving to reach that state, in other words, will benefit all biota, complex or simple, and will promote the well-being of every locale selected for colonization; subsequent adaptation and development to meet human needs will be ecologically and socially integrated in and adapted to place.

Creative Creation

Confronted with the beauty and immensity of the cosmos, perhaps most strikingly on a starry night, people in cultures around the world have constructed over millennia diverse and distinctive creation narratives, incorporating into them their respective and initially differing concepts of a Creator (which might become integrated if the parties involved incorporate elements of each one's beliefs and understandings[1]). Over time in some cultures, story became history in the minds of those who heard it, and then became the immutable description of divine being(s) at work. In Western cultures, as biblical criticism by biblical scholars examined creation and other stories in the Hebrew Scriptures—the Bible for Jews, called the Old Testament by many Christians—scientist and biblical scholars alike determined that the

---

1. Thomas Berry and Brian Swimme have proposed that a science-based common story, evolving to incorporate new knowledge, would serve this purpose well. See Swimme and T. Berry, *Universe Story*.

narratives were neither literally historical nor scientifically possible. Christian scholars and Church leaders came to understand and to teach that the Bible is neither a history book nor a science text: it is a sacred book whose culturally based writers developed stories or interpreted actual historical events in such a way as to teach religious truth to the best of their ability, given their contextual limits of culture, language, time- and tools-limited scientific understanding, and prevailing doctrinal constraints. Those Christians who adapted, by contrast, a continuing interpretation of Sacred Scripture as the literal recording of divine teachings as dictated by God to biblical writers, became increasingly conflictive with scientists, biblical scholars, and clergy and laity of other Christian faiths. As a result of biblical literalism, eventually some scientists (especially but not exclusively atheists) came to deride not only Christian fundamentalism, but Christianity in general and Christians' Bible. In response, fundamentalist Christians developed "creation science" and its political and legal form, Intelligent Design.

Through millennia during and after these theological developments, Christian thinkers and leaders continued to ponder cosmic beginnings and ongoing processes, and God's role in and with them. Key questions focused on whether or not God created from any material existents at the cosmic creative moment, since it had been believed that no matter existed prior to the appearance of the universe: *creatio ex nihilo*, creation out of nothing; or, God created from previously existing material (a corollary speculation: whether this was or was not an eternal materiality, external to and co-existing with God): this appeared to be suggested by a literal reading of Genesis 1, where the Spirit moves over the waters (some disagree with this literalism, and suggest that primeval "waters" are solely symbolic, meaning a state of "chaos," a cosmos without order until God brings order through divine creativity): *creatio ex materia externa*; or, God created by birthing cosmic origins from God alone (a new understanding, perhaps, of the teaching of Maximus the Confessor who taught that the *Logos* is intimately related to every creature, all of whom are *logoi* who have emerged from divine creativity): *creatio ex Dei*. John 1:1–3, 14, which incorporates an early Christian hymn, celebrates in words paralleling Genesis 1 the eternal Logos who entered history in human form: "In the beginning was the Word [*Logos*], and the Word was God. He was in the beginning with God. All things came into being through him, and without him not one thing came into being.... And the Word became flesh and lived among us, and we have seen his glory...." Perhaps the *Logos* immanented (presenced) the cosmos within the primordial cosmic elements of the "primeval atom": indeed, "without him not one thing came into being" in the birth of the universe—Spirit was immanent from the beginning, in elements of the divine that were the bases for the Spirit's creativity, a material immanence of no like kind and still ongoing and complexifying.

The fullest expression of divine creativity is not *creatio ex nihilo* or *creatio ex materia externa*, but *creatio ex Dei*: there is no previously co-existing cosmic matter, as symbolized for some, by the Gen 1 "waters"; nor is there "nothing" at all since creation emerges from elements of the Spirit by a unique creative act of divine partial *kenosis*, a selective "self-emptying" or emanation of elements of divinity or of divine creative power about which Paul teaches regarding the Incarnation (Phil 2:5–7). All creation is sacred because it is infused with divine being; Incarnation in Jesus is a special divine mode of being in a particular material and visible form in a particular cosmic locale (though perhaps not just one visible human form in that Earth context, nor only one material form in only one locale at one moment of time in a vast universe that is more than three times as old as Earth).

Complementarily, a *machina in Dei* and a *deus in machina* (a machine in which a god is inherent/indwelling) would not be considered seriously as possibilities by peoples of faith traditions; and a *Deus in nihilo* or *nihilo in Dei* are intellectually and spiritually absurd concepts.

The mechanistic view of the universe that was stimulated by Newtonian physics is, effectively, a *machina ex Dei*, a machine that emerged from God. By contrast, Steven Hawking's proposal for a spacecraft to carry representative humans to the moon and Mars is, effectively, a literal *deus ex machina*, a "god of the machine" (a deity borne by mechanical means, operated by a stage crew behind the curtain) who arrives just in time to rescue someone in danger (a technique developed initially for and used in early Greek drama, millennia ago). In Hawking's proposed rescue scheme, the spacecraft is both the machine that rescues and the god it carries (in this case, the human-authored anthropocentric and materialist ideology that underlies the technology developed and used).

In regard to *creatio ex Dei*, might it be possible that since matter and energy are integrated and interchangeable, spirit and materiality are analogously inherently integrated? To some extent, this might be intellectually and spiritually expressed in the doctrine of *transubstantiation* developed in the Catholic tradition to describe how the substance of bread and wine are transformed into the body and blood of Christ, while the appearance ("accidents") remain the same. Such an understanding would enable *creatio ex Dei* to be more comprehensible and perhaps acceptable in some religious traditions.

It should be borne in mind in the formulation and sharing of all religious doctrines that humankind cannot comprehend, define, or limit divine being, although humans often anthropocentrically and anthropomorphically project what they believe or want divine being to be. Sometimes people come to understand aspects of divinity in deliberative thought based on sociohistorically and culturally developed doctrinal speculation, as evidenced by perceived historical acts and incidents, or as received in the spiritually intimate engagement of spiritual seers with the transcendent-immanent Spirit.

The cosmos, then, does not come into being as a *creatio ex machina*, a *creatio ex nihilo*, a *machina ex Dei*, or a *creatio ex materia eterna*. The cosmos comes into being as a *creatio ex Dei*.

Creation cannot be reduced to a machine or to a constricting design. It is a dynamic, evolving, complex aspect of integral being. It includes complementary and perhaps even seemingly contradictory or conflictive aspects, all in dynamic unity. The Creator brings forth—births—creation because of divine love, and imbues it with dynamism, creativity, and freedom and such inherent, autopoietic processes and characteristics as might, over time, be discovered by intelligent life and called "laws." (Enlightenment scientists and their successors, several of whom were Christian clergy or monks, declared that they were "thinking God's thoughts after Him.") God is not just in "heaven" or "the heavens," but in all creation, on Earth as in (the) heaven(s); in God we live and move and have our being—and the creativity and responsible freedom to be integrated co-creatively within cosmic dynamics and within biotic evolution. People share in the Spirit's creative power and creative processes when they are open to the Spirit's guiding Presence and strive to promote and sustain ecological well-being while meeting their sustenance needs. People reject the Spirit's Presence, creative power, and creative processes when they seek solely or even primarily to use their freedom to control creation to satisfy their wants without considering ecological impacts on their abiotic context and their extended biotic community.

Creation versus "Intelligent Design"

Toward the end of the twentieth century, some conservative Christians developed a theory of "Intelligent Design (ID)." The ID proposal presents creation as a *machina ex Dei*. It embodies a certain *mathematica ex Dei*, reducing creative freedom—God's, intelligent beings', and the universe's—to mathematical formulas in an effort to put a scientific "face" or veneer on "creation science," itself an expression of literal interpretations of Bible texts; it is biblically-related (not biblically-based) creationism. In ID thought, creation is seen not for the dynamic, creative, free integral being that it is, but as the predetermined development of a rigid divine plan, under initial and ongoing divine control. ID proponents expected their belief to be accepted as a legitimate "science" that should be taught as an option in US public school science classrooms, on a par with the natural sciences, whenever cosmogenesis and cosmic dynamics, and biogenesis and evolution, are presented in science courses. Their quest was thwarted in federal court, where judges detected and rejected ID's fundamentally religious, not scientific, nature.

ID is not *cosmology*; it is similar to *cosmetology*. In beauty salons, the "beautician" uses cosmetics to alter the appearance of their client, and thereby to alter to some extent that person's inherent sense of who they are, or inherent projection of whom they would like to be or at least appear to be; it is a human imposition on existing materiality, physiology, and psychology. In like manner, the ID advocate seeks to shape humans' perception of the materiality and energies of continuing, dynamic creation according to particular limited and limiting current religio-cultural human conceptions that flow from a particular religious base and rigid dogmas. ID strives to put a cosmetic "mask" or human "face" on a divine creation to make the cosmos appear to be that which it is not. ID masks creation's nature, its dynamic, divinely imparted creative freedom, under an imposed cosmetic design just as is done to a woman or man in a beauty salon on Earth, but on a much larger scale. ID hides, beneath a human-constructed and human-satisfying veneer, what has naturally been formed and continues to be forming in creation.

Creationism in itself and in its ID appearance is not wholly erroneous; it is a belief system which its adherents have a right to hold and seek to convince others to embrace. While not scientific (and with anti-science theories and beliefs), and while it denies that God gave freedom to creation to develop cosmic dynamics and biotic evolution, at least it teaches that divine creativity is immanent in the cosmos. ID attempts to limit that creativity, however, to a specific formulation expressed in Genesis creation stories that were not intended to narrate history or science, written in an era in which the ancient Hebrews had no scientific instruments to study astronomy and had a limited understanding of the cosmos. All materiality that exists was, for them, entirely contained in Earth and Earth's atmosphere, which included a "firmament," a hard ceiling that separated waters above and waters below, a firm structure to which heavenly bodies were attached and on which they followed their assigned tracks to orbit Earth.

The Creator Spirit does not premeditatedly specify a prescribed "design" that all that comes to exist after the singular "Day Without Yesterday" (in the terminology of priest and theoretical scientist Georges LeMaître) is predetermined to follow. Rather, the loving and creatively free intelligent Spirit initiates, in giving birth to original existents from divine Being, a mode of creative freedom that enables abiotic cosmic dynamics and biotic evolution. In reflecting on the process of biotic evolution, after observing that "natural selection primarily promotes adaptation rather than evolution,"[2] evolutionary biologist Francisco J. Ayala states:

> Natural selection is simply a consequence of the differential survival and reproduction of living beings. It has some appearance of purposefulness because it is conditioned by the environment: which organisms survive and reproduce more effectively depends on which variations they happen to possess that are useful in the place and at the time the organisms live.
>
> But natural selection does not anticipate the environments of the future; drastic environmental changes may be insuperable to organisms that were previously thriving. Species extinction is a common outcome of the evolutionary process.[3]

ID as a belief system and not a science is best taught, as are other religious faiths and doctrines, within its places of worship or its own teaching facilities—but, like other religious faiths or their particular doctrines (which include diverse creation stories), should not be imposed by governments or public education institutions on those who do not choose to accept ID's particular view of divine action and mode of creativity.

Ecclesia: A Spirit Transcendent and Immanent

Over millennia, since its origins in the life and teachings of Jesus and his followers ("disciples"), the Christian Church has focused primarily on *God-Transcendent*: a divine Being distinct and separate from creation and cosmos, unchanging and unchangeable, eternally the same. Despite the principal ecclesial dogma of divine Incarnation in Christ Jesus, a complementary understanding of the Spirit has been minimally acknowledged or accepted until (and ever since) the twentieth century—*God-Immanent*: a divine Being-Becoming who is unaltered in essence but suffers with and transforms the suffering of those wounded and killed as biologically and socially evolving creation unfolds. The subsequent overarching theological and doctrinal understanding did not emerge in depth until approximately the last decade of the last century: God is neither God-Transcendent nor God-Immanent but both, *God Transcendent-Immanent*: a divine Being-Becoming who permeates, is lovingly solicitous about, and engages with creation in love, mercy, and justice but is simultaneously distinct from creation and cosmos. Creation is, effectively, simultaneously God's first "incarnation," the first moment in cosmic history when God assumed a material presence. In

---

2. Ayala, *Darwin and Intelligent Design*, 55.
3. Ibid., 68.

this instance and instant, the material manifestation of the Creating Spirit occurs when the Spirit assumes materiality in part and within divine being. John 1:14 teaches that the "Word became flesh and lived among us ," but the Logos already was immanent in the cosmos, since the Logos-logoi dialogic relationship continued since its beginning in God's primordial creative self-giving act. All biota and abiota are *logoi* related to the eternal *Logos*. This insight was expressed originally by the monk, mystic, and martyr Maximus (580–662 CE). An elaboration of the insight would be that all creatures are, in a spiritual perspective, word-beings spoken by the creating *Logos*, the Word-voice; each in turn expresses the divine Voice in their being and in their diverse voices, and expresses themselves in combination and concert with other *logoi*. All *logoi* are sparks of a divine *Logos* flame as they express words from the Word.

The concept and term *panentheism*, which indicates that creation is dialogically *in* God, replaced both the distant God "in heaven" or "in the heavens," and the God identified with and confined within creation, the latter as designated by the concept and term *pantheism*. A new ecclesial, ethical, theological, and philosophical realization has been formulated and developed: pan*en*theism better expresses the divine mode of Being-Becoming in the cosmos, which through its abiotic dynamics and biotic evolution is integral being-becoming. Panentheism is based on in-depth consideration of the ramifications of Paul's teaching about God's *kenosis* ("self-emptying") in Philippians 2:5–8, in which God's care for the world meant that "Christ Jesus, who, though he was in the form of God, did not regard equality with God as something to be exploited, but emptied himself, taking the form of a slave, being born in human likeness. And being found in human form, he humbled himself and became obedient to the point of death—even death on a cross."

Similarly, in the birth and dynamic development of the cosmos: the divine Spirit did not coercively design and determine exactly what the universe would be and become after the singular, initial divine act of creativity eons ago when Spirit gave birth to the primordial cosmos and provided it with its "genes"—its abiotic cosmic parameters and potential, and guiding inherent laws to develop over eons; the cosmic means and "tools," in time and over time, to creatively and gradually be and become in immanent divine Presence. It is possible that God self-humbled, and self-emptied some elements of infinite divine Being-Becoming, by creating the cosmos with (meaning together with and as part of) them and taking a material form in the "primeval atom" and existing in and sacralizing, while remaining distinct from (as Logos-logoi) material-energy manifestation in the cosmic energy-matter singularity. Through time, then, the Logos suffers with those who and that which suffers distinct from the Logos, and empathizes with the sufferers and transforms and incorporates that suffering, in the spiritual dimension of reality.

The Spirit, therefore, did not enter cosmic history initially and solely when becoming incarnate in Jesus to teach and guide humanity directly and specifically, but was already present in and sacralizing integral being.

Considerations of a Cosmic Commons, by adherents of Christian faith in particular, are ecclesially and ethically clarified by reflection on Philippians and other biblical texts—both the scriptural texts discussed elsewhere regarding the initial and ongoing divine creation of the sacred cosmos, and verses describing responsible intelligent beings' care for creation, as evidenced and imaged by God; and by exploration of core teachings about divine essence, existence, and action. These are expressed or modeled in 1 John 4:7-8, 11-12, which teaches, "Beloved, let us love one another, because love is from God; everyone who loves is born of God and knows God. Whoever does not love does not know God, for God is love. . . . Beloved, since God loved us so much, we also ought to love one another. No one has ever seen God; if we love one another, God lives in us, and his love is perfected in us"; 1 Corinthians 13:8, 13, which observes concerning core Christian virtues, "Love never ends. . . . And now faith, hope, and love abide, these three; and the greatest of these is love"; and Matthew 22:36-39, in which, when asked which of the commandments is the greatest, Jesus replies, "'You shall love the Lord your God with all your heart, and with all your soul, and with all your mind.' This is the greatest and first commandment. And a second is like it: 'You shall love your neighbor as yourself.'"

Christian Testament teachings seriously question and contradict "only faith" or "only faith and Scripture" ideologies and theologies. In Matthew 7:21 Jesus states, "Not everyone who says to me, 'Lord, Lord,' will enter the kingdom of heaven, but only the one who does the will of my Father in heaven." James 2:14-17 declares, "What good is it, my brothers and sisters, if you say you have faith but do not have works? Can faith save you? If a brother or sister is naked and lacks daily food, and one of you says to them, 'Go in peace; keep warm and eat your fill,' and yet you do not supply their bodily needs, what is the good of that? So faith by itself, if it has no works, is dead." First John 3:17-18 teaches, "How does God's love abide in anyone who has the world's goods and sees a brother or sister in need and yet refuses help? Little children, let us love, not in word or speech, but in truth and action."

In these considerations, Incarnation is not interpreted as being intended to lead to sacrificial atonement. Rather, as understood by Irenaeus (ca. 130-202 CE) in the second century, the Logos came among us to teach and exemplify the way in which we should *be*: in relation to God and to each other (and, it should be added, to all biota and to Earth). Irenaeus believed that Christ would have come whether or not humans had sinned. Since they had sinned, Christ came as their Savior, as a new Adam. In Jesus, the self-humbled Logos suffered materially and visibly, to the point of death, in conscious material creation. In the spirit of Irenaeus, a response to Thomas Paine's questions about and rejection of divine Incarnation and multiple sacrifice in a plurality of worlds could be that the sacrifice was not a divine intention in this or any other world. God among us does not want innocent suffering, and the innocent should not suffer on behalf of the guilty. Jesus, in this understanding, was not sent or mandated to be crucified, but sacrificed divine power to teach humanity how to walk with God and care for each other, and as a consequence was executed by those with power who rejected his teaching and example.

Natural Goods

One important conceptual and linguistic change that would alter both how humans view creation and the integral being of the cosmos, and how they interact with their Earth home, would be to realize that Earth provides natural *goods*, not natural "resources." A "resource" is something regarded as awaiting human extraction or exploitation in order to meet human needs and wants. A river might be regarded principally or solely as a source of hydroelectric power, to be dammed to enable agricultural irrigation, provide urban residents with water to drink, wash, or refresh lawns; its benefits are limited to being an instrumental, anthropocentric provider of something beneficial to and used by the human community, no matter what the impacts would be on fish populations which provide needed nourishment and livelihood for human fishers, and food for bears and eagles. In this view, too, a forest would be viewed as merely standing lumber, awaiting logging, sawmills, and carpenters and cabinet makers; a mountain might be seen solely as an immense source of gold, silver, or coal, awaiting mining operations that would blast off its crown and tunnel deep into its core.

"Goods," by contrast, should be understood to be natural places—water, soil, and air, and minerals and other natural features or Earth components and contents—that are available where they are and often best left in place to meet the essential needs of biota other than humans, as well as for humankind. Particular types of natural goods, such as those mentioned, might be needed by humankind even more over time for survival, and should be carefully used in the present so that a sufficiency might remain intergenerationally for those to come.

Corporeal Common Ground: Commons Good, Commons Goods

Earth is a commons. It is the abiotic context of biotic existence. In it, diverse and complex types of life coexist and coevolve when Earth provides the elementary necessities of life for all life in its being and becoming.

While Earth is a commons, the cosmos is a commons of commonses, the "mother of all commonses." Earth itself is born from and in divine Being-Becoming: Earth's existence first became possible when the cosmos emerged in the singularity, the primeval explosive inflation with which it began, from which it grows and matures from its initial state in the "day without yesterday," and as a result of which it has become 14.82 billion Earth years of days since yesterday, and now gives birth itself to every commons that has come to be. In every historical moment, humans' and other intelligent beings' then-currently available science and technology limit the extent to which any intelligent life can know, comprehend, and interact with the vast cosmos, let alone any parallel dimensions of integral being that might exist. Earth's scientists and humanists alike can explore to no greater extent, with their current data, theories, and tools, than that permitted by what is presently visible (through telescopes and microscopes) or audible (through radio telescopes and other technologies that detect sound transmitted in ocean depths, on Earth's surface and in its subsurface [by radioactive matter, for example]), from near and far.[4]

Every habitable place that is inhabited is a commons. It has emerged and developed in a way that enables ongoing generation of life—biota that have over time experienced and adapted to geological changes and challenges (such as earthquakes and tsunamis, both born of plate tectonics), metereological changes and challenges (such as alterations of climate, powerful storms, and fierce winds), and biological changes and challenges (such as predator-prey relationships, entry into new places occupied already by other biota, achieving balance with or successfully rejecting biota entering its place, disappearance of plant and other food supplies, adapting to new niches). After all of this, intelligent biota have evolved, and adapted to and from altering and altered contexts.

Throughout the cosmos, in every commons and among commonses, intelligent beings become a consciousness of the local and universal

---

4. Some scientists, such as astrophysicist Jacques Vallee, suggest that "extraterrestrial intelligence," ETI, might rather be or also be "extradimensional intelligence," EDI: intelligent beings who can cross dimensions, appear to humankind, linger, and then disappear suddenly (see chapter 7). Should this indeed be the case, then there are locales and scientific data and technologies that enable transdimensional travel—knowledge and tools that humans currently lack (and might continue to lack for another thousand or more years).

commons; as participants and part of a commons, they are that commons and the cosmic commons reflecting on itself. Terrestrial and extraterrestrial intelligent life, therefore, has or could have communion and community on the common ground they share: both the shared intellectual and spiritual space, and the material places in which they come or will come into Contact. A pre-Contact potential for respect, interaction, and communion with intelligent others awaits its realization in a positive engagement of terrestrial and extraterrestrial intelligent life.

The *commons good* on the material common ground shared by intelligent life is the well-being of their home, the integrity of the place where they live. The *commons goods* are those natural places that provide for the needs of their visitors or inhabitants by their specific setting, geophysical characteristics, or the natural goods embedded within them which are intended to be equitably shared to meet the needs of each and all.

The Holy Qur'an, the sacred book of Islam, expresses, without using the words, the concepts of a commons in which all life is interrelated and of natural goods by which they are enabled to provide for their sustenance. This state of being on Earth flows from Allah's solicitude for all creatures. Allah calls humankind to be responsible for divine creation, and to have and express in acts of compassion a like regard, expressed by fulfilling their responsibilities for each other and all life:

> "O ye who believe! Spend out of (the bounties) We have provided for you" (S. II, Section 34, v. 254). According to scholar Abdullah Yusufali, "spend" means "give away in charity, or employ in good works, but do not hoard. Good works would in Islam include everything that advances the good of one that is in need whether a neighbor or a stranger, or that advances the good of the community, or even the good of the person himself to whom God has given the bounty. . . . The bounties include mental and spiritual gifts as well as wealth and material gifts."[5]

> "And the earth We have spread out (like a carpet); set thereon Mountains firm and immovable; and produced therein all kinds of things in due balance" (S. xv, vv. 19–20). Yusufali explains that Allah says, "We provide sustenance of every kind, physical, mental, spiritual, etc., for you (i.e., for mankind). But We do more: We provide for everyone of Our creatures. And there are those of which mankind is not even cognisant. We provide for them also. There are those who may at first sight appear hostile to man, or whom man may consider hostile, such as wild and

5. *Holy Qur'an*, 102.

noxious animals. They are Our creatures, and We provide for them also, as they are Our creatures."[6]

"He has created the heavens and the earth for just ends" (S. XVI, v. 3).[7]

The ideas presented can have a significant impact on terrestrial transformation. They can catalyze a cosmic consciousness and a contextual conscientiousness that might be static (remaining only in the realm of knowledge) or dynamic (ranging outward from knowledge of possibilities, to visions of what might be, to projects to realize knowledge and visions materially in the commons, not merely conserve them solely philosophically, theologically, intellectually, and ideologically).

International Visions and Projects for Terrestrial Transformation

While an abundance of documents exists that have originated in and been developed by major social institutions, consideration for terrestrial transformation will be limited to and will focus on international political-ethical-economic-ecological statements that involved widespread consultation before promulgation: United Nations documents (developed by representatives of the respective member states), and the Earth Charter (developed by representatives from six continents, numerous nations, distinct cultures and ethnicities, multiple religious commitments, and diverse levels of socioeconomic status). The basic rationale for this source limitation is that just as the UN documents needed to be approved by diverse peoples and nations before being issued as binding international instruments, any documents offered by humankind extraterrestrially (whether for solely human use or as bases for terrestrial-extraterrestrial agreements) should be dynamic, potentially collaborative working documents. A document developed by a secular give-and-take that involves diverse cultures has as its precedent the "practice" of formulating UN documents. A complementary terrestrial-extraterrestrial collaboration in context would be established more readily.

Religious bodies, too, including the Vatican, the World Council of Churches, the National Council of Churches, the National Association of Evangelicals, the Coalition on Environment and Jewish Life, the Shalom Center, the Muslim World League, and other international, national, and regional religion-based groups have strongly addressed ecological problems that harm creation and oppress human communities (especially the

---

6. Ibid., 640.
7. Ibid., 656.

vulnerable poor and indigenous populations, at home and abroad). All the religions cited require human remorse and responsibility for past and current adverse actions against creation, and human resolve to make restitution for social harm and to remedy ecological harm to shape a better human and planetary future. Religions' involvement in socioecological concerns is well documented among their respective constituents and beyond, and has been influential in partially addressing pressing crises afflicting Earth.

The focus in these pages will be on non-religious statements. While religion-based documents are particularly fruitful internally in their own tradition and among their own members, and at times effective for influencing the ideas and actions of open-minded people external to particular traditions, this is not universally the case: some people reject religions' social statements precisely because they emerge from religious faith, while some adherents of a particular faith reject statements developed by other faiths. Secular statements, by contrast, have global appeal for a broader public, including most religions' followers.

United Nations international documents related to terrestrial socioecological transformation include the Universal Declaration of Human Rights (1948), the Antarctic Treaty (1959), the World Charter for Nature (1982), the Millennium Development Goals (2000), and the Declaration on the Rights of Indigenous Peoples (2007). The complementary and also internationally developed and distributed Earth Charter (2000) originated with United Nations personnel, Canada's Maurice Strong in particular. These significant instruments related to socioecological conditions and concerns have neither emerged from nor express a particular faith tradition. Their faith-neutral character enables them to provide insights and principles more globally acceptable, without alienating one or another segment of the population because of a perceived bias religious statements have in favor of a particular faith-based worldview.

Concern about Earth events that have been catastrophic socioecologically in particular times and places because of their impacts on human communities around the world, prompted member states of the United Nations, and environmentally conscious and concerned organizations and individuals without an official political status, to seek solutions that include mitigation of harms present now, and prevention of scientifically projected future harms.

In the international documents that follow, numerous and diverse topics and issues are explored. Consideration of the documents will necessarily be space-restricted. Discussion will include presentation of selected segments of the documents, with affirmation of positive elements (which comprise the majority of the documents' content), and critical analysis of, and suggestions for amending other parts. This will help stimulate development and evolution of future documents that promote the well-being of Earth and Earth's biota, and of celestial contexts and their biotic communities, and prompt projects to promote realization of the documents' positive hopes for future Earth, which would have a positive impact, if realized, on humankind residing elsewhere in the future cosmos.

UN Terrestrial Visions and Principles

In a world rife with confrontation and combat, collaborative development of internationally accepted statements is often an elusive task. The general consensus at the United Nations and among its member states (at least publicly) is that amid all the strife afflicting nations and peoples, ecological well-being and human rights are areas in which substantial agreement can be reached. The final versions of the documents are a tribute to the collective insights, tenacity, creativity, and foresight of their authors and the government leaders and bodies that supported them and are committed to implement them.

Universal Declaration of Human Rights

The human rights declaration (DHR) attempted, in the aftermath of World War II and the Holocaust, to promote peace among nations and peoples, and to establish principles to which there would be global assent and commitment, even if they were not immediately implemented. The principles reflected historical contexts and realities and affirmed common aspirations for a better global future for all peoples.

In its Preamble, the DHR declares that global freedom, justice, and peace are founded on recognition of the dignity and "inalienable rights" of all humanity (par. 1). Disregard for human rights led to "barbarous acts" (par. 2). If peoples' only perceived recourse in response to these acts is not to be rebellion, human rights must be safeguarded by laws (par. 3), and by fostering "friendly relations" among nations (par. 4); UN members' citizens reaffirm faith in human rights, including "equal rights of men and women" (par. 5). There must be "universal respect" for human rights (par. 6), based on a "common understanding" of what they are (par. 7). In the Preamble's concluding paragraph, the UN General Assembly proclaims the DHR to be a "common standard of achievement for all peoples and nations."

The thirty Articles that follow present human rights thinking and principles upon which the signatories agree. These include: "all human beings are born free and equal in dignity and rights" (Art. 1); all people are entitled to all the DHR's rights and freedoms, regardless of distinctions of "race, color, sex, language, religion, political or other opinion, national or social origin, property, birth or other status," or based on any geopolitical differences (Art. 2); rejection of slavery (Art. 4) and torture (Art. 5); equal status in legal contexts (Art. 6, 7); freedom within borders (Art. 13) and to seek asylum externally (Art. 14); right of property ownership, individually or "in association with others" (Art. 17); freedom of "thought, conscience, and religion" (Art. 18); right to participate in government, "directly or through freely chosen representatives," whose authority depends on the "will of the people" (Art. 21); right to social security (Art. 22); right to work, to choose their job, have good working conditions and "protection against unemployment," to "equal pay for equal work," to "just and favourable remuneration," and to "form and join trade unions for protection of interests" (Art. 23); and the right to an adequate "standard of living" to provide self and family with essentials needed for "health and well-being," which include food, clothing, housing, medical care, necessary social services, and the right to security in the event of unemployment, sickness, disability, widowhood, old age or other "lack of livelihood in circumstances" over which they have no control (Art. 25). The rights require that each person fulfill "duties to the community" (Art. 29).[8]

Antarctic Treaty

On December 1, 1959, after several years of negotiation, the Antarctic Treaty (AT) was promulgated by the governments of twelve nations. Prior to the treaty, the respective countries had been laying claim to segments of Antarctica and maneuvering to control them. Each nation had some basis for their claim, usually related to elements of the unnamed Discovery Doctrine. Previous to the Treaty, scientific teams from many of the nations had worked collaboratively during the 1957–58 International Geophysical Year; some were still in the field.

The signatories agreed in the Preamble that it is "in the interest of all mankind that Antarctica shall continue forever to be used exclusively for peaceful purposes" (par. 2), that substantial scientific contributions resulted from previous cooperation (par. 3), that ongoing cooperation in science is in accord with the "interests of science" and will promote human progress (par. 4), and that the Treaty will facilitate continuing use of the continent for "peaceful purposes only" and advance UN Charter "purposes and principles" (par. 5).

The Articles that follow state that nations' use of Antarctica can only be for "peaceful purposes": no military bases, maneuvers, or weapons testing,

---

8. As will be seen in chapter 11, "Cosmic Charter," these principles, and those that follow from other UN documents for Earth and outer space consciousness and conduct, are readily adaptable to human rights and responsibilities in exoEarth contexts.

and members of the military present and the equipment they have must be intended only for "scientific research" and other "peaceful purposes" (Art. 1); that "scientific investigation" and "cooperation" shall continue (Art. II); that to sustain scientific cooperation, the Governing Parties should exchange "scientific programs plans," "scientific personnel," and "scientific findings," and establish "cooperative working relations" with organizations, including UN agencies (Art. III); that no interpretation of the Treaty can mean that a "Contracting Party" has renounced rights of or claims to territorial sovereignty in Antarctica, nor any basis of its claims to territorial sovereignty, or that any acts while the Treaty is in force can be a "basis for asserting, supporting or denying a claim to territorial sovereignty, or create any rights of sovereignty (Art. IV); that nuclear explosions and nuclear wastes' disposal are prohibited, while any laws or policies regarding the foregoing that have been developed external to Antarctica will be applicable to Antarctica (Art. V); that the Treaty's provisions do not apply to the high seas in proximity to Antarctica (Art. VI); that the Contracting Parties noted in Article IX may designate their own observers who will have access to any nation's Antarctic areas or facilities to inspect stations, installations, equipment, and ships or airplanes bringing in or taking out personnel and equipment, and that every nation involved will have the right to aerial reconnaissance over all areas at any time, and the Parties will provide notice in advance of their exploratory and other expeditions on the Antarctic surface, and of military personnel or equipment (as allowed by Art. 1.2) that are taken to Antarctica (Art. VI); that Contracting Parties alone have authority over their observers, and any jurisdictional disputes should be resolved immediately into a "mutually acceptable solution" by Contracting Parties (Art. VIII); that Parties shall meet in Canberra to exchange data, consult on "matters of common interest pertaining to Antarctica," and propose measures to further Treaty "principles and objectives," including that Antarctica be used solely for "peaceful purposes," that scientific research and international scientific cooperation be facilitated among Parties, and to ensure "preservation and conservation of living resources in Antarctica" (Art. IX; the phrase "living resources" is jarring and provocative, and should be altered: it callously reduces biota with intrinsic rights to things that have solely an instrumental function, to be used as humans choose); Contracting Parties in conflict should resolve disputes among themselves, and the Parties involved should consent to refer any irresolvable dispute to the "International Court of Justice for settlement"; failing that, peaceful means of resolution should be pursued (Art. XI); the Antarctic Treaty may be amended or modified by unanimous consent of the Contracting Parties (Art. XII); and any UN State or other State invited may accede to the Treaty if all of the Contracting Parties consent (Art. XIII).

The Antarctic Treaty and its international acceptance should stimulate serious discussion not only about how states should act in space among themselves, but also how they should interact with other intelligent beings upon Contact and thereafter, on Earth and in the heavens.[9]

World Charter for Nature

In the latter decades of the twentieth century, United Nations' member states became increasingly concerned about geophysical, biological, and ecological implications of humans' utilization, exploitation, and depletion of Earth's natural goods. Consequently, in response to these and other pressing and potential ecological crisis issues and concerns, the United Nations General Assembly selected representatives of member states who were respected experts to collaborate in the development of a World Charter for Nature (WCN). In the WCN, promulgated in 1982, the UN responded to the devastation of Earth's air, water, and earth on a global scale, recognizing that ecological relationships mean that no form of pollution is solely locally situated and limited. Previously, the document declared, the General Assembly had stated (1980) that "the benefits which could be obtained from nature depended on the maintenance of natural processes and on the diversity of life forms"; these benefits were being "jeopardized by the excessive exploitation and the destruction of natural habitats"; collaborative efforts were necessary to protect Earth. Member states should take into account, in the process of altering or removing natural goods within their borders, "the supreme importance of protecting natural systems, maintaining the balance and quality of nature and conserving natural resources, in the interests of present and future generations."

In its opening paragraphs the WCN declares that "natural processes" and a "diversity of life forms" are being jeopardized by human actions. The document advocates maintaining nature's own rhythms and quality, and protecting "natural resources" through international cooperation to ensure nature's intergenerational sustainability, and the ongoing availability of natural

---

9. The AT provides a "road map" for how Earth nations might resolve similar disputes over territory or the natural goods therein on Earth and in space. Currently, there is conflict over contradictory claims to areas opened up because of global warming at Earth's other pole, in the Arctic. Commercially and militarily beneficial sea routes are becoming available, as are natural goods below the waters that have replaced what was once impermeable ice. Ironically, a major international dispute has arisen over nations' rights to petroleum that might be beneath the Arctic Sea: global warming caused by humankind, particularly by the burning of fossil fuels, has catalyzed the Arctic meltdown that now makes available additional fossil fuels, whose extraction and use will effect still more global warming, with rising seas threatening urban and agricultural areas elsewhere—and the lives and livelihoods of their residents.

goods for humans and other biota. The WCN states, too, that humankind is part of nature and depends on natural systems, and that civilization itself has originated in and is sustained by nature. In this assertion, the document breaks away from anthropocentric, androcentric, and hierarchic understandings of humans' relationship with nature that have been expressed in philosophical, political, religious, and economic ideologies for millennia. Instead, humans should see themselves as integrated with and related to other biota in coherent ecosystems; they are responsible to and for nature as a whole and in its rhythms and processes. Humans are not situated over and superior to an externalized, inferior nature; they are immersed in and mutually engaged with nature—or should be—and are to respect natural biological processes and biotic diversity.

Similarly, the WCN notes the uniqueness of all biota, and advocates respect for organisms; this respect is to be guided by a "moral code of action," to be operative whether or not specific biota are considered to have instrumental value for humanity. It deplores adverse impacts on nature by "excessive consumption and misuse of natural resources" and the lack of "an appropriate economic order" in the local and international human community—thereby displaying concern about issues surrounding both the use and the distribution of natural goods. (Since capitalism is the dominant and dominating "economic order" globally, this statement presents a critique, however indirect, that global capitalism, whose ideology and activities are adversely impacting global nature and human societies within nature, is not an appropriate system. Recognizing that environmental degradation has not been confined solely to capitalist countries, however, General Assembly delegates did not endorse competing economic systems such as socialism or communism.)

The WCN proposes a series of principles intended to ensure conservation of nature in the present and future. These principles include: respect nature and protect its "essential processes"; maintain genetic viability, including through safeguarding population levels sufficient to enable species to survive, and ensuring sufficient habitats to provide for their needs, with special attention paid to unique areas and ecosystems; secure nature against impacts caused by war; ensure that social and economic efforts include nature conservation (expressing thereby human self-interest and self-assurance, since conservation is advocated to provide for human subsistence, and may be enhanced by science and technology); plan development of Earth's areas in such a way that biota and aesthetic beauty are protected; use carefully "natural resources" [sic]; safeguard biota: "*living resources* shall not be used in excess of their capacity for regeneration"; maintain soil productivity and water quality and supply; and take care that non-renewal resources are "*exploited* with restraint," with such "*exploitation*" to be compatible with ecosystem functions (emphases added). An anthropocentric ideology is apparent in these principles, which are intended to safeguard natural goods and biota not for themselves or because they have intrinsic value, but because they have instrumental value for humankind. *Living resources* and *exploitation* are particularly troubling, as they run counter to the document's title and stated intent to be a World Charter for Nature: the phrases highlighted urge conservation of Nature because it is in human self-interest to do so.

The startling and provocative phrase *living resources* (as noted earlier in AT analysis) should be substantially altered; it is intricately intertwined with "exploitation." In both, human hegemony over and superiority to other biota is advocated, or at least asserted, by this (mis)understanding of the interdependent relationship and ecosystemic integration of humans within the biotic community and their shared common ground, their Earth context. Even a softening of "living natural resources" by replacing it with "living natural goods" would not help much, if at all, to ameliorate the attitude and actions expressed; even though it at least would implicitly indicate the intrinsic value or inherent worth of nonhuman biota, its use in this context reduces that value to instrumental value: land, natural goods, and natural living goods are all awaiting human "exploitation," and thus care for and conservation of these goods is exhorted primarily so that they might provide human benefit, not for what they are in themselves as part of the natural world.

The word *exploitation*, too, should be replaced, in this and other contexts. Unfortunately, it has become the common word used generically in regard to extraction and use of natural goods, without sufficient thought given to its full meaning. If one were to talk analogously about "exploiting" the abilities of working people, particularly ethnic minorities, for example, such a proposal would not be well-received because of historical (including in present history) abuse of segments of the human population by dominant ideologies and the groups and governments espousing and exercising them. Lamentably, in the contemporary socioecological reality, "exploitation" expresses well current human consciousness and conduct vis-à-vis Earth and Earth's natural goods and biota. As with the use of the word *resources*, which indicates something awaiting human exploitation and use, rather than "natural goods" that are available to humans, other biota, and operative geodynamics, *exploitation* enables and exacerbates human ecological and economic excesses—causing the very problems which the WCN hopes to ameliorate or avoid. A better term than *exploitation* would be "responsible development of natural goods to meet human needs"—lengthier, but a stimulus to thought about how humankind should interact with its Earth home and the broader biotic community.

The WCN advocates other practices to safeguard nature and natural systems through responsible human assessment of proposed courses of development of natural goods; responsible decision-making; and minimal impacts on ecosystem dynamics.[10]

Declaration on the Rights of Indigenous Peoples

The United Nations approved the Declaration on the Rights of Indigenous Peoples (DRIP) in December 2007. The document was in the making for some thirty years, and succeeded in receiving passage by the UN General Assembly because of efforts initiated and led by the International Indian Treaty Council (IITC), which in 1975 became the first native peoples' organization to receive nongovernmental organization (NGO) status from the UN. The DRIP supports the efforts of perhaps the most ignored ethnic and cultural group—indigenous peoples—to secure human rights in territories and on lands where they have lived for millennia, and to seek redress for historical violations of their human rights. As noted previously, indigenous peoples have been victims of and oppressed by the Discovery Doctrine since the fifteenth century.

The DRIP states that indigenous peoples "have the right of self-determination," and therefore they are empowered to "freely determine their political status and freely pursue their economic, social, and cultural development" (Art. 3); they have a "collective right to live in freedom, peace and security as distinct peoples" (Art. 7); national states should establish "effective mechanisms" in order to provide redress for any past action which had "the aim or effect of dispossessing them of their lands, territories, and resources," and to prevent such action in the future (Art. 8); the right to "maintain and develop their political, economic and social systems or institutions," and to securely exercise traditional subsistence and development practices and economic activities (Art. 20); the right to ongoing spiritual relationships with their traditional areas and in regard to the natural goods ("resources") existing there, intergenerationally (Art. 25); the "right to the lands, territories and resources which they have traditionally owned, occupied, or otherwise used or acquired," and should receive from states recognition of this right; the recognition should respect indigenous peoples' "customs, traditions and land tenure systems" (Art. 26); states should provide an open and transparent

---

10. The international WCN principles, policies, and proposed practices will prove helpful as bases for developing an intergalactic or interstellar interspecies instrument, a covenant between distinct intelligent species that becomes an ongoing and evolving negotiation of mutually accommodating and beneficial interrelationships. (That task will be undertaken in chapter 11.)

process, which should include indigenous peoples' representation and participation, to implement the rights cited (Art. 27); the right to redress violations of rights cited "by means that can include restitution" or at least a "just, fair and equitable compensation" for traditional lands and natural goods that were "confiscated, taken, occupied, used or damaged without their free, prior and informed consent," and if indigenous peoples' consent, the compensation may be monetary or made with "lands, territories and resources equal in quality, size and legal status" (Art. 28); the right to decide and put into place their own priorities for use of their land and natural goods (Art. 32); the right to receive from states and their successors recognition and enforcement of, and honor and respect for, treaties and other agreements and arrangements that have been made (Art. 37); the UN and its agencies shall assist in the realization of the DRIP, including by "financial cooperation and technical assistance," and establishing means to ensure that indigenous peoples participate in resolution of issues (Art. 41); the UN will promote "full application" of the DRIP and follow the extent to which its provisions are being effective (Art. 42); the rights recognized in the DRIP are only "minimum standards" needed to ensure that indigenous peoples not only survive, but have "dignity and well-being" (Art. 43); and all rights recognized in the DRIP are "equally guaranteed to male and female indigenous individuals" (Art. 44).

In 2012, the UN Human Rights Council (UNHRC) appointed a Special Rapporteur (UN parlance for "Special Investigator") to determine the extent to which nations that were signatories to the DRIP were fulfilling their responsibilities to the indigenous peoples within their borders. The UNHRC selected James Anaya, Professor of Law at the University of Arizona School of Law, a graduate of the Harvard University Law School, a long-time worker for human rights in the UN system, and an Apache, to be the Special Rapporteur. Anaya held seven Consultations in the United States, at locations selected to enable maximum participation by regional Indian communities' members.[11]

---

11. William Means, Board Member of the International Indian Treaty Council, invited the author to testify at the final and longest Consultation, held for two days (May 1–2, 2012) at the Sinte Gleska University, on the Rosebud Reservation, Mission, South Dakota. Means asked that I address the topic "Discovery Doctrine and the Catholic Church." Preparation for my presentation provided me with my first in-depth and in-breadth engagement with Discovery. After my oral presentation to Special Rapporteur Anaya in the plenary session, and submission of my lengthier document to him, it occurred to me that this would be an excellent addition to and frame for my consideration of impacts on indigenous populations by Earth explorers in space. I had included this issue already, but now considered it more extensively as I perceived elements of Discovery present in US actions in and attitudes toward space voyages, the location and exploitation (literally) of natural goods, and possible encounters with simple and complex forms of extraterrestrial life. These imperialistic acts and plans began to be replicated among other nations, and a

The Declaration on the Rights of Indigenous Peoples provides a direct counterpoint to the Discovery Doctrine, and is intended to halt its continued use in national and international political and juridical arenas.

The Earth Charter

The internationally developed Earth Charter[12] originated in a document prepared for the United Nations Conference on Environment and Development (UNCED), popularly known as the Earth Summit, which was held in Rio de Janeiro in 1992, a decade after the UN promulgated the WCN. The EC represents a significant nongovernmental international effort to promote global ecological well-being and justice for human communities.

The Preamble calls people to respect and recognize their responsibility to promote the inherent worth and dignity of each other and of all life. The four narrative paragraphs that follow discuss the themes "Earth, our home" (humans live as part of an extensive biotic community in an evolutionary cosmos); "the global situation" (human production and consumption are devastating the environment and depleting needed natural goods; the rich-poor gap is widening in the midst of injustice and violent conflict; human overpopulation is straining Earth's ability to provide for life); "the challenges ahead" (people can choose to work together globally to care for Earth, each another, and all life); and "universal responsibility" (people, as citizens of the world, should have a sense of global responsibility). A shared vision and values would "provide an ethical foundation for the emerging world community."

The introductory paragraphs are followed by interdependent principles that outline and call attention to human needs, broader biotic community needs, and Earth needs, and urge people to take action to meet those needs—now and in the future. The principles are elaborated in four parts, each of which has four major principles, and sixty-one supporting subprinciples, and provide a holistic vision for Earth's well-being.[13]

Earth Charter principles include Part I: respect Earth and life in all its diversity; care for the community of life with understanding, compassion, and love; build democratic societies that are just, participatory, sustainable,

---

new kind of "space race" seemed to be developing among the more powerful Earth states.

12. The Earth Charter is available in both print form and as a free download; it can be printed and disseminated, with appropriate attribution, from www.earthcharter.org. The Earth Charter was placed in the public domain, without copyright and in multiple languages, to facilitate its widespread publication, broad dissemination, and local community discussion and utilization.

13. The EC principles are discussed, too, in Hart, *Encountering ETI*.

and peaceful; and secure Earth's bounty and beauty for present and future generations; Part II: protect and restore the integrity of Earth's ecological systems, with special concern for biological diversity and the natural processes that sustain life; prevent harm as the best method of environmental protection and, when knowledge is limited, apply a precautionary approach; adopt patterns of production, consumption, and reproduction that safeguard Earth's regenerative capacities, human rights, and community well-being; and advance the study of ecological sustainability and promote the open exchange and wide application of the knowledge acquired; Part III: eradicate poverty as an ethical, social, and environmental imperative; ensure that economic activities and institutions at all levels promote human development in an equitable and sustainable manner; affirm gender equality and equity as prerequisites to sustainable development and ensure universal access to education, health care, and economic opportunity; and uphold the right of all, without discrimination, to a natural and social environment supportive of human dignity, bodily health, and spiritual well-being, with special attention to the rights of indigenous peoples and minorities; and Part IV: strengthen democratic institutions at all levels, and provide transparency and accountability in governance, inclusive participation in decision making, and access to justice; integrate into formal education and life-long learning the knowledge, values, and skills needed for a sustainable way of life; treat all living beings with respect and consideration; and promote a culture of tolerance, nonviolence, and peace.

The EC's conclusion, "The Way Forward," suggests that a "new beginning" is needed, to be initiated through implementing EC principles. This would necessitate "a change of mind and heart," with "a new sense of global interdependence and universal responsibility."

Earth Charter principles would be beneficial guides for humankind in exoEarth settings, too. The vision and proposals they present would be helpful in space voyages and colonization.[14]

Restoring Common Ground in the Earth Commons

The international documents discussed provide a strong foundation for ecological restoration. Since *ecology* has to do with relationships, the restoration would be in terms of how humans relate to each other, individually and in discrete social communities; how humans relate to other members of the biotic community, in which they have responsibilities incumbent upon them as the most complex species, and requisite for them as the species that has

---

14. Chapter 11 provides an amended version of EC principles that helps clarify humankind's role and responsibility in space.

most damaged other living creatures and their shared areas on Earth's commons and common ground, and most diminished natural goods present in or on the commons; and how humans relate to the Earth they inhabit, and to "nature" in a wider sense. As an integral, evolving part of nature, humankind must ponder—scientifically, social-scientifically, and spiritually—what is and might be its niche on Earth, and project, using presently available knowledge from the fields mentioned, what might be their niche elsewhere in the cosmos.

The human species must take seriously the fact that it is the most complex and knowledgeable reflective consciousness of Earth, and might well encounter extraterrestrial beings who have a complementary consciousness (and perhaps conscientious conduct related to it), in terms not only of their relationship to their home planet but of their sense of cosmic place developed, possibly, over billions of years prior to their encounter with humankind.

International documents such as the World Charter for Nature, the Declaration on the Rights of Indigenous Peoples, and the Earth Charter could all contribute to a closer approximation of humankind's shared utopian vision and hope on Earth, and eventually in other places in cosmic space.

PART II
TERRA CONSCIENTIA

4

Extraterrestrial Discovery

Cosmos Community Contexts: In Caelis *and in* Caelo *as on* Terra?

The human venture into space advanced considerably in the decades that spanned the twentieth and early twenty-first centuries. An ever-progressive trajectory is evident in a stunning succession of events that have unfolded rapidly over little more than a half-century: the USSR launch of Sputnik (1957); the US Apollo missions (1968–72) that included Apollo 11, which provided the first lunar landing and moonwalk, featuring Neil Armstrong (1969); the joint US-USSR Apollo-Soyuz orbital mission (1975); the US Voyager 1 and 2 missions (launched: 1977; by June 2012: 11.1 billion and 9.1 billion miles away from the sun, respectively, and still transmitting data); Space Shuttles (1981–2011; 135 missions); Mars Pathfinder (1997)[1] and Mars Exploration Rovers Spirit and Opportunity (landed: 2004) and Curiosity (landed: 2012); and, currently, the International Space Station (ISS: constructed 1998–2011) orbiting Earth and gathering scientific data about Earth and space.

The ISS is the most visible of these efforts, and can be viewed nightly (barring cloud cover) with the naked eye as a bright "star" in the sky that outshines all celestial bodies except for Earth's sun and moon. It is also a signal of possible ongoing collaboration among Earth's most powerful nations at the moment—Russia and the United States—as well as another thirteen nations, whose astronauts have constituted its resident and rotating scientists and construction crews.[2] The ISS is serviced in part by commercial

---

1. The Mars Pathfinder included the Carl Sagan Memorial Station lander and the Sojourner Mars Rover, named for Sojourner Truth (1787–1883), African American former slave, abolitionist, and suffragette.

2. The International Space Station has been continuously occupied since 2000. Its occupants have represented fifteen countries; it has been serviced by five space agencies and by several private companies.

companies' staffed or unstaffed shuttles; they carry needed materials to the ISS and return with scientific data to a safe splashdown on Earth. In the United States private industry involvement replaced government (National Aeronautics and Space Administration [NASA]) involvement in the supply phase of station support, and was a portent of future commercial freight and even civilian passenger trips into near and distant space.

The most significant scientific discoveries to come from all exploration efforts to date might well be those announced in the near future by NASA scientists working with Curiosity. The Rover continues to gather samples of rocks and soils seeking evidence of past or present life on Mars, and of possible natural goods that would benefit humankind.

The ISS, an ongoing NASA scientists' project,[3] and the projected or planned pursuit of a permanent lunar base and a colony on Mars go well beyond, in a relatively short time, Neil Armstrong's declaration: "That's one small step for a man, one giant leap for mankind."[4] They might indicate, too, a potential continuation of the fifteenth-century Doctrine of Discovery. As discussed previously, Discovery was and continues to be permeated with elements of international competition and imperial expansion. It was first conceived and developed by Europeans as a rationale and self-interestedly "legal" basis for them to claim as their own any land or other newly found places whose occupants did not have two requirements of a "superior civilization": first, a culture congruent with or very complementary to the self-designated "superior" European culture with its social and economic structures (including private property ownership); and second, the only "true" religion, Christianity, and thereby participating in the Christ-mandated "salvific" mission of the Christian church. Euroamerican political and judicial systems and later, other nations around Earth adopted Discovery and have continued its injustices for centuries, particularly against indigenous peoples and cultures.

Current space achievements catalyze caution that there might be conflictive strains of nationalism and political ideologies operative in future space exploration, and when humans make a "discovery" in space. Plans for exploration of the solar system, and exploratory operations of complex rovers on exo-Earth planetary terrain transmitting their finds to scientists and government agencies' personnel in their country of origin, will likely lack transparency in terms of a nation's or nations' actual intentions. (A parallel: already, satellites orbiting Earth for ostensibly meteorological or other benign purposes have been used militarily.) As programs proceed for planetary

---

3. NASA provides a continually updated site with news about the Space Station, including scenes of prior construction and past and current resident scientists at work: www.nasa.gov/topics/shuttle_station/index.html.

4. Armstrong et al., *First on the Moon*, 268.

landings and for orbital flights around distant planets by scientists voyaging in space vehicles, some of the spacecraft likely would depart from the space station. Previously, space voyages have historically been intended (as have past ocean voyages made by exploring sailing ships), overtly or covertly, for territorial acquisition and other military, industrial, and commercial purposes—as much as or even more than for scientific purposes.

When the history of Discovery is kept in mind, the use of the particular word *discovery*, when apparently solely denoting finding new *"terra" firma*[5] for possible human settlement or "terrigenous" ("earth/land born," earth/land-produced) natural goods acquisition and extraction, should evoke careful analysis of the meanings actually attached to that term. Over the past five centuries, the "discovery" concept has been used historically (including today) to establish or to reinforce nationalistic or private interests' claims to land and goods: such might be the meaning and intention of future use, so that an even apparently casual mention of the word *discovery* implies, even if it is not verbally stated, a "discovery" claim that will be exercised by customary rituals and symbolic actions.

"Discovery" of the Moon

In 1968–69, as preparations were made in the United States to put the first member of the human species on the moon as part of humanity's initial venture into space, humanists, scientists, and religious leaders all over Earth suggested and even pleaded that a United Nations flag be placed on the moon when the lunar landing occurred. Their proposal was that landing on the moon was a common human achievement, not the accomplishment of a single nation; a United Nations flag on the moon would recognize humanity's scientific and technological progress, and still honor Armstrong and the United States for having placed it there. The gesture would have been appropriate, too, because the United States itself is a polyglot of humankind; its population is comprised of people of all races and ethnicities, the result of the presence of indigenous native peoples who had arrived millennia ago, and the intermingling immigrations, over the five most recent centuries, by people from Europe, Africa, Asia, Australia, and Latin America. In fact, the technologies that would enable a lunar landing were invented by scientists who were members of diverse ancestral ethnicities.

---

5. Since this familiar term usually means literally "solid Earth" or "solid earth/ground," *terra* is not literally or technically appropriate for other planets or celestial bodies that might be habitable or that have needed or wanted goods; a secondary meaning of the phrase might be analogously appropriate: "mainland," as distinct from "island."

The US government, in a defiant display of narrow nationalism, decided to proceed with the now customary practice of a nation's representatives placing on newly trod territory, in this case the moon, a flag from their particular homeland—rather than from the culturally interrelated, politically integrated Earth homeland from which they came.

Despite the pleas, then, the US government decided to make a nationalistic statement. Consequently, Armstrong's act belied or at least contradicted his words that he was in the midst of a making "giant step for *mankind*." His claim that all humankind was making a major advance in space exploration would have been better reinforced had the UN flag, an obvious symbol and firm statement and celebration of the common technological and scientific achievement of humankind as a whole, been placed on Earth's natural satellite instead of (or, in addition to) a national flag. The banner of the nations of the world, organized to promote peace and mutual understanding and benefit, would have been an obvious international and intercultural symbol that indicated recognition of the interrelated and interdependent nature of scientific research and technological development on Earth, and of a common human interest and share in what separate but integrated segments of humanity have achieved and will achieve.

The evident nationalism in the US decision provides a reality check against optimism about possible future collaboration in the cosmos among diverse humans, or between humankind and other intelligent species. Even if space voyages come to be collaborative ventures among nations, as perhaps signaled by the International Space Station, the conduct of humanity as a whole—given the present military-industrial-commercial consciousness of governments and transnational corporations—in distant "discovered" places might be destructive of local ecologies, extinct[6] species, and devastate indigenous lives and communities.

Although the word, concept, and "doctrine" of Discovery was not mentioned in the public exchange of ideas regarding the proper flag to plant on Earth's nearest "celestial body," its underlying ideology was a dominant factor. (More recently along these lines, the crew of a Russian submarine along with a Russian government official succeeded in placing a flag beneath Arctic waters on precisely the location of the North Pole in order to make a claim on it, on the petroleum reserves presumed to be beneath it, and on the northernmost water passage opening above it, as part of Russian national

---

6. As noted previously, a species that perishes because it loses its evolutionary niche due to natural evolutionary processes (such as altered habitat, predator conduct, entry into its territory of a competing species better fit for its context, or some other factor) "goes extinct." If a species that would have continued to evolve and adapt through natural biological processes perishes because of harmful human intrusion into its environment, it is "extincted."

territory because of its contiguity and continuity, under water, with Russia.) Beyond the flag planting, the United States brought back "moon rocks" to solidify (literally) its moon claim back on Earth. In subsequent years, other scientific analyses of the moon's natural goods were made.

The acts of planting the US flag and other events related to the moon were also a symbolic, direct contradiction to the spirit and even the letter of the 1967 Outer Space Treaty, which states clearly that space territories and natural goods are humans' common heritage—precisely to prevent such nationalistic displays and their commercial, industrial, and military implications. The Mars explorations and Mars materials analyses have represented similar quests not only for scientific knowledge but also for potential industrial development of natural goods discovered on the Red Planet (so named because of its red appearance, a result of the iron oxide so prevalent on its surface). Mars the planet, if its surface or subsurface contains human-desired natural goods, might prove to have been appropriately named after the Roman god of war if there is no significant change in human consciousness and conduct prior to the discovery of such goods, and nations' rivalry to exploit them for themselves rather than acquire them to be shared among all peoples, for mutual benefit.

Careful consideration of the moon landing event should give pause to reasoning and reasonable people. Those familiar with the Doctrine of Discovery should have particular cause for concern, if not alarm. The Space Treaty and the Antarctic Treaty both provide for international sharing of territory and natural goods that come under human control as humanity explores for and finds new lands or parts of lands, on Earth or in the heavens. Is the US lunar flag planting the opening salvo of a battle that could eventually erupt not only over lunar control (for example: as the site of a moon base to launch voyages of discovery and Discovery ever deeper into space; or as an industrial site where minerals are mined for commercial benefit), but for seizure of territory and natural goods on other celestial bodies? Here is presented dramatically what was discussed previously only theoretically: human consciousness and conduct, if humans are to survive on Earth and in the heavens, must be transformed prior to forays into the near and distant cosmos.

Recognition of the pervasiveness and continued use of outdated and unjust "discovery" attitudes and laws should stimulate anew consideration of what "discovery" meant and continues to mean. The twofold focus of the present discussion will be analysis of original, terrestrially formulated Discovery Doctrine ideology, tenets, and symbols, and their use in extraterrestrial contexts. The latter will be linked to potential ongoing implications and impacts of Discovery on distant celestial bodies' native populations and places, including local environments (abiotic places) and ecologies (biota's relationships to and in abiotic places, and with other biota sharing places). Initial and ongoing responsible conduct of explorers is crucial; it will determine what will be the long range consequences of human forays into the contexts in which they will find themselves. This will be especially significant where indigenous, physically and culturally diverse intelligent beings live, when the natives first discover the newcomers arriving on their home planet.

Discovery Doctrine and Christian Doctrine

The origins of religious and political doctrines of Discovery are both *religious* and *political*, having been intertwined as discussed previously. Both popes and monarchs—and their respective theology scholars and legal advisors—claimed that Discovery Doctrine and its implementation were mandated by God, and that therefore God was on their side. But is a divine attitude and intent truly present in imperial ventures, whether the latter are religious or political, on Earth or in the heavens? Are divine thinking and planning really behind, and an impetus for continuing, an insidious centuries-old doctrine whose specific originally Eurocentric perspective, plan, and projects on Earth decimated native populations, destroyed native cultures, and devastated native lands? Does God promote genocide? Has the Discovery Doctrine been, and should it be so understood in the future to be, a divine mandate understood in Heaven as it is professed on Earth? Or, in a perverse conversion of the Lord's Prayer, is Discovery merely a narrow, culturally born and borne human belief and hope that "God's will in heaven is being done on Earth," and even, though not so directly and arrogantly expressed, "God's will be done in heaven as it is on Earth," which calls on God, not humans, to change behavior? The latter two expressions, of course, directly exemplify Ludwig Feuerbach's idea that humanity creates divinity in its own image: the God in whom people have faith reflects human (mis)understandings of divine being; in effect, humans worship a projection of themselves, a high form of nationalistic narcissistic idolatry, to justify their actions.

While the religion-based aspect of Discovery ideology would not be directly expressed in space exploration, given that a diverse majority of nations lack an official, state-recognized religious creed, it might be covertly held by a significant number of people within one nation such as the United States, even though the United States constitutionally provides for freedom of religion. (Even after the ratification of the federal Constitution and Bill of Rights, in the policies and practices of the new federal government and of new states individually, "freedom of religion" often and at times exclusively meant "freedom of Christianity," "freedom of Protestant Christianity," and even "freedom of my denomination's Protestant Christianity." This was especially, but not exclusively, the religious rationale for persecuting "unbelievers," who included whatever native peoples were encountered.)

The Lord's Prayer: God's Will Be Done in Heaven and the Heavens as on Earth?

The Lord's Prayer, or "Our Father" in the Christian Scriptures (Christians' "New Testament"), praises God in *heaven* and prays that God's will be done "on Earth as it is in *the heavens*." The Latin translation of the text is helpful to illustrate this:

> *Pater noster qui es in* caelis, *sanctificetur nomen tuum* (Our Father who is in heaven, holy be your name).... *Fiat voluntas tua, sicut in* caelo *et in terra* (Your will be done on Earth as it is in the heavens) [literal translation].

In the prayer, *caelis* means "heaven," and *caelo* means "heavens." A similar distinction between the singular and the plural is made in the Greek version of the prayer. (Greek was the original biblical language for all twenty-seven books of the specifically Christian Scriptures, with the possible exception of Matthew's gospel: some scholars think its original language might have been Hebrew or Aramaic.) In some cultures, the singular, *caelo*, might have been understood to refer specifically to the divine dwelling place high above, while the plural, *caelis*, referred to an intermediate area above Earth and below the exclusive divine region, possibly also a divine abode but also the locus of divine intent and action, and the place in which different types of weather develop and emerge. In the Hebrew Scriptures, the plural "heavens" is always used, never the singular.

In the Christian Scriptures, English translations of the Hebrew language use both singular and plural. In the first verse of the first biblical book, Genesis, the translation is literal: "In the beginning, God created the heavens and the Earth" (1:1).[7] In Exodus the manna story in the original Hebrew says that God gave Israelites in the wilderness bread/manna from "the heavens" (16:4); the English version uses "heavens," in reference to the wilderness experience of the Hebrews; and the Gospel according to John in the Greek original states that God "gave them bread from the heavens"—but this becomes "bread from heaven" in English. In the last book of the

---

7. An interesting twist to both the Exodus and John verses is present in the differences between the Catholic Church's New American Bible (1970 translation of the Old Testament/Hebrew Scriptures; 1986 translation of the New Testament/Christian Scriptures), and *Nelson's Complete Concordance of the New American Bible* (1977 publication, also copyrighted by the Catholic Church). In the Bible, "heaven" is used; the Concordance has "heavens."

Christian Bible, Revelation, the plural form exists in the original Greek and Latin texts—"Then I saw new heavens and a new Earth. The former heavens and the former Earth had passed away" (21:1)—but the English rendition replaces "heavens" with "heaven" each time it occurs.[8]

If at Contact extraterrestrial intelligent species have superior technology and a much older evolved civilization (remember that the universe is about 14.82 billion years old, and Earth is only about 4.5 billion), might they make the case that Christians' own Lord's Prayer empowers ETI, coming from the "heavens" that are "above" Earth, to claim a divine mandate that an "inferior" human civilization on Earth must submit, once discovered by ETI, to the "superior" ETI civilization in the heavens? ETI could claim that when humans pray as Jesus taught his followers (Matt 6:9–13 is the version used in Christian churches; the version in Luke 11:2–4 is shorter), ETI is a response, by divine command, to humans' petition that God's will be done "on Earth as it is in the heavens": ETI, coming from "the heavens," has been doing God's will, and will propose for humanity or impose on Earth their spiritual consciousness and social conduct.

The preceding paragraph is not presented to suggest that ETI would have their own Discovery Doctrine, or would appropriate and invoke terrestrial humans' religious doctrine and practice to use a political, Europe-based Discovery Doctrine as justification for cultural, economic, and political imperial imposition. Rather, it is intended to highlight the culturally and religiously arrogant and absurd premise used since the fifteenth century to seize already inhabited lands from their indigenous residents—all on the basis that European civilization (and its genealogical and ideological descendants) and European Christianity were superior to whatever other culture or religion the explorers encountered and "discovered." Just as indigenous peoples around the world objected to Europe-based (geographically and ideologically) political, economic, and religious imperialism centuries ago, humankind in the present or the future would object to the imposition of an alien culture, and the seizure of Earth's earth and other natural, terrigenous goods to benefit primarily that culture. Neither Europeans nor ETI should, in justice, claim that their own "true religion" must be propagated; neither should any intelligent life throughout or in parts of the cosmos submit to another's politically based and culturally biased doctrine of discovery.

---

8. Alteration of the Revelation verse is seen also, as above, in the differences between the New American Bible and *Nelson's Complete Concordance of the New American Bible*. While the Bible has "new heaven and Earth," and "former heaven and Earth," the concordance lists "heavens" as the word used twice in 21:1. (Earth is not capitalized in current biblical translations, as it is in this author's and others' ordinary usage.)

Although the US scientific community is decidedly secular, and US Christian churches are decidedly more split—including internally—on peace and war issues than was the case from the period of US history spanning colonization through the end of the nineteenth century, vestiges of the "God on our side" ideology spring to the fore on occasion. As of this writing, the most recent example was US President George W. Bush's call for a "crusade" against Iraq and, by extension, against Islam. If some people can be convinced to view US wars as a "crusade" against non-believers of Christian doctrine who live on the same planet Earth, how much more might they be committed to use force to spread Christianity to other worlds if it is deemed "necessary" because existing cultures reject humans' alien religious claims and religions? When "national security" is thrown in the mix to support "religious security," the combination can be convincing for some citizens. Cultural arrogance has become evident currently, too, in some US politicians' and even academics' claims that they are defending "Western civilization" against other nations' or religions' peoples who resist US political, economic, or cultural imperialism.

Clever obfuscation of the true, economic purposes of war declared by national governments (which lurk behind the scenes and are surreptitiously pushed and pursued by transnational corporations and their wealthy shareholders) might be attempted by using a purported religious purpose for war. Corporations are practiced in careful manipulation of public opinion through advertising, carefully planted "news" stories in selected media, and excessive campaign contributions—*at home* to candidates willing to speak for them in the guise of "ordinary citizens," and *abroad* to political leaders or economic elites who seek to convince ordinary citizens that people with a different religious faith commitment than their own must be displaced and even killed in order to protect the integrity of the "true" faith of which they are dedicated devotees. In the twentieth century the US government, political leaders around the nation, military recruiters, and even church officials often stated that the struggle for hegemony between the US and the USSR was a US fight against "atheistic communism," implying that the system supported was "Christian capitalism"; this implied in turn that the Cold War (and a potential heated military conflict) was not only a political struggle but an economic and religious struggle. Today, Muslims and Muslim nations resent and resist efforts to impose on them political and economic structures they view as Western Christian ideologies. They and other religious cultures and nations became even more concerned about the United States exporting and imposing a so-called democracy when they saw its flaws revealed in efforts by a major political party in the US 2012 presidential campaign to suppress voters' constitutional right to vote, and to buy the election through biased campaign propaganda aired on television. It was obvious that not all US citizens accepted democracy as their political system, and usually it was such as these who claimed to want other nations to have democracy as their system. (Similarly, in Latin America: for more than a century the CIA and US military deposed democratically elected leaders, a practice which General Smedley Butler exposed and opposed during his retirement in the 1930s.)

As expeditions are organized for space exploration within the solar system in which Earth orbits (and then beyond), General Smedley Butler's revelations both during and after the Great Depression (discussed in chapter 2)—that as a marine he fought under the guise of "national security" for US banking interests and corporations—present an ever-contemporary warning to counter government-expressed claims that such voyages are solely for scientific purposes. "Science" has displaced "Christianity" as the basic rationale for discovery voyages, but the economic motives might, in reality, remain the same: acquisition of territory and natural goods. If the human species is not to do in the heavens what some of its members have done on Earth, a change of consciousness and conduct is needed beforehand.

Discovery in Space

Discovery Doctrine on Earth, beginning in the fifteenth century, has wrought political, economic, ecological, and religious harm globally. It has especially but not exclusively been devastating to indigenous peoples, cultures, territories, and natural goods. The ideology and practice of Discovery might well be replicated in exoEarth milieus.

Prologue (and Prophecy?): Avatar *and Discovery*

John Cameron's 2009 film *Avatar* presents a fictional representation of and warning about space ventures for human commercial financial gain that are supported by military personnel. Through its obvious parallels with *Dances With Wolves*—even to the extent of using several of the same actors, including accomplished Indian actor Wes Studi—the film does not hide its message that what might happen in space already has happened on Earth, to indigenous peoples of the Americas. *Avatar* illustrates analogously (and intentionally), in a futuristic context and on a world, Pandora, "far, far, away," the consciousness and conduct of European explorers, rulers, and colonizers during the formulation, elaboration, and continuing application of the "Doctrine of Discovery," and the Discovery ideology's probable future use and consequent impacts on intelligent life on other worlds. The planet Pandora's Na'vi people re-present the injustices perpetrated on indigenous peoples in the past (and perpetuated in the present), and represent all indigenous peoples who likely will suffer similar injustices in the future, barring changes in human consciousness and conduct. The alien humans' greed to acquire natural goods that already have for the Na'vi *intrinsic value*[9] in place and *instrumental value*[10] when required to meet the needs of local populations prompts the humans, as invading extra-Pandorans (to particularize "extra-terrestrial" in this extra-Earth setting), to attack the native population. As ordinarily occurs in such contexts, the extra-Pandorans employ cultural denigration and disparagement as a first step (as did Europeans when they relentlessly referred to Indians as "savages," and claimed to have a "superior" civilization and religion), that would lead inevitably to physical and cultural genocide and ecological destruction unless thwarted.

The film's strength, then, is that it presents a forecast of and caution about the possible continuing application of the "Discovery Doctrine" in the cosmos as humankind expands among distant celestial bodies. It provides, therefore, both a possible (and unfortunate) fictional prologue to space exploration, natural goods exploitation, and colonization by aliens originating on Earth; and, a factual prophecy (in the full sense of the word, as will be discussed) about what would result from irresponsible conduct in space. (*Avatar* will be explored more in depth in *Encountering ETI*.)

Prophetic Purpose

The dual role of the biblical Hebrew prophets was to deplore and decry the then-current consciousness and conduct of their people, which included worship of false gods and oppression of, rather than compassion for, the most vulnerable people in society: the "widow and the orphan" and the resident alien; and, to warn about what the consequences would be if they continued to violate God's laws. The prophet was God's messenger, and the mediator between the divine and the human. The principal temporal concern of the prophet was the present, not the future: people were acting in ways that

---

9. Value that something has in itself, inherent to that being.

10. Value that something has that is assigned to it by others, to meet others' needs or wants. Natural systems and their component parts can have instrumental value in place, such as a flowing river enabling both habitat for fish species or water to slake the thirst of animals or nearby trees drinking from it in their respective ways, or for animals to have fish for food: grizzly bears or golden eagles fishing for diverse salmon species, for example. Something with instrumental value can be beneficial to meet human needs, too, but its integrity in place and instrumental value for other species is eliminated when it is exploited to meet human wants and greed, such as extracting gold and mining coal irresponsibly to meet insatiable corporate and individual citizen desires for wealth and energy generation.

affronted and angered God, and should be altered. The prophet brought into public consciousness a divine reprimand and threat about consequences of continued misconduct. Over time, however, most people focused on the future aspect of the prophetic message: a "prophet" came to mean principally, in popular thinking, someone who can foretell the future. More recently, however, beginning in the twentieth century, the original understanding was recovered: people who stood over against and denounced rulers, governments, and prevailing ideologies (such as racism, classism, and sexism) were said to be "prophetic"; Martin Luther King, Jr., Dorothy Day, César Chavez, and Daniel Berrigan particularly stood out in the United States for their focus areas: racism, classism, and militarism.

Avatar can be prophetic in both senses. First, *Avatar* deplores, by implication and symbolism, past and current oppression of native peoples (as initiated and sustained by the Discovery Doctrine discussed in these pages); Earth's devastation by industrial corporations; imperialism exercised by any nation or combination of nations against cultures that are less advanced in military technology; economic exploitation of peoples of whatever ethnic identity; abuse of "aliens" and "alien" cultures when, in fact, the human arrivals on Pandora from Earth were the real aliens (much as had been the case when the Europeans approached and landed upon the shores of native peoples' native territories, and were discovered there by the native inhabitants). Second, *Avatar* decries the continued activities of the military-industrial complex as it, and the governments that represent its interests, map the course for and calculate expected profits from humans' ventures into space. It is highly probable that the Curiosity land rover sent by the United States to Mars is seeking more than just the past or present presence of life as it bores into the surface of Mars, and its laser smashes rocks or mountain sides: NASA has a team of geologists examining the results; little mention has been made, as of this writing, of biologists and astrobiologists having had as significant a role in analysis of Mars fragments.

It would be most unfortunate, and even catastrophic, if *Avatar* proved to be a prescient prediction of or predilection about Earth explorers' entry into exoterrestrial environments, and encounters with extraterrestrials. Here, again, Stephen Hawking's push for human bases and colonies to "save" humanity from the consequences of its prior and ongoing actions toward indigenous populations and terrestrial environments proves to be, ultimately, fruitless if there is no change in consciousness and conduct prior to space voyages. The classic television series *Star Trek* offers a vision that contrasts sharply with that presented in *Avatar*: the Earth explorers aboard the *Enterprise* represent diverse cultures and ethnicities, and they are part of an interspecies planetary federation that accommodates distinctions in the service of common interests—much along the lines suggested by UN space documents.

Discovery Dangers in Space

The "elements of discovery" described by Robert Miller[11] that have characterized European and Euroamerican attitudes on terrestrial places provide insights into considering, and a subconscious warning against, potential seeds of conflict in the cosmos. The conflict could occur because of competitive claims on territory desired for settlement and for bases for further space exploration, or claims on simultaneously sought natural terrigenous goods: among Earth nations, and between terrestrials and extraterrestrials (for whom humankind is the "extraterrestrial," "extraPandoran," etc.). This would preclude collaboration in cosmic space as humankind explores diverse regions, seeks to colonize celestial bodies, and considers types of terraforming or otherwise altering distant planets and other celestial bodies to meet humans' needs—and wants. If human consciousness and conduct are not transformed prior to space adventures and ventures, Earth's nations might begin such conflict on Earth prior to or during voyages to distant bodies, much in the vein of European nations that at times went to war for control of parts of the distant "new world"—even, as described earlier, numerous places which were already inhabited, and had agricultural operations, herbal and spiritual medical practices, science (including astronomy), and, in places, elaborate and extensive urban planning and structures.

Attorney and legal scholar Robert Miller summarizes the impacts of the Discovery Doctrine on Indian peoples:

> As a consequence [of the Discovery Doctrine], American Indians lost valuable natural-law rights of self-determination, sovereignty, and real-property ownership without their knowledge or consent. The confiscation of these rights by Euro-Americans was not justified by any rational, legal, free exchange of rights, but was just presumed because of the ethnocentric assumption of the "superior genius" of Europeans.[12]

Steven Newcomb complementarily analyzes biblical roots, expressed through European interpretations of both the Bible and Christian doctrine, which Europeans believed to support their Discovery claims:

---

11. See chapter 2.
12. Miller, *Native America*, 174.

> During the fifteenth, sixteenth, and later centuries, the monarchies and nations of Christendom lifted the Old Testament narrative of the chosen people and the promised land from the geographical context of the Middle East and began carrying it over to the rest of the globe. Genesis 1:28's directive to subdue the earth and exercise dominion over all living things, for example, and Psalms 2:8's mention of the "uttermost parts of the earth" provided a cognitive basis for the globalization of the Chosen People-Promised Land model during the Age of Discovery. The monarchs of Christendom and their seafaring subjects imaginatively projected themselves into the Old Testament narrative of the chosen people and the promised land. Accordingly, they conceived of themselves as having been commanded by God to take possession of the "uttermost parts of the earth."
> ... The presumption by Christian potentates that they had the divine right to take possession of heathen lands (lands not possessed by any Christian prince or people) was a direct result of their belief that God had previously commanded the Hebrews to take possession of Canaan and that they, as Christians, had "become" God's "new chosen people."[13]

Discovery Doctrine elements must be confronted directly and considered analytically in regard to the potential for and manner of their utilization in space. Substantive discussion of an alternative ideology and alternate human behavior must be undertaken, too. It should be complemented by human resolve, on an international scale, to embrace and embody alternative ways of thinking and acting for and in space, if devastation is not to be inflicted upon other planets' biota, including on their intelligent beings, and on their abiota—including their terrigenous natural goods that serve ETI or other species' needs where and how they exist in place, or as they have been extracted and modified by ETI prior to humans' arrival.

Elements of Cosmic Discovery

Humans have left their "fingerprint" already in space, not only via exploratory vehicles sent forth but by radio signals, nuclear explosions, and the activities of US Mars Rovers gathering data on Mars and transmitting it to Earth. Humans likely have been or will be detected by intelligent species that use diverse technologies to probe the cosmos. The human presence in the cosmos will be evidenced to a greater extent if nations compete to arrive first at potentially lucrative planets or specifically promising regions thereof, and

---

13. Newcomb, *Pagans in the Promised Land*, 43.

engage in conflict thereafter because of their respective claims to places and goods. The Hubble and other telescopes already have detected hundreds of Earth-like planets on which life, and perhaps even ETI, might have evolved. A signal that the US government is interested in focused efforts to detect ETI is that the USAF, which once openly developed Project Blue Book to investigate UFOs and then claimed it had no interest in them, provided grants to support the work of a renewed SETI program which has begun scanning areas in which promising planets have been detected by NASA telescopes. The USAF claimed that it wanted the telescopes to provide data on space debris that has resulted from Earth nations' launches of rockets into Earth's atmosphere and beyond.

Discovery: In the Heavens as on Earth?

Miller[14] analyzed ten key elements of the Discovery Doctrine developed and enforced by European popes and monarchs. Humankind as a whole or in its nationalistic parts might seek to use these elements in space if current consciousness and conduct continue, and are exported into space. In an age replete with technologies far advanced from what they had been in colonial times centuries ago, the original elements would have to be altered if, overtly or covertly, Earth nations exploring space did so with the same imperial intentions as those of fifteenth- and sixteenth-century Europeans. Thinking about ways in which nations, individually or collectively, might attempt to assert Discovery claims is useful for our ongoing thought experiment: if humankind's consciousness and conduct have not been substantially altered prior to or soon after intensive space journeys, then nations and private transplanetary, transgalactic, or transstellar corporate enterprises might well seek to do in the heavens what has been done on Earth, to the detriment of intelligent beings in all places.

The first element of the Discovery Doctrine, *first discovery*, could be invoked from afar while humans are voyaging in space, or even beforehand as a result of data compiled by Earth-based telescopes, Earth-launched satellites and space stations, lunar and planetary bases currently proposed, and, eventually, colonies within the solar system. Likely habitable or potentially goods-laden celestial bodies might be located through finely tuned optical or radio telescopes and with advanced technology, including spectroscopy. Nations with the most advanced technologies and financial resources might attempt to make such claims decades or even generations before arriving on site. (This extensive time for journeys assumes use of vehicles and technologies currently available; other means of travel might be developed that

14. See chapter 2.

expedite space voyages in visible space or via an accessible, invisible other dimension of the cosmos.) As happened in Antarctica because of land claims based on a "discovery" ideology, nations intending space journeys could, by agreement, apportion areas of the sky for their respective exploration and development—or, as happened in earlier centuries, award desirable bodies to those who first arrive.

A provocative new aspect of the first discovery element might emerge, given transnational corporations' increasing power and influence: a private company, funded by affluent investors, might build and operate a telescope to scan the skies and seek planets that are human-compatible and laden with desirable terrigenous goods that previously were available solely terrestrially. The 'discovery' by private corporations or individuals seeking personal benefit and profit might be accorded the same rights as would be accorded to discovery by a nation's scientists on behalf of their country. "First discovery" claims, then, when humankind uses optical and radio telescopes and infrared detection devices from vast distances, would be especially problematic to evaluate and verify in order to determine justly "ownership" and use of territory and goods (in the event there is no evolved intelligent life already *in situ*).

The second Discovery element, *occupancy and possession*, would only be satisfied when a nation or group of nations was able to establish a base or settlement on a planet or celestial body newly found by humans. This element renders almost useless a claim made solely on the basis of long distance telescopic sight or sound. A difficulty could emerge should space vessels with different origins—terrestrial nations and extraterrestrial beings—converge upon a large planet from opposite sides, establish their respective bases, and initiate scientific research or mineral extraction prior to coming into Contact with each other.

The third element, *preemption*, might be attempted by individual Earth nations when they encounter peoples who are not as technologically advanced as they are, and who have territory or natural goods desired by humans, in order to secure for a particular nation the sole right to trade with ETI and use the desired goods. Here again, if human competing claims are mediated to maintain peace, the places claimed could still catalyze conflict should ETI arrive on the scene.

The fourth element, *native title*, would mean that native inhabitants would lose territorial rights, and be allowed only to occupy and use their traditional lands. Here, too, consciousness about and a will not to replicate injustices perpetrated in the Americas and elsewhere during European expansion should impel human explorers, in good conscience, not to harm native intelligent beings in the heavens as indigenous peoples had been harmed on Earth. Earth nations should be bound by the United Nations Declaration on the Rights of Indigenous Peoples, which, although formulated to rectify injustices perpetrated on Earth and to prevent their recurrence on Earth, might be invoked to prevent replication of injustices against native inhabitants in space.

The fifth element, *indigenous nations' limited sovereign and commercial rights*, whereby European nations took away native peoples' customary rights to diplomacy or to trade with any peoples they chose, and arbitrarily decided that they could only do so with the European nation that had "discovered" them, would be implemented only with great difficulty, if at all. Surely humankind, especially given its integrated cultural, racial, and ethnic reality five centuries after European exploratory voyages, would not attempt to similarly control extraterrestrial intelligent beings (unless, of course, species-centric ideologies become dominant and operative in space).

The sixth element, *contiguity*, might be difficult for any nation to assert because of a planet's terrain. On Mars, for example, where Curiosity currently roams and records, there are no apparent rivers whose mouth would enable a "discovering" nation to claim all territory contiguous to it. (An interesting new possibility: subsurface rivers and aquifers, should they exist on Mars or elsewhere, might be used, given technological detection capabilities currently available, as bases for claims to surface lands.)

The seventh element, *terra nullius*, could be a source of contention: when indigenous populations are in place, the UN and many (if not all) terrestrial nations (though not necessarily private industrial or commercial corporations) would object to any single nation or a collaborative bloc staking a claim based on this Discovery element.

The eighth element, *Christianity*, would no longer be a factor because in the contemporary era and into the foreseeable future, diverse religions and nontheistic perspectives are dominant in distinct Earth nations, and Christianity is not globally, nor would it be universally, recognized as the only "true religion."

The ninth element, *civilization*, would no longer hold because multiple advanced civilizations from diverse continents exist on Earth now, many of which have complementary political, economic, and military power, and none of which would dare claim to be superior to all the others and have a divine mandate to subjugate the others; and, ETI might have a more advanced civilization and spirituality.

The tenth element, *conquest*, would be as unjust in space as it has been on Earth when less technologically developed peoples came into contact with more technologically advanced people; and it would not be feasible if, as might well be the case, the inhabitants of a place are far advanced beyond humans in technological achievements and weaponry.

Mars: Curiosity about Curiosity

A significant US space accomplishment in 2012 provides an example for the potential practice of Discovery on a planet: the successful landing (using an innovative and complex procedure) of the Mars explorer Curiosity.

Previously, the United States had successfully landed two Mars rovers to explore the Martian terrain and transmit data back to US scientists and other NASA personnel on Earth. The stated scientific mission was to seek evidence that life exists or had existed on the red planet. In the latter aspect of the rovers' quest, the mission appears to have been successful. A possible clue to biotic existence, in the past and possibly still in the present, was found: the presence of ice. (Incredibly, in November 2012, ice was discovered on Mercury, the planet nearest to the sun; previously it had been assumed that water could not be present there because of intense solar heat.) Ice indicates that there might still be subterranean water, as yet not found, and therefore possibly life. The "proof of life" aspect of Curiosity's mission remains, as of this writing, unfulfilled.

The landing and operation of Curiosity (whose name exemplifies interest in what might be found on Mars in terms of biota and abiotic benefits), which contains the most complex technological equipment landed to date on exoEarth territory, provides the United States, if it were so to choose, the opportunity to invoke Discovery in ways not previously possible or permissible to any Earth nation. The United States, while not having a "settlement" on Mars, would have operative "explorers" that are simultaneously national symbols of the US presence; although not an actual human "occupation" of Mars, their technological capabilities, communication with the nation that sent them, and transmission of data that can indicate the types of subsurface minerals present enable them to engage in operations that previously could have been done only on planet Earth, and there solely by human explorers who were seeking to stake a firm claim of Discovery ownership on behalf of their monarchs and nations, using customary and appropriate symbols to do so. At this stage of space exploration, the United States appears to have the only technological and financial means (now with the assistance of private corporate interests) to place an actual enduring human presence on Mars.

Two possible scenarios for Mars present themselves: first, that life, possibly only very primitive life, is found on Mars; and second, that conditions are not present that would permit life to exist currently or in the future. Both scenarios have the potential for elements of Discovery to become operative and create conflict.

In the first case, *primitive or intelligent life is found*. Most elements of Discovery might be invoked if life is intelligent, and many more might be invoked if life is primitive. The eighth element, *Christianity*, would be problematic principally because there is no religious dimension at all inherent in the mission of Curiosity, and because even if there were, Christianity is not the US's state religion, nor would it be acceptable as a Discovery element by non-Christian nations. Qualifiers would be added to element six, *contiguity*, because while Curiosity freely roams the land within the Gale Crater, it seemingly has no capability to exit it; therefore, any Earth nation (including the United States) that is able to land outside the crater an exploratory vehicle that is internationally recognized as the equivalent of Curiosity could lay claim to territory contiguous to its explorations; and, should ETI become a new arrival (assuming it has presently no presence) with superior technology, it could claim any and all territory and all goods therein. Element nine, *superior civilization*, would have to be reinterpreted in case there is an as yet undiscovered civilization that is superior to humankind in some way: not necessarily in technology, but by being in a more advanced evolutionary stage (perhaps a stage past technology, or one of having never used humanly recognizable technology such as was developed on Earth). Element ten, *conquest*, would be difficult to invoke unless nations did it collaboratively for mutual benefit; however, an existing civilization might possess powers—not necessarily technology-based—that are superior to those of humankind and would thwart conquest.

In the second case, *life is not found nor can life be projected to emerge and evolve*. The United States might attempt to invoke all possible elements of Discovery prior to other nations' arrival on the red planet. The first three are obvious: the United States is the first to find and explore, initially from afar, the parameters of the crater; to occupy the crater by the technological equivalent of machines that do what human explorers did in ages past; and to preempt other nations' exercise of the first two elements. The seventh through tenth elements are automatic: the land lacks living beings and is now and will be in the future truly "vacant" of life; there need be no claims that intelligent biota present are not recognized to exist, as happened in the Americas and elsewhere on Earth, because their civilization and religion were unilaterally declared to be inferior, and thus since no religion or civilization exists, conquest is unnecessary. In the second scenario as in the first, the arrival of new Earth nations renders obsolete the first three, the sixth, and the last two elements invoked by the United States in its Discovery claims; and the new arrival of ETI would render all Discovery elements and claims null and void, and obsolete. In fact, ETI could use superior technology to enforce its claims, or benevolently collaborate to promote benefits for both TI and ETI.

As it explores, Curiosity has the capability to drill exploratory holes or shallow pits in the Mars terrain, or to use a laser to shatter segments of crater walls, and in both practices to examine surface terrain and subterraneous areas, collect Mars fragments, and perform data analyses that can be transmitted to Earth—all purportedly to seek life, or evidence that it now exists or once existed. Through Curiosity activities, it is possible that US scientists, and thereby the US government and interested corporations, would uncover evidence of minerals needed or desired by Earth nations or industry, or even able to provide for the human common good, such as gold, uranium, or platinum.

Under both scenarios—life is found; no life is found—the United States could attempt, as the only nation present on Mars in some form, to use Curiosity and the elements of Discovery to make a national and nationalistic exclusive claim to Mars territory and natural goods. (A fascinating possible additional twist to Curiosity's endeavors: if the United States does have remnants of an alien spacecraft that landed at Roswell in 1947, and Curiosity finds the same minerals used for the craft's exterior, as they are or as amalgamated with other materials, would the United States keep such a discovery and its related data secret?)

A new type of "space race" might well develop. This time, it would not be for the purpose of putting into Earth's atmosphere a satellite or a vehicle with a human passenger aboard; this has been accomplished. Rather, it would be to invent and manufacture technologically advanced vehicles, with or without humans aboard, that would have the capability to mine and transport back to Earth raw materials needed to benefit nations' citizens or to profit commercial enterprises. Such vehicles would simultaneously provide diverse nations (and private enterprises?) with a justification for invoking Discovery in their own behalf, as they would embody the second and third Discovery elements, and potentially, depending on nations' power, be used to make claims under the ninth and tenth elements. Such a space race could result in military conflicts—on Earth, as in space, as on Earth . . . —in an ongoing cycle of violence, an unending dialectic that has no synthesis or resolution.

An alternative attitude and consequent conduct would be for the United States to be faithful to the Articles of the UN Outer Space Treaty and other UN documents in which extraterrestrial territory and goods are declared to be the common province of all humankind, and are to be shared and distributed in such a manner as to meet first the needs of underdeveloped nations and economically poor citizens in all nations, and only thereafter to benefit all nations, in a way upon which there is mutual international accord. Although UN instruments do not note or consider the possibility of an ETI presence or ETI claims on the moon and "other celestial bodies," direct Contact with ETI would alter human thinking about solely human practices of Discovery, or solely human claims to that which is "discovered" by humankind.

If humankind contemplates and initiates space exploration, territorial acquisition, and natural goods extraction while employing a "Discovery" ideology, people should also consider a parallel possibility: that ETI has developed, in its own history, a Discovery Doctrine of its own, similar to the European and Euro-American, Euro-Canadian, Euro-Australian, and Euro-New Zealander perspectives, and continues to operate under its ideology. Perhaps such a theory and its practice had been used on the extraterrestrials' own home planet when more technologically developed peoples found hitherto unknown territories and their indigenous inhabitants, and now will be used by ETI to judge humankind as similarly culturally and technologically inferior,[15] and subject to conquest.

Invoking their theory, the ETI would justify conquering, removing, and resettling native residents—or even exterminating them, as per what Thomas Jefferson and other US revolutionary leaders suggested, and upon which they periodically acted.[16] In ETI thinking all of humankind would be perceived as Earth's native peoples in such a scenario; all must be subdued and perhaps eliminated in order that ETI might make a permanent place for themselves on Earth. An unsettling complementary thought is that ETI will develop such an understanding only as they analyze human thought and actions and encounter this still-operative perspective of Europeans and European descendants vis-à-vis indigenous peoples, and decide to utilize it themselves as a useful tool for expansion and acquisition, beginning soon after Contact and engagement with humans on Earth or in the heavens.

In contrast, if the historical realities and current impacts of Discovery are not faced head-on today by humankind and the elements of Discovery rejected, then indigenous and less technologically developed ETI on distant

---

15. In such a case, *Avatar* suggests caution regarding not only future human conduct, but also in terms of an uncritical acceptance of ETI statements about their own past conduct. The complementarity of *Dances With Wolves* and *Avatar*, and their respective descriptions of the impacts of human exploration on indigenous populations and places, on Earth and in the heavens, might then illustrate how humans would be replicating ways in which ETI oppressed native populations on their own planet before departure, a type of consciousness and conduct that continued as ETI voyaged into and colonized (or only exploited the natural goods of) other celestial bodies. The issues are explored in depth in Hart, "Cosmic Commons," and Hart, *Encountering ETI*.

16. See Miller, *Native America*, especially chapters 2–4.

planets and celestial bodies might well suffer the physical or, minimally, cultural genocide experienced by Indians of the Americas.

On Earth, repentance for the past, response to present issues (including restitution of illegally and unethically seized land and terrigenous goods), and resolve not to continue "business as usual" toward native peoples and their territorial holdings would halt the Discovery trajectory. It would alter, too, the types of relationships possible between humankind and ETI. If humankind and ETI have rejected or will reject Discovery, however perceived in their respective worldviews and wherever and whenever practiced on the worlds they find, consequent Contact and mutually respectful communication would enable all to live in integrated, interdependent, and interrelated communities, and collaborate on common enterprises on common ground.

Stephen Hawking and ETI

Aliens' impacts on indigenous intelligent life were partially discussed by theoretical scientist Stephen Hawking when, after he advocated human-inhabited sites on the moon and Mars and agreed that there was a very high probability that there were other intelligent beings living in and possibly exploring the cosmos, he advised against efforts to make Contact with ETI. Hawking warned that intelligent explorers likely would be aggressively seeking places in which to live, or at least from which to extract natural goods, and would malevolently act against humankind. He projected anthropomorphically that ETI's conduct would be similar to that of Europeans who went to the New World five centuries ago, where Contact between Europeans and native peoples had catastrophic impacts on the latter. Hawking declared that the aftermath of post-Columbus 1492 Contact "didn't turn out very well for the Native Americans."[17] Interestingly, even though Hawking warned against Contact with potentially invading aliens who as extraterrestrial intelligent life would colonize human habitations, he did not argue against similar human abuse, when humans are parallel invaders landing as intelligent extraterrestrials on other celestial bodies. He did not discuss how upon Contact, humans might cause environmental degradation and devastate resident life, from primitive cells through intelligent beings. His recollection of European conduct in the Americas did not prompt a plea for attempts to effect pacific human interaction with extra-Earth indigenous intelligent life.

At their core, Hawking's contradictory or at least incomplete statements are emphatically anthropomorphic and anthropocentric. Hawking anthropomorphically projects human conduct onto extraterrestrial intelligent life,

---

17. Hawking expressed these sentiments in his 2010 BBC series *Into the Universe with Stephen Hawking*. See the Introduction, especially footnote 12.

which he expects to behave against humans just as terrestrial intelligent life has done in intrahuman encounters where land and natural goods have been sought. Hawking anthropocentrically asserts that intelligent humans should save themselves from being dominated by extraterrestrial intelligent life; he romanticizes his own species *a priori* as having a culture that should be preserved. Effectively, he asserts that the human species and culture are superior to other intelligent beings and cultures that might be encountered on Earth or in the heavens, and thereby privileged to subjugate rather than be subjugated. In so doing, perhaps he is merely stating a perceived biological evolutionary imperative and instinct to survive, which also expresses an anthropocentric, rather than biocentric or terracentric, attitude about humans. Given Earth's devastation by humans which, ironically, he sees as impelling at least a select group of humans to relocate to distant places in space, Hawking's position is inconsistent. Finally and also ironically, while Hawking declares that humans should colonize space to save themselves from their conduct on Earth, he does not suggest that humankind should alter its conduct and respect extraterrestrial places (which might be pristine) and resident intelligent life (which might have a superior, non-conflictive, ecologically integrated, and communal rather than individualistic culture).

In these and other respects, Hawking inadvertently advocates for exercise of a human Discovery Doctrine when human explorers encounter other "peoples"[18] and explore newfound places. Such an ideological underpinning would promote conduct by humankind as a whole on celestial bodies that would mirror the very type of European ideology and action that he deplores (or, perhaps he merely notes in passing in order to make his case for non-Contact with alien intelligence) because of its impacts on indigenous peoples.

Stephen Hawking's evocation of events surrounding European entry into the Americas, then, and potential outer space parallels in the twenty-first century as humankind ventures into extraterrestrial exploration, is reminiscent of elements inherent in the Doctrine of Discovery, as amended and refined over centuries. Its tenets and practices must be eliminated to prevent catastrophic Contact in the cosmos.

Contact and Contextual Socioecological Ethics

Considerations about Contact, and Contact itself, can catalyze peoples' theoretical and practical engagements with socioethical issues surrounding human impacts not only on habitable places in space, but also on their

---

18. Although humans do not know beforehand the physical form that ETI will have, using the word *people* for them indicates beings with intelligence, creativity, and culture, however alien these might be to specifically human experience and expression.

home planet. An idea expressed by Richard Randolph, Margaret Race, and Chris McKay in "Reconsidering the Theological and Ethical Implications of Extraterrestrial Life" (1997) is particularly pertinent here. They declared that theological and ethical discussion of extraterrestrial modification "may help us see our own environmental issues more clearly from this distant perspective."[19] This phrasing should be amended, to some extent. It should include engagement between the present and the future, and between terrestrial and extraterrestrial experiences, reflections, and projections (which would be a corrective for Stephen Hawking's romantic idealism about human conduct in space, as noted previously). Terrestrial socioecological ethics should be in conversation with extraterrestrial socioecological ethics in a dynamic, dialectical, and dialogic process wherein each informs and even alters the other. What humans do on other worlds will, initially, reflect what they do on Earth, so humans' consciousness and conduct must change on their home planet as they seek to project what they will do elsewhere; what humans learn as they adapt to other worlds' contexts can in turn, then, impact their attitudes, values, and practices on Earth.

While scientists might be interested in adding to scientific knowledge, and experiencing the adventure of exploration and discovery, those who finance space explorations often are more interested in the commercial or military benefits that would accrue from such efforts. In the light of Earth issues such as global climate change or mining-caused pollution, scientists (and politicians and entrepreneurs) as well as ethicists should reflect on impacts of humans' interference in an emergent evolutionary process on another planet–whether the developing life be in its early stages or intelligent–through mining or other heavy industrial production, agriculture, forestry, or residential construction, when any or all of these might pollute the land or water, or alter climatological processes. In turn, reflection on extraterrestrial human impacts on biota or abiotic contexts should initiate new consideration of parallel terrestrial impacts. Human cosmic consciousness might flow into cosmic commitments, while complementarily Earth consciousness might flow into Earth commitments; these complementary considerations and their contextual implementation would dialogically catalyze or influence a human sense of responsibility for humans' conduct in both terrestrial and extraterrestrial contexts.

The human hope must be, in the light of the preceding, that not only would humans' theory and practices change, but that intelligent beings encountered in space or on Earth before or during the time when humans travel further into space must have overcome and discarded on their home planet both international, intercultural, and interracial conflict and the ideologies

19. Randolph et al., "Reconsidering," 6.

that fueled them, and any remnant thereof that would become operative upon Contact with inhabitants of "new worlds" they "discover."

Dismissing "Discovery," Inculcating Interrelationship

An effort to eliminate export of false and anthropocentric expansion of Earth "discovery" doctrines into the heavens would be successful most expeditiously with the support of United Nations member nations and agencies, through their consensus development of international, enforceable covenants, prior to humans' space voyages; and, of faith traditions, whether theist or atheist, mutually, cooperatively, and collaboratively refusing to engage in any missionary proselytizing efforts. Among Earth nations, visionary politicians, entrepreneurs, and military personnel would not seek personal or planetary profit via domination of newly-engaged cultures and newly found natural goods; among faith adherents, scholars and leaders would have to support and embrace not only interreligious ecumenical association on Earth, but also interreligious or inter-philosophical cross-fertilization with ETI.

It would be to the detriment of current and future human communities if Discovery were to continue in the present, whether blatantly, indirectly, or subconsciously. Discovery constructs a formidable barrier to cultural engagement and respect. In its mutated Euroamerican form, adapted by transnational corporations and the political leaders and economically well-to-do who benefit from its ongoing implementation, it imperialistically decrees that all other cultures, political systems, and economic arrangements must conform to its model. Thus, for example, imposition of a "Western-style democracy" and a capitalist economy, following a US model, continues to be an objective of US foreign policy, despite its failings at home. Cultures and nations that believe in communal well-being and compassion rather than individualism and self-centeredness as cultural values, and whose political structures embody those values, are urged or coerced (politically, economically, and militarily) to alter their values and practices. This is done despite the fact that their values better express Christian principles and practices that most US citizens claim to believe and follow.

Absent significant change of mind and substantial conversion of heart, the "in the heavens as on Earth" perspective today, which has been preceded by and would be modeled on Discovery as it was formulated, developed, and practiced in Europe's colonization of other continents, would be asserted anew. It would replicate Discovery elsewhere such that the current "in the world as in Euroamerica" consciousness and conduct would impede acceptance in cosmic contexts of alternate understandings of social responsibility and commitments, and of beneficial ecological relationships.

If Euroamericans and others influenced by their ideology do not respect other members of the terrestrial human race, who share their general physical characteristics and mental capacity but who do not share their cultural background or embrace their economic system or replicate their political systems, neither would they respect and adapt to extraterrestrial intelligent beings who have distinct physical characteristics, cultures, economic arrangements, and political structures. Discovery continues to denigrate terrestrial ethnicities, cultures, and societal institutions whose roots are planted deeply in Earth. Discovery in space would denigrate, covertly or overtly, the histories, cultural characteristics, and ethical thought and practices of extraterrestrial intelligent beings, however advanced their own civilization, that did not correspond to Euroamerican perspectives and practices. Discovery in space would catalyze interplanetary intercultural conflict, in which the technologically more advanced intelligent beings would prevail. Discovery would disrupt the potential for harmony and cooperation that exists within cosmic integral being.

United Nations documents focused terrestrially, and aspects of their extraterrestrial implications for space exploration were discussed earlier and extensively. The conflict between UN ideals and hopes, and the current lingering use of the Discovery Doctrine on Earth in political and legal consciousness and contexts, requires reflection, revision, and removal in order to resolve actual and projected conflicts between humanist theory and political-military-industrial-corporate practice.

The particularly pressing questions that emerge from the issues presented in this chapter are whether or not humankind will enforce in the heavens the European-Euroamerican Discovery Doctrine that originated, developed, and was imposed on Earth, especially against native peoples, and whether or not ETI will have developed somewhere in the heavens their own Discovery Doctrine that they will attempt to implement on Earth.

Congenial and collaborative cosmic Contact requires departure from and the disappearance, on Earth and in the heavens, of the corruptive consciousness, culture, and conduct embedded in Discovery doctrines—whether Discovery originates on Earth or in other cosmic contexts. A new *topia* ("place"), currently envisioned as a *utopia* ("no place"; a hoped-for place that does not yet exist), then could replace the current *dystopia* ("extremely oppressive place") experienced in different degrees and places by many Earth citizens, and resolve the experience of dis-placement humans have experienced historically when groundbreaking scientific discoveries have shaken their existing cultural and religious beliefs.

5

Spirit and Science

Consciousness and Collaboration

In the mid-second millennium CE, the complementary theories of Nicolaus Copernicus (1473–1543), Galileo Galilei (1564–1642), and Johannes Kepler (1571–1630) shook religious, theological, philosophical, and scientific foundations in Europe. Previously, Europeans (and, globally, people from other geographic regions, cultures, and religions) believed that Earth was the center of the universe. Their belief (which they assumed to be a fact) was based on empirical observation (the sun and moon appeared to "rise" on the eastern horizon, traverse the sky, and "fall" on the western horizon; stars appeared to circle Earth). This belief was supported by the limited scientific knowledge and scientific instruments then available, and influenced or was influenced by prevailing religious doctrines and philosophical speculation. It was apparent to each and all, merely by looking at the sun in the sky during the day, and the moon and other celestial bodies at night, that all the lights in the sky orbited Earth. The "world" in the West and elsewhere included Earth—the abiotic context, habitat, and common ground of all biota; invisible air; and visible stellar bodies. The depth and breadth of Earth "below" and cosmos "above" were not perceived or even suspected by humankind.

Since the time of ancient Israel until the sixteenth century, religious adherents of the faiths of Judaism, Christianity, and Islam, as they emerged historically, accepted the apparent "evidence" of their eyes and the interpretation of its meaning as expressed in their respective traditional sacred texts. They believed that they were a unique, special creation of Yahweh/God/Allah, whose ancestors were placed on Earth in a primordial Garden of Eden. Christians added to that belief the assertion that people were special because God had become Incarnate (enfleshed; embodied) in one form and one place in the cosmos—as a human being on Earth—in order, as Christ Jesus, to "save" humankind for eternal life, and renew the whole "world," understood to be the entire cosmos.

Humans Displaced from Place

Humankind's belief in and self-designation of their lofty position of dominion on Earth—above all creatures—and in space—above all beings, in a cosmos that was created solely for "man"—was diminished and eventually lost as major scientific discoveries, based in part upon research made with previously unavailable instruments and technologies, provided new data not only about the human condition and place (in terms of both location and status) but about the cosmic milieu in which Earth was situated.

First Human Dis-placement: Earth Is Not the Cosmic Center

When Copernicus first proposed a heliocentric system, a theory published posthumously because of his fear of Church and societal reactions, his ideas went largely unnoticed. Galileo, by contrast, a more assertive and egotistic personality, at times dramatically and even flamboyantly announced discoveries he had made based on his observations through telescopes he invented and on his associated mathematical computations. As had Copernicus and Kepler, he stated that Earth (humans' home) and other planets existed in a solar system, a heliocentric entity surrounded by far-distant stars: humans' planet (and, by extension, humanity) was not the center of the universe after all, these scientists proclaimed. In doing so, they contradicted literal interpretations of several biblical passages and their associated Christian doctrines.

People in the West aware of such statements experienced a profound sense of loss of their customary religiously and ideologically developed place in the world. They experienced dis-placement from their perceived, self-declared, and self-interested status as creation's highest creature which, as preeminent "images of God" lived on the planetary center of the cosmos.

Galileo ran afoul of the Catholic Church, and in his "plea bargain" with Church officials of the Inquisition (to avoid threatened torture and execution) he "recanted" his scientific cosmological assertions, and affirmed that Earth was indeed the cosmic center—even though he knew that that essentially creedal statement was untrue. The sense of human dis-placement from preeminence in the universe was, on the surface, seemingly eliminated by Galileo's "conversion" to the traditional cosmological view.

Kepler, for his part, had corrected the theories of Copernicus and Galileo and the speculations of Aristotle through his discovery that planets travel around the sun in orbits that are parabolic, not circular. Unlike his predecessors, he did not "fudge" what he saw, since he was not rigidly bound by prior thinking that assumed and asserted that since the circle was a perfect and elegant figure, circular orbits were a given for celestial bodies in God's creation.

The Church's condemnation of Galileo, and of his scientific data and theories because they contradicted traditional dogmas and biblical texts, cast a pall upon scientific research and release of significant scientific data for centuries to come. Until science was able to definitively prove otherwise, most people could slip back into their comfort zone: Earth and humans were still at the cosmic center after all. The "conflict between science and religion" was born.

Gradually, however, science prevailed in the terracentric *vs.* heliocentric debate. Earth was proven beyond doubt to be part of a solar system, and accepted as such. Simultaneously, new theories and practices emerged regarding biblical scholarship: the field of "biblical criticism" (which meant a critical, not a superficial or faith-blinded, interpretation of biblical texts) prompted biblical analysts and interpreters to reexamine the historicity, internal integrity, and scientific compatibility of Bible narratives. Gradually, a consensus evolved: the Bible was neither history book nor science text, but a faith-based interpretation and even embellishment of stories designed to teach religious truths, not scientific or historical facts. Some Christians experienced a sense of theological displacement from comfortable, "certain" dogmas as a result of this knowledge. (Augustine [354–430 CE] provided small comfort for Christians who became aware of a comment he made a millennium earlier regarding Genesis stories: when science and the Bible are in conflict, the Bible should be interpreted allegorically.)

Second Dis-placement: Humans Are Not Independent Divine Creations

In the nineteenth century, Charles Darwin provided the next great jolt to human global and universal understandings. Once again leaders of religious institutions—and their faith tradition's adherents—had to consider setting aside, or at least rethinking in part or in whole, notions that humankind was the center of the cosmos, a unique, special divine creation unlike and superior to, and placed as ruler over—or at least manager or "steward" of— all other life and planet Earth. Instead of being a distinct creature created independently from other creatures, humans' corporeality resulted materially, Darwin stated, through a long process through which they emerged as offspring of primates.

The seeds of Darwin's *On the Origin of Species* (1859) were planted during his voyage as a naturalist on HMS *Beagle* and his concurrent scientific studies, particularly of Galapagos biota. He proposed subsequently that biota were not individually created in a six-day period, as described in Genesis 1, but had gradually emerged and evolved from simple organisms over eons of time. A subsequent work, *The Descent of Man* (1871), reiterated evolutionary theory but with a special focus on the human species. The idea that humans had emerged from primates as part of a long evolutionary trajectory rather than instantly at God's creative Word, as described eloquently in the first two chapters of Genesis, was psychologically, philosophically, and religiously threatening and even abhorrent to many clergy and laity—and scientifically stimulating and socially exciting to members of the scientific community and of science-amenable members of the general population.

Some clergy of Darwin's time accepted his ideas, although usually with the caveat that humans were the exception to evolutionary processes and were instead a special creation of God, at least to the extent that God instilled in them a spiritual, eternal soul; most clergy felt threatened religiously, and some professionally, as their long-held dogmas, rituals, and moral structures, as well as their personal and professional status in their ecclesial role, seemed to be under attack. Darwin, his evolutionary theory, and his writings narrowly avoided condemnation by the Catholic Church, primarily because reasonable churchmen observed that when Galileo had expressed scientific truth centuries earlier, he had been assailed unreasonably by Church officials; Darwin's clerical contemporaries declared that a similar Church error in interpreting scientific data—in this case, evolution—would become apparent over time, to the detriment of church credibility.[1]

The debate over evolution continues to rage in some religious circles (and evolution has been coercively suppressed in others) as people, particularly religiously fundamentalist clergy and laity, continue to resist what they view as humans' dis-placement from their status of preeminence amid creatures and creation. Religious "creationism" and its political variant, "Intelligent Design," provide ideological opposition to scientific knowledge and theories. The theory of evolution, of course, has continued to evolve as new data becomes available, particular in the area of genetics; and "mainline" Christian churches officially accept its tenets as scientifically established (some, however, still allow for human exceptionalism: God directly created humans, body and soul; or, God provided a human soul at an appropriate moment during primate evolution).

    1. In late November 2012, conservative evangelist Pat Robertson, founder and host of the *700 Club*, made a similar statement on national TV. He declared that evangelicals should accept that dinosaurs existed long before humans, the Bible, and the Genesis stories, and that Earth was more than just thousands of years old. He stated further that Christian churches would lose their younger members if they continued to assert and teach otherwise.

Third Human Dis-placement: Humans Are Not the Sole Cosmic Intelligent Life

Twenty-first century scientific discoveries and technological advancements—those already made and those yet to come—have the potential to diminish even further humans' self-understanding and sense of self-worth. Humanity has begun to explore the near and distant heavens with more powerful optical and radio telescopes, scientific instrument-laden space exploring unstaffed probes, artificial satellites, space shuttles and, where feasible, sophisticated all-terrain robotic vehicles. Concurrently, scientists supported by national governments, universities (including those receiving periodic and declining federal research grants), and private commercial and industrial interests have embarked on a more dedicated search for extraterrestrial life (ET) and extraterrestrial intelligence (ETI). Concurrently too, governments and industrial entrepreneurs have begun preparations for a more extensive and intensive exploration to locate and eventually extract extraterrestrial natural goods (the latter downgraded in human consciousness and speech to "resources" available for human exploitation in place, or for extraction for relocation). Already, civilians have begun paying substantial travel fees to accompany scientists on shuttle expeditions, and commercial civilian space rides in Earth's vicinity are in the works.

Theoretical developments in physics, and technologically advanced optical and radio telescopes in astronomy, have provided data that indicate the likelihood that millions and perhaps even billions of planets much like Earth exist in our own Milky Way Galaxy; extrapolation of this data suggests that innumerable other such planets exist in near and distant galaxies. A small percentage of these planets have conditions necessary for the existence of life as we know it; on an even smaller percentage, intelligent life might have evolved. Recent space explorations have discovered evidence of water seemingly capable of sustaining (or of having previously sustained) living organisms; the Mars rovers have been the most public examples of data collection vehicles that have provided such information, but in 2012 scientists discovered ice on Mercury, the planet closest to the sun. A new field of astrobiology has emerged. Studies of and on other celestial bodies[2] will continue to be made, thereby increasing the possibility of finding simple or complex forms of life that exist (or have existed) elsewhere in the cosmos.

The debate is on regarding the nature and ongoing impacts of the combination of time-matter-energy, intertwined in an inflationary universe which has been a dynamic outcome and continuation of the explosive cosmic singularity from which the universe began. Some scientists and religionists

---

2. "Celestial bodies" is the generic term used by NASA, the UN, and others to denote extra-Earth bodies such as planets, satellites (including Earth's moon), stars, etc.

continue to affirm that life evolved in the cosmos only on Earth: for the former, as a chance event, and for the latter, as a divine creation; others state that life emerged in several places through natural, autopoietic planetary dynamics and biotic evolution; yet others declare that life exists elsewhere in space as part of divine creativity.

The primordial cosmic event through which all that exists in the universe came into being was termed the "Day without Yesterday" by Belgian priest-scientist Georges Lemaître in 1927, when he first theorized that an initial explosion, whose resulting debris was expanded into space, was the origin of the cosmos—a theory and concept derisively dismissed by British astronomer Fred Hoyle as a "Big Bang" theory, the name by which it continues to be known, though no longer in a derogatory or disparaging way. Scientists currently discuss whether the emerging universe is so constituted as to make inevitable the beginning and evolution of life, or if the emergence of life—and, even more, the evolution of intelligent life—is a chance event, which perhaps occurred only once, on Earth, or on a handful of Earth-like planets.

Biblical Blocks to Spirit-Science Communication and Collaboration

A stumbling block to congenial science-religion consociation and collaboration is Christian fundamentalists' interpretation of the Bible. They are certainly free to hold their beliefs, even when they contradict reason and biblical scholarship and analysis. However, conflicts erupt when they strive to impose on others—not only other Christians, but humanists and people of other faiths—their ahistorical, implausible, anti-science, and biblically incorrect misinterpretations of biblical texts. In the United States, there is no state religion; people are free to hold and to live according to their beliefs, to the extent that they do not evolve into practices harmful or potentially harmful to others' well-being. Enthusiasm for a particular religion, denomination, or sect can be internally beneficial in peoples' religious beliefs, places of worship, and education. But some seek to use the external power of government to *coerce* people with different beliefs, rituals, and ethical practices to follow them, when they are unable to *convince* others, through moral suasion and doctrinal exhortation, to accept beliefs presented to them. When creationists' internal institutional power is useless in a nation having religious freedom, and their expression of their ideas fails to convince others to accept them voluntarily, they aggressively pursue governmental support to implement them. This is unacceptable and even morally reprehensible in a religiously pluralistic society that protects its religious diversity, including in its Constitution.

Biblical literalist Christians are challenged by established historical data and scientific knowledge and theory to rethink traditional interpretations of verses in the Bible, from both the Hebrew Scriptures and related Christian Scriptures. Biblical interpretation is substantially related to believers' understandings of "inspiration," "interpretation," and, in narratives that embody both, "myth."

Biblical Inspiration

Literalist Christians (and likeminded proponents of religious fundamentalism in Islam, Judaism, and other religions) believe that to say a sacred text is "inspired" means that it is literally, historically, and even scientifically accurate and "true" because its verses were divinely given. Among Christians, verses of the Bible are always literally "true" historically (e.g., 'Adam' and 'Eve' and all biota—including dinosaurs—were individually created by Yahweh several thousand years ago; Samson killed a thousand Philistines with the jawbone of an ass because of the power in his long hair), and scientifically (the Genesis 1 'days' of creation were literally of twenty-four hours' duration, even though the sun and Earth were not yet created when the 'days' began; and that the Flood story describes a historical global flood that destroyed all Earth's fauna except for the reproducing pair of every species herded by Noah and his family into an immense boat capable of holding them). A qualifier on biblical literalism accepted by even the most ardent creationist is that Earth is part of a heliocentric (sun-centered), not terracentric (Earth-centered) planetary system, and stars do not circle around Earth—even though these very obvious scientific truths contradict visual "evidence" and several biblical passages, and thereby conflict with a literal interpretation of such passages.

A different understanding of religious inspiration would be that *inspiration* is a faith tradition-related insight that is received by an historical person in a particular time, context, and culture; is expressed in their own language with its meanings, symbols, synonymous possibilities, and ambiguities; is intended to teach people of a specific culture in a given place and time; and which might teach future generations if they carefully discern its fullness of meaning and express this thoughtfully in their own language(s) to convey complementary meanings and symbols. People should recognize, in this process of acquiring knowledge and spiritual insight, that cultural and linguistic expressions used in the past, present, or future might cloud textual meanings for succeeding generations. Further, the complexity and fullness of a message might be understood only over time and after exegesis in different places by members of diverse cultures. Sometimes, as new information becomes available, seemingly unalterable texts are found not to express objective truth, even if apparently presented as historical facts; other texts do not state eternal laws promulgated by God, even if stated by a biblical writer to be a direct verbal command from God to a known, even heroic, biblical figure.

Biblical Interpretation

Once people realize that "inspiration" does not mean divine dictation they are better prepared to engage communally, interculturally, and collaboratively in interpretation of texts. The process of *interpretation* provides a way to seek the deeper meaning of sacred writings. It helps people discern which passages might be (*a*) *teachings from God* for all people, times, and places; (*b*) *teachings from God* for a particular people, for a particular time or all time, and in whatever place(s) they live; or (*c*) *teachings not from God*, which are advocated as if by divine command or instruction because of misunderstandings about God; misunderstandings about God's expectations of humanity; political or social considerations; understandings limited by historical time or place, including pre-scientific perspectives; literary expressions customary in a particular era or locale which might, in a culturally accepted practice of that time and place, augment historical data regarding events or people; or cultural biases such as nationalism, patriarchy, sexism, classism, rigid religious ideology, and ethnocentrism.

Biblical teachings from the Hebrew Scriptures—particularly those believed to be expressing directly divine commands—are sometimes hotly debated today because people from Jewish and Christian traditions interpret them in markedly diverse ways. Specific commandments that illustrate such interpretive conflicts are those regarding the Sabbath day; the prohibition against eating pork and shellfish; and the "ban" (*herem*) which was, essentially, biocide against all life in a particular place—and therefore included genocide against its human inhabitants. The representative texts and verses which follow illustrate these commandments, though not exclusively.

The Sabbath day is promulgated in Exodus 20:8–11. The sixth day of the week is to be dedicated to God. Sabbath practices include rest from labor, reflection on religious teachings, and particular Sabbath rituals. Is this an (*a*) type of commandment—a commandment from Yahweh for all peoples, times, and places—or a (*b*) type of commandment—a commandment from Yahweh for a particular people and their contextual places?

The prohibition against eating pork and shellfish (and consumption of other types of biota) is promulgated in Leviticus 11:6–8. Is this an (*a*) or (*b*) type of commandment from God, or a (*c*) type of commandment *not* from God? If the latter, what might it have been intended to teach, and why?

The *herem* is advocated in Joshua 6:15–21 (and elsewhere). The attack against Jericho is dedicated to God: the warriors are engaging the enemy to give glory to Yahweh, not to gain the usual spoils of war such as livestock, slaves, wives, etc.; precious metals are not to be distributed but dedicated to Yahweh and melted down to make ritual articles that will be used to worship Yahweh. Every living being in the city of Jericho is to be killed by Joshua and the attacking Israelites. Is this an (*a*), (*b*), or (*c*) type of commandment? If type (*a*), to whom does it apply today? If type (*b*), in what circumstances should it apply today? If type (*c*), why was it ever applied?

Over millennia, Orthodox Jews have continued to observe the Sabbath on the sixth day, Saturday, while most Christians since the first century CE have observed the Sabbath on the seventh day, Sunday (except for Jewish Christians who observed both Sabbaths). Orthodox Jews do not eat pork or shellfish, while most Christians do eat them, even regularly (as do self-described "secular Jews"). Neither Jews nor Christians advocate publicly that the *ban* should be invoked against an adversarial nation or people (although some Jews in modern Israel did advocate a *ban* against Palestinians years ago, and Christian officials in the US government and military effectively exercised a *ban* against the cities of Hiroshima and Nagasaki).

Biblical Myths

Biblical myths have provided intellectual obstacles that thwart science-religion cooperation when they are believed to be literal history and not literary stories. A *myth* is a narrative expressing what is perceived to be true, which expresses or reinforces a religious belief, cultural practice, or a cultural understanding of the acts of an historical person or the import of a purported historical event. "Perceived" is key here: the myth might actually express a historical occurrence, but what is perceived as truth cannot be verified as such, or the myth is not about a historical event or person, but is a story conveying religious doctrines or advocating moral behavior. The Bible includes four basic types of myth:

(1) myth as *factual* truth: the myth describes a historical event or person not (yet) independently corroborated by sources external to the sacred text other than those that elaborate religious doctrine;

(2) myth as *idealized* truth: the myth affirms, elaborates on, or even exaggerates what transpired at a historical event or the accomplishments of people participating in it, going beyond the factual data to express cultural religious, political, or economic understandings, needs, wants, or biases (e.g., doctrinal or ethnic affirmation or prioritization, or exhortation to moral conduct);

(3) myth as *fiction*: the myth expresses or describes beliefs or values, in order to teach people of a given culture and era by means of symbolic figures or events (which in a later historical period might be viewed to have actually occurred, as if the myths described exactly historical persons and events); and

(4) myth as *aspiration*: the myth promotes an individual's or people's vision of what they hope will come—even if the future hoped for is expressed as if it had been a historical reality at some time in the past—or of what they hope they will be as a person or as a people.

In the Samson stories in Judges, all four types of myths are present; the fourth is evident, too, in the Genesis 2 creation story—a source of some science-religion conflicts—among other places.

In terms of the first type, a myth with an historical basis, it might be accepted that Samson was an actual historic person to some extent (cf. Judg 13–16). For a period during its pre-nationhood history, Israel was governed by local or regional figures called "judges" (hence the title of the biblical book in which their lives and exploits are narrated). They usually gained prominence because some pressing problem plagued the people, often in the form of ethnic invaders or despots: "Then the LORD raised up judges, who delivered them out of the power of those who plundered them" (Judg 2:16). The Judges included Deborah, also a prophetess, who routed the Canannites; Gideon who defeated the Midianites; Jephthah who bested the Ammonites; and Samson who confronted and periodically defeated the Philistines. Samson, then, is a type of warrior-judge.

A religious practice of the time, one that lasted at least through part of the first century CE, was for a person to take (or to have been obligated to fulfill because of a parental pledge) a multifaceted Nazirite vow. The Nazirites were viewed as particularly chosen by or at least singularly dedicated to Yahweh, second only to the prophets in that regard. They were not allowed to cut their hair; they were not allowed to drink wine or any alcoholic beverage, or even to eat grapes (in case the grapes had begun to deteriorate on the vine, and were being transformed through fermentation into wine on the vine); and they were not to be in physical contact with deceased people.

The Samson myths of the first type, then, might describe a historical judge who had taken a Nazirite vow. In a way complementary to the story of Daniel and other youths in the court of Nebuchadnezzar who are healthier when they follow biblical dietary laws and refuse to eat Babylonian food, Samson is unbeatable as a warrior and hunter when he is faithful to his Nazirite vow.

Regarding the second type of myth, an elaboration and even exaggeration of a person's attributes or of a historical event, Samson's extraordinary physical strength enables him to kill, barehanded, a lion who attacks him (Judg 14:6), and uses the jawbone of a dead and decaying donkey to kill a thousand Philistine warriors who had come to capture him (Judg 15:9–19).

The third type of myth, a fictional narrative, is present in the Samson story when it states that Samson's physical strength is in his hair and that because of his great strength (which he lost when he had been captured and shaved, but had grown anew while he was a prisoner) Samson was able to pull the columns of the Philistine temple to which he was chained, until they collapsed, the temple fell, and all within it were killed (Judg 13:4–5, 14; 16; 17).

A myth of the fourth type, a narrative expressing a people's aspirations, is evident in the Samson narrative where the Nazarite vow is prominent. The vow and its fulfillment embody the hope of the ancient Israelites that they will faithfully observe Yahweh's commandments. The Nazarite vow, while it has a literal meaning, also signifies fidelity to Yahweh; long hair is a sign of this fidelity. When the Nazarite's hair is cut, that means that he has fulfilled his vow, if it has had a specified time period, or that he has not been faithful to his vow in some way. The Sampson narrative is an affirmation of tradition and a warning against infidelity to the tradition with its particularities. The strength is not literally in the hair; it lies in faithfulness to what the uncut hair symbolizes: the tradition, not the hair, is the strength of the people as a community and of the person individually. Similarly, the Israelite tradition is a gift of Yahweh; being strong in maintaining the tradition means that it will be more powerful than the traditions of the ethnically and religiously other, in this case the Philistines: the power of Yahweh destroys the temple and kills followers of false deities. (Samson's marriage to Delilah, a Philistine who betrays him for silver [Judg 16:4–21], has a similar meaning: the culturally and religiously alien "other" can seduce people away from following Yahweh's teachings and their traditions, to their detriment.)

The story of "Adam" and "Eve" in Genesis 2 might also be understood as a myth expressing the aspiration to be faithful to Yahweh: to do what one ought to do, and not to do what one ought not to do. Garden Eden is planet Earth. Adam and Eve might then be viewed, in today's evolutionary understanding, as "rising apes" rather than as "fallen angels" (they have angel-like—angelic—conduct in Eden). The ancient writers, of course, would not have been aware of human evolution from primates; however, religious and scientific perspectives converge in analysis of the story in their respective understandings that life emerged from basic elements: biblically, the soil or clay; scientifically, the primeval sea.

Extraterrestrial Myths?

Perhaps, like humans, extraterrestrial intelligent beings carry with them myths developed or retained over millennia. The myths, if they are the types just described, will have elements of fact and fiction. The integration of the best myths, those most universally able to be understood and become means for interspecies communication and conduct, or at least of those that have a common denominator, would serve well as a means for terrestrial and extraterrestrial intelligent beings to uncover mutually shared values and visions. As humans continue communication after initial Contact with other intelligent life, they might come to understand, in an exchange of ideas, that some of their scientific and spiritual "dogmas" were, in fact, types of myths. (Some scientific reductionists express a fundamentalist scientism, stating emphatically that spiritual reality and a belief in a divine Spirit are mere "myths" that a more advanced extraterrestrial civilization will have already rejected: such scientists have projected onto other intelligences, *a priori* and anthropomorphically, the scientists' own atheist myth, expanded into cosmic "truth.")

In the cosmic commons, intelligent beings need new, universal myths that would embody shared understandings and, especially for intercultural peace and well-being, aspirations. While shared historical myths might be difficult to develop, the exception might be "myths" about the origins of the cosmos (which, however true, can never be objectively replicated in a scientific laboratory) that could be mutually accepted as intelligent beings share their data and theories. The possibility for an ongoing terrestrial-extraterrestrial intelligent beings' benevolent post-Contact relationship would be enhanced and encouraged by the development of a new, yet old, myth of the fourth type: congruent, or even complementary existing myths from each culture might be rethought and reworded to express, based on shared scientific understandings, stable social constructs and shared socioecological ethics. The latter would provide an intellectual milieu in which they would have common ground upon which to build a concrete foundation for cosmic community in the cosmic commons.

As people consider human history and observe current events, they note how on their Earth home and habitat members of diverse faith traditions have not learned to accept each other. Consequently, hope for transcosmic shared myths might seem extremely "utopian" in the negative sense of the word: an unrealistic aspiration that the "real world" will crush. However, just as a certain ecumenical spirit began to take hold in the twentieth century among Christian churches (embodied, among other ways, in the World Council of Churches and diverse national Councils of Churches), and as some bodies began to incorporate Jewish and Muslim representatives of likeminded organizations (in the United States, the National Religious Partnership for the Environment is such an association), it is possible that particular intelligent species' aspirations are congruent or complementary and, in a spirit of cultural ecumenicity, could be conjointly integrated within the cosmic commons community).

It has generally been the case that diverse religions and diverse constituencies within a particular religion have been able most often to find compatibility and a working relationship not by attempting to reconcile or at least make compatible their respective doctrines, but by discerning and jointly promoting shared socioecological values, principles, and beliefs. Movements for social justice, particularly for the poor and the racially or ethnically other (in the United States, this has been evident in the civil rights movement and, periodically, movements for workers' rights and dignity, adequate housing, and universal health care, and at times for women, which have been most effective), civil rights and human rights policies have been pursued, and laws promoting social justice have been enacted. Associations of religious bodies working to protect the environment, locally and globally (including by promoting eco-justice to confront eco-racism), have helped to raise public consciousness about such issues; consequently, political support and civil legislation to safeguard Earth and equitably distribute Earth's common natural goods, and individual, corporate, and voluntary association support and activities, have followed: not only within particular religious constituencies, but also in the broader secular arena (and have complemented the longtime environmental conservation efforts of environmental organizations generally comprised of secular humanists and secular humanist scientists).

When thinking creatively and cooperatively, intelligent beings will have a transformed and transforming *conscientia*—consciousness of cosmic realities—and *consciencia*—conscientiousness in relating to cosmic others with compassion.

Bearing in mind the preceding, it might well be possible, on a cosmic and eventually truly universal level, to construct shared myths and then to integrate shared principles and the socioecological consciousness, commitment, and conduct that would put them into practice. This would help to realize (make "real") utopia (*u-topia:* "no place") in the positive sense of the concept: a place that does not yet exist, but can exist if people hold on to their vision and work in every present to make it a real *topia* for intelligent beings, biota in whatever form, and abiotic places in transcosmic contexts.

Tradition in Transition and Transformed

Catholicism has been, and is still viewed by some, as anti-scientific (particularly in the scientific community; on some issues: erroneously). However, representative visionary priest-scholars from that tradition—often in tension with their Church—have provided alternative perspectives to church dogmatic rigidity, as they affirmed newly emerging scientific data and theories. Despite personal and professional tensions with their Church, they offered innovative insights and concrete work that have benefitted both science and religion.

In order to highlight the gradually developing new stance of one of the traditional adversaries of scientific thought in past eras, the focus here will be on developments in the Catholic tradition. Certainly, similarly courageous and visionary figures have emerged in other traditions; however, a narrow focus here will provide an initial entry into evolving Christian thought. Three priests who have interacted in diverse ways with science will serve as representatives of transitions in and even transformations of tradition: Pierre Teilhard de Chardin (1881–1955; France); Georges Lemaître (1894–1966; Belgium); and Thomas Berry (1914–2009; United States). Coincidentally, Teilhard (French Army) and Lemaître (Belgian Army) were decorated veterans of World War I. Coincidentally, too, their respective groundbreaking scientific discoveries occurred just two years apart—Lemaître's cosmological/physics theory of a "Day Without Yesterday" in 1927, and Teilhard's participation as a scientist (primarily as a geologist) integrally involved in the Chinese scientific expedition that unearthed Peking Man in 1929.

As evidence for Darwin's theory of evolution by natural selection became ever more widely accepted in scientific circles and beyond in the latter part of the nineteenth century and early years of the twentieth century—including among adherents of religious traditions—some backlash resulted from conservative religious leaders and laity. Contrasts between a literal interpretation of their textual traditions and an in-depth consideration of the theory of evolution prompted people to ponder implications of evolutionary theory for Christian doctrines and rituals. Religious leaders noted that contradictions were evident between scientific statements regarding the chronological emergence and evolution of life, and a literal interpretation of the biblical creation stories in Genesis 1–2 and the flood story in Genesis 6–9. Fundamentalist Christians and the Roman Catholic hierarchy, in particular, became especially concerned not only about the possibility and consequences of attempts to relate the Bible to then current scientific knowledge. They also worried about implications of that knowledge, especially its apparent irreconcilability with key theological teachings, particularly regarding an "original sin" resulting from the "Fall" of "Adam" and "Eve," and the related Redemption from sin wrought by the salvific crucifixion and death of Jesus; teachings from Paul and Augustine, among others, which had been transmitted for almost two millennia, seemed threatened. In this historical context, the Vatican I Council (1869–70) briefly considered condemning the teachings of Charles Darwin much as it had condemned Galileo Galilei centuries before. However, caution guided their reflections as some Church leaders wondered whether or not, if they condemned Darwin, they might make a religious-doctrine-over-scientific-fact error of the type that led to Galileo's conviction in a Church court. Although in that case Galileo was forced to withdraw and even deny his theory (while knowing it was fact) in order to avoid ecclesial execution, eventually the Church had to withdraw its condemnation of him in order to avoid continuing embarrassment; it had become obvious that Earth did indeed orbit the sun. Hostility toward evolutionary theory and data continued however, as the Church continued to prioritize religious dogma over scientific knowledge.

In this social, intellectual, and ecclesial milieu Pierre Teilhard de Chardin was born. His keen interest in science, particularly in geology, biology, and paleontology, complemented by his concurrent interest in theology and his responsibilities as an ordained priest, led him to thoughtful and respectful efforts to integrate current science and traditional church doctrine. His work, however, resulted in his being forbidden by the Vatican and the Jesuit order to publicly speak or write about his integrative ideas, and consequently catalyzed his exile from France, his homeland. His public silencing, however, did not prevent his personal thinking and writing, and posthumous publication of his books and essays on the topic.

Pierre Teilhard de Chardin, SJ

Early in the twentieth century, Pierre Teilhard de Chardin, SJ, a French priest-geologist-paleontologist, pioneered in efforts to integrate Spirit and science. He was an ardent advocate of the theory of evolution and of its reconcilability with Christian faith and thought (more privately and less publicly as his position was met with alarm and persecution by Vatican and Jesuit officials, and eventually catalyzed open conflict). His scientific journeys, both intellectual and geographical, led to an impressive body of writings, international lectures, and participation in significant scientific expeditions as a geologist.

Teilhard was born to theologically conservative Catholic parents on May 1, 1881 in Auvergne, France, and baptized Marie-Joseph-Pierre Teilhard de Chardin. As he grew up in this bucolic setting and developed a strong appreciation for the beauty he saw in nature, he sought a sense of permanence:

he wanted to find and believe in things that endured. After several disappointments when he saw seemingly permanent objects perish, he settled on rocks as the most enduring, and his specific interest in geology was born. This closeness to nature contributed to his construction of concepts later in life that fused, philosophically, his understandings of God and creation. In the words of his major biographer, Claude Cuénot, Teilhard's entire life became "a steady progress towards an all-embracing synthesis that integrated man with the planet on which he lives, with the universe in which they both evolved, and with the God to whom he proceeds."[3]

Teilhard entered a Jesuit novitiate in 1900, and was ordained a priest in 1911 in Hastings, England. From 1912 to 1914, he pursued scientific research in Paris at the Institut Catholique, Collège de France, Paris Museum, and Sorbonne. In December 1914, he was called to military service to fight in World War I: clergy in France were not exempt from the military. At his own request, a month later he became a stretcher bearer at the frontlines, serving in a Moroccan regiment—the fourth combined light infantry and Zouaves—which was comprised principally of Muslim troops. His comrades admired his courage under fire; his officers commended his heroism with medals. He participated in several major battles: Ypres in Flanders (1915), Verdun (1916), second battle of the Marne (1918), and the Kehl bridge into Germany (1919). Even while in the military, he continued to write scientific articles and collected field specimens of pre-humans on the battlefield. As he pondered the meaning of the universe in this context, he wrote *La vie cosmique* (*Cosmic Life*) in 1916.

During battle, Teilhard would duck and weave among hails of bullets to rescue wounded soldiers and to retrieve bodies of the fallen. He was promoted to corporal; served as *ad hoc* chaplain; and was popular with and highly respected by officers and soldiers across ethnicities and religious backgrounds. Although General Guyot wanted to promote him to captain and to be chaplain for the 38th division, Teilhard declined; he wanted to stay "among the men" on the battlefield. His military citations include the Croix de Guerre; Médaille Militaire; and Chevalier of the Legion of Honor.[4] After the war, he professed his Jesuit vows in 1918. The war had had a profound impact on him as he pondered life's deeper meaning amid the carnage around him.

During the next several years, Teilhard renewed his Parisian studies in science. He did research on mammals of the Eocene Period in France, including through participating in field excavations, for his doctoral thesis in geology, which he successfully defended in 1922. During 1920–23, while

3. Cuénot, *Teilhard de Chardin*, 1.
4. Ibid., 22–25.

writing the thesis and for the year after graduation, he held the chair in geology at the Institut Catholique. As a Jesuit priest and geologist, Teilhard integrated his studies in theology and science, particularly geology and biology. He came to reject any mind-body or body-soul dualism, accept evolution, and develop a dynamic theory of cosmogenesis.

In 1923–24, he went on his first expedition to China, to Tianjin, where he took part in site excavations. He returned to Paris in 1924–26, where his sermons, and his lectures at the Institut Catholique, aroused Vatican and Jesuit interest and displeasure—but were popular among seminarians, parishioners, and the general public who were interested in the relationship of evolution to Catholic dogma, particularly doctrines concerning the "Fall" of Adam and Eve and "original sin." His Jesuit superiors ordered him not to speak or write on theological subjects, only on scientific matters. He was forced to resign from the Institut Catholique and return to Tientsin for 1926–27. Subsequently, after a brief period in France and Ethiopia, he returned to China.

Cuénot comments, regarding Teilhard's work, that he was "first of all a geologist, secondly a palaeontologist, specializing in mammals, and only thirdly a prehistorian and anthropologist. Of foremost interest in his palaeontological and anthropological discoveries, therefore, is the fact that they always had a solid, or in any case rarely questionable, geological basis."[5]

Teilhard worked with the Geological Service, and took part in the major excavations at Choukoutien. In 1928, fragments of the jawbone of *Sinanthropus pekinensis*, Peking Man, were found; in 1929, with Teilhard present as the expedition's geologist, Pei Wen-Chung and his team found an adult skull. Tools, too, were found at the site, and evidence of fire. These discoveries indicated that *Sinanthropus pekinensi* was *Homo faber*, and at the time became the oldest human remains that had been found.

His scientific work reinforced Teilhard's affirmation of evolution, and his dissatisfaction with the rigidity of the Catholic hierarchy and official Church teachings. Yet, he never mentioned that he was considering or ever had considered separating himself from either the Jesuits or the Catholic Church, as he continued to hope that both would come to recognize the ever-emerging scientific evidence, and appreciate his efforts to reconcile Church doctrine and scientific developments. Gradually, while yearning to return to his native France, he became resigned to exile, and plunged himself into his work. As a priest, devoted Catholic, and developing Christian mystic he felt isolated from both his family—he was not allowed to return to France on the occasions of the deaths of his parents and three siblings—and his religious sentiments, which were fostered initially in childhood by his devout mother,

5. Ibid., 91.

and subsequently developed more fully during his time as a soldier, in his Jesuit education, and through his mystical experiences. However, his situation enabled him to develop in more depth his integration of religion and science. His scientific research and expeditions, and his theological and philosophical speculations on their significance, mutually stimulated each other and catalyzed his creativity in this regard. While he attempted to formulate a vision for the future, ever-informed by his and others' scientific work, many people preferred to look backward rather than forward, as he laments in the collection *Letters from a Traveler*: "[T]here are still so many people who behave as though the past were interesting in itself, and treat it as only the future deserves to be treated."[6] Along these lines, he writes that "the essence of Christianity is nothing more nor less than belief in the unification of the world in God through the Incarnation. . . . In order to be Alpha and Omega, Christ must, without losing his human determination, become coextensive with the physical immensities of time and space. To reign upon earth, he must extend his quickening power through the world";[7] and, "I call the cosmic sense that more or less conscious affinity which binds us psychologically to the whole in which we are enveloped. . . . In the cosmos as I have described it here, it becomes possible for us, however surprising the expression may be, *to love the universe*. And it is in fact in that act alone that love can develop with unlimited clarity and power."[8]

Teilhard's exile from France, then, which was imposed and enforced by the hierarchy of the institutional Catholic Church in the Vatican, by bishops in France, and by the top leadership of the Jesuits in Rome and France, provided him with ample opportunity to extend his scientific explorations into the fields of biology and paleontology. A fortuitous result of his exile—unanticipated by church leaders—was the invitation he received to be part of the team of scientists searching for human origins: they discovered Peking Man. Ironically, Teilhard was exiled because of his advocacy of evolution, and found further, significant evidence of human and other biotic evolution during his exile. For Teilhard, in-depth reflection on paleontological data regarding human origins and their relation to biotic evolution was stimulated and strengthened consequent to the discovery of Peking Man. The China find did not limit his scientific thinking: his ongoing research led him to speculate that humans had first evolved in Africa, but he never had an opportunity to travel on that continent. (Discoveries later in the twentieth century would validate his theory that humankind had African origins.) He continued, too, to reconcile evolutionary theory with both Catholic doctrine,

6. Ibid., 146, citing Teilhard, *Letters*, 209.
7. Ibid., 194
8. Ibid., 196.

including biblical texts, and his experiential sense of a Creator God who was both transcendent to and immanent in creation. His work was not accepted by church officials until well after his death, in New York City, on Easter Sunday, 1955. Catholic historians and theologians consider his voice and his writings to have been significant stimuli for, and a major undercurrent in, the Second Vatican Council.

Teilhard's ideas on science-religion complementarity, engagement, and integration, particularly as expressed in evolutionary theory and data related to official church dogmas and his own scientific and religious intuitions and mystical experiences, were published posthumously in books and essays that he had given to friends to ensure that they did not disappear. Most of the latter were printed and made available in this way because of Church concerns bordering on condemnation, and the Church's consequent coercive constraints. His publications on science-religion themes have included *The Phenomenon of Man* (1959), a scientific book focused on human origins and evolution, and future possibilities for humankind in a dynamic cosmos in which divine presence is immanent and influential (the manuscript he had submitted to Roman and Jesuit censors was not allowed public dissemination and publication); in it, he sought to reconcile and integrate scientific theories and data with theological doctrine and ethics; *The Divine Milieu* (1960; *Le Milieu Divin*, 1957), a collection of essays reflecting on ways in which creation was permeated by divine presence); and *Christianity and Evolution* (1974), a collection of essays focused on the compatibility and integration of Christian doctrine and evolutionary theory.

Teilhard, Extraterrestrial Worlds, and Extraterrestrial Intelligent Life

In mid-1953, while in New York, Teilhard wrote "A Sequel to the Problem of Human Origins: The Plurality of Inhabited Worlds," which was not published during his lifetime.[9] His scientific knowledge regarding biotic evolution on Earth (reinforced by paleontological data from his field expeditions), the age of Earth and the universe (reinforced by geological data from his expeditions), and his understanding of the cosmic time required for Earth and life to have eventually come into existence, stimulated him to consider similar possibilities elsewhere in the cosmos. His ideas here were, as was the case with his thinking about the relationship of evolution and theology, well ahead of his time and place. They were written just six years after news of a crash of an alien craft near Roswell, New Mexico which had been followed by other UFO reports and scientific and theological speculations about

---

9. Teilhard, "Sequel to the Problem of Human Origins," in *Christianity and Evolution*, 229–36.

extraterrestrial life possibilities—including in a *Time* magazine article (September 15, 1952) which he cites.

Teilhard observes in his essay that "there are millions of galaxies in the universe, in each of which matter has the same general composition and is going through essentially the same evolution as that inside our own Milky Way."[10] This leads him to speculate that *if* proteins appear as early as possible and wherever possible, and *if* life, once existing, evolves as high as possible, including "up to 'hominization' if it can," and if there exist billions of solar systems where life can emerge and become "hominized," then "our minds cannot resist the inevitable conclusion" that if some form of technology could somehow detect the "radiation of the 'noospheres' scattered throughout space," then "it would be *practically certain* that what we saw registered . . . would be a cloud of thinking stars."[11] He states further that what once could only be imagined "is seen by us in the twentieth century to be *by a long way the most probable* alternative,"[12] and adds: "In other words, considering what we now know about the number of 'worlds' and their internal evolution, the idea of *a single* hominized *planet* in the universe has already become in fact (without our generally realizing it) almost as *inconceivable* as that of a man who appeared with no genetic relationship to the rest of the earth's animal population."[13]

Teilhard speculates along these lines that "at an average of (at least) one human race per galaxy, that makes a total of millions of human races dotted all over the heavens."[14] However, he offers a caution: "No matter how great a probability may be, we must be careful not to treat it as a certainty—that is obvious. . . . There is no question, then, of having to begin work on a theology for these unknown worlds. We must at least, however, endeavour to make our classical theology open to (I was on the point of saying 'blossom into') the possibility (a positive possibility) of their existence and their presence."[15]

Venturing into theology, Teilhard suggests in a subsequent essay, "The God of Evolution," that "in a universe in which we can no longer seriously entertain the idea that thought is an exclusively terrestrial phenomenon, Christ must no longer be *constitutionally* restricted in his operation to a mere 'redemption' of our planet."[16] In his theological thinking, Teilhard expresses an unrecognized anthropocentrism, based on his traditional Catholic

---

10. Ibid., 230.
11. Ibid., 230-31 (Teilhard's italics).
12. Ibid., 231.
13. Ibid.
14. Ibid., 232.
15. Ibid., 233-34.
16. Teilhard, "God of Evolution," in *Christianity and Evolution*, 237-43, here 241.

theology and on his vision of Christ as the Omega Point, when he assumes that God became incarnate solely on Earth for all the cosmos. Extraterrestrial intelligent life, he assumed, would benefit from this earthly incarnation: "For Christ's work of divinization to spread over the universe, it is sufficient to assume that God has raised up on each thinking planet (and continues to do so until the end) prophets and priests to whom knowledge of the redemptive Incarnation has been revealed and its grace communicated.... For the universe is so perfectly one that the Son of God has only to enter into it once to occupy and permeate it in its entirety.... By taking a human nature, the Word was 'cosmified'. He had to be born but once of the Virgin Mary to make his own and divinize the whole of creation. Just as Christ's birth is cosmic, so are his passion and death."[17] Teilhard wrestled with understanding the interaction of divine transcendence and divine immanence, and speculated: "How are we to succeed in apprehending the presence of the divine current beneath the continuous web of phenomena—the creative transcendence through evolutive immanence? ... Considered objectively, material facts *have in them something of the Divine.*"[18] In these words, his thoughts correspond well with those of Maximus the Confessor, who spoke of the relationship between *Logos* and *logoi*.

In theological thinking that results in anthropomorphic speculation Teilhard assumes, projecting from human historical conduct, that other "hominised" life will need redemption just as was necessary for humankind. Teilhard's understanding that all intelligent life harms itself is very like Stephen Hawking's assumption that evolved intelligent life inevitably tends toward self-destruction—and unlike the ideas of eighteenth century evangelist Thomas Chalmers, who declared in his sermons and stated in his essays that some intelligent beings might not have sinned at all. A certain "cosmic anthropomorphism" is present in both Hawking and Teilhard, but emerges independently from their contrasting philosophical bases. (Hawking, of course, does not accept the theological idea of "redemption" by an incarnate deity.)[19]

17. Teilhard, "Sequel," 235 n. 11.
18. Teilhard, "Note on the Modes of Divine Action in the Universe," in *Christianity and Evolution*, 25–35, here 29.
19. At the present historical moment humankind, without knowing or directly communicating with other representatives of intelligent life in the cosmos, inevitably projects onto theorized others its own conduct and shortcomings. It remains to be seen, at this stage of the cosmic journey, whether such be the case universally (in both senses: throughout the cosmos; characteristic of everyone), which would indicate that apparent anthropomorphism is actually a human manifestation of a universal intelligent life failing, and Teilhard's understanding is therefore accurate and realistic rather than pessimistic.

Teilhard had a slightly different take on the issue of Incarnation and Redemption in a previously unpublished essay decades earlier: "Fall, Redemption, and Geocentrism" (1920). He suggested at that time, too, in words similar to those above, that it seemed likely that intelligent life exists elsewhere in the cosmos: "[I]t is almost impossible to conceive that, among the millions of Milky Ways which whirl in space, there is not one which has known, or is going to know, conscious life—and that evil, the same evil as that which is such a blemish on earth, is not contaminating all of them. . . ."[20] Here he affirms the likely possibility of intelligent beings dwelling elsewhere, and speculates that evil is "contaminating" all of them. In a sharp contrast with his "prophets and priests" proposal cited above, however, he states that "the idea of an earth chosen *arbitrarily* from countless others as the focus of Redemption is one that I cannot accept; and on the other hand the hypothesis of a special revelation, in some millions of centuries to come, teaching the inhabitants of the system of Andromeda that the Word was incarnate on earth, is just ridiculous. All that I can entertain is the possibility of a multi-aspect Redemption which would be realized, as one and the same Redemption, on all the stars . . ."[21] He holds fast here to the idea of a single redemptive act, but states that it was not needed in every setting where life appeared; and, since "all the worlds do not coincide in time," it is necessary to introduce "a relativity into time."[22]

Teilhard's thoughtful theological speculation about ETI is contrary to both some Protestant evangelicals' assertions that ETI must be baptized on Contact to "save their souls" (which he does not mention), and to the statement by a Catholic theologian (who is contemporary to him, and whom he does mention) about conflictive Contact with ETI, which he cites, somewhat with astonishment: "It is embarrassing (unless it was meant as a joke) to read in *Time* [15 September, 1952] the advice given by a teacher of theology [Fr. Francis J. Connell, a Dean of Theology] to be wary of pilots of 'flying saucers': if they landed from a planet not affected by original sin, they would be *unkillable*."[23]

Teilhard's pioneering work on biotic origins and evolution on Earth is complemented by his fellow priest-scientist George Lemaître's work on abiotic origins of and dynamics in the cosmos.

20. Teilhard, "Fall, Redemption, and Geocentrism," in *Christianity and Evolution*, 36–44, here 38.
21. Ibid., 44.
22. Ibid.
23. Teilhard, "Sequel," 233 n. 7.

Georges Lemaître

Georges Lemaître, like Teilhard de Chardin, was a priest-scientist who attempted in his own area of expertise to promote the independence and the compatibility of science and spirituality. A *New York Times* article in 1933 featured a photo of Einstein and Lemaître, beneath which was the caption, "They have a profound respect and admiration for each other." In the accompanying article the author observed, "'There is no conflict between religion and science,' Lemaître has been telling audiences over and over again in this country.... His view is interesting and important not because he is a Catholic priest, not because he is one of the leading mathematical physicists of our time, but because he is both."[24]

Lemaître was born in 1894 in Charleroi, Belgium. In 1911, at the age of seventeen, he enrolled in the school of engineering at the Catholic University of Louvain (Université catholique de Louvain [UCL]), but left three years later to volunteer in the Belgian Army to fight in World War I as an artillery officer and engineer. His heroism under fire earned him accolades and several medals, including the Croix de guerre avec palmes. He would state later that "after 53 months of war ordeals and military camps, he lost interest in a professional career and decided to become a priest."[25] His postwar studies shifted from engineering to mathematics at the UCL and he earned his PhD in mathematics there in 1920 after he completed his thesis "Approximation of functions of many real variables" under the direction of Professor Charles de la Vallée Poussin. He subsequently entered the Malines Seminary and was ordained a priest by Cardinal Mercier in 1923.[26] Unlike both Teilhard and Thomas Berry, he did not belong to a religious order in the Catholic Church. However, he was a member of a fraternity of priests, the Friends of Jesus, whose members professed religious vows and provided mutual support as they strove to be dedicated to their calling.

Following his ordination, Lemaître received a fellowship to study in Cambridge University, England with noted astronomer Arthur Stanley Eddington, who initiated his studies in contemporary stellar astronomy and related mathematics. In 1924, he studied at the Harvard College Observatory of Cambridge (in the United States) under the direction of another renowned astronomer, Harlow Shapley, noted for his work on nebulae. He received his Doctorate in Science at the Massachusetts Institute of Technology in 1927;

---

24. Cited in O'Connor and Robertson, "Lemaître," pars. 7 and 8.

25. R. N. Tiwari, cited in ibid., 4.

26. Biographical data has been compiled from three sources: O'Connor and Robertson, "Lemaître"; the Web site of the Georges Lemaître Centre for Earth and Climate Research: https://www.uclouvain.be/en-316446.html, 1–5; and Farrell, *Day Without Yesterday*.

his thesis focused on "The gravitational field in a fluid sphere of uniform invariant density according to the theory of relativity." He had spent 1924-25 in Belgium as associate professor at the UCL while he worked on his dissertation; when it was accepted, he was promoted to professor.

In 1927 he published in the *Annales de la Société scientifique de Bruxelles* an article that would establish his reputation later, when it was more widely disseminated: "A homogenous Universe with constant mass and increasing radius explaining the radial velocity of extragalactic nebulae," in which he elaborated his groundbreaking concept of the physical expansion of the universe. When he discussed the article with Einstein at the Solvay Conference of physicists in Brussels, Einstein told Lemaître that "Your calculations are correct, but your grasp of physics is abominable," principally because Lemaître used Einstein's theory of a cosmological constant, which Einstein had discarded, and proposed an expanding universe, which Einstein rejected in favor of a Steady State universe. In 1929, astronomer Edwin Hubble discovered and published astronomical data that seemed to provide evidence of an expanding universe, unaware of Lemaître's published paper. Lemaître sent his paper to Eddington who had it translated into English and published. His ideas were at first ridiculed by those who thought that as a priest he was trying to use scientific theories to justify Genesis 1, but when astronomical observations reinforced his ideas they became more widely accepted and disseminated. He would call his theory of a singular origin of the now expanding universe "The Day Without Yesterday." In his "Primeval Atom Hypothesis," he stated that from a singular explosive burst of a quantum the entire universe eventually emerged, and continued to expand. In a May, 1931 article in *Nature* he explained:

> If the world has begun with a single quantum, the notions of space and time would altogether fail to have any meaning at the beginning; they would only begin to have a sensible meaning when the original quantum had been divided into a sufficient number of quanta. If this suggestion is correct, the beginning of the world happened a little before the beginning of space and time.[27]

Lemaître and Einstein kept their conversations going for several years, often in person. Einstein and Vallée nominated him for Belgium's highest scientific award, the Francqui Prize, which he received in 1934 from King Leopold III. He was elected to membership in the Royal Academy of Sciences and Arts of Belgium in 1941. He was honored by Pope Pius IX in 1936 when he was invited to be one of the founding scientists of the Pontifical

---

27. O'Connor and Robertson, "Lemaître," 2.

Academy of Sciences, and later was its president, from 1960–66. In 1950 the Belgium government honored him and his work during the 1930s with the "Prix décennal des sciences appliquées pour la période 1933–1942," a decadal prize for applied sciences, and in 1951 he was selected by the Royal Astronomical Society to be the first recipient of the Eddington Medal.

In his later life, Lemaître continued teaching at the UCL until he was awarded emeritus status in 1964, after which he retired. In subsequent years he became absorbed in computers, computer development, computational theory, program languages, and program writing, an interest he had begun to develop in 1930. Shortly before he died in Louvain in 1966, Lemaître was told that Bell Laboratories physicists Arno Penzias and Robert Wilson accidentally had discovered background interference when doing radiometer tests studying stellar radio sources. They were contacted by Princeton scientists Robert H. Dicke and P. J. E. Peebles, who with their team had done research on background radiation that should have resulted from a fireball origin of the universe, and had predicted this "cosmic microwave background radiation" would be found. After communicating, both teams published separately, in the same scientific journal, their complementary findings: the theory of cosmic background radiation developed by Dicke and Peebles et al., and the tests by Penzias and Wilson. This was, of course, the radiation residue that Lemaître, in his Primeval Atom "fireworks" aftermath, had theorized should be present in space. In 1978, Penzias and Wilson received the Nobel Prize for Physics for their "discovery." Neither Lemaître nor Dicke and Peebles received equivalent recognition for the work they did to enable Penzias and Wilson to realize that the "noise" which interfered with their quest for radio signals from the stars was radiation residue from the primeval singularity.[28]

As noted earlier, most of Lemaître's scientist contemporaries initially dismissed and even disparaged his theory about a "Day Without Yesterday" because he was a Catholic priest and because they thought that he was trying to provide a scientific justification for the Genesis 1 creation story. (According to Fred Hoyle in a 1950 radio program, Lemaître's idea was nothing more than a "Big Bang" theory; that dismissive designation continues today, although now as a popular phrase it is used positively to indicate approximately what Lemaître had proposed.) Lemaître, however, always made a careful and even sharp distinction between his theological beliefs and his scientific theories, to the extent of pointedly asking Catholic Church officials who linked publicly his scientific accomplishments and Genesis narratives

---

28. Elaborated in Farrell, *Day Without Yesterday*, 137–40. Farrell observes (138) that anyone today who turns a cable TV to a blank channel will be able to see what some writers have called the "echo of creation": some of the "snow" on the screen results from cosmic microwave background radiation.

to cease doing so. He advocated autonomy for both of the "two ways," as he called them, science and revelation.

In a 1933 interview in California in which he was asked how he could reconcile faith and physics, Lemaître responded:

> The writers of the Bible were illuminated more or less—some more than others—on the question of salvation. On other questions they were as wise or as ignorant as their generation. Hence it is utterly unimportant that errors of historic or scientific fact should be found in the Bible, especially if errors relate to events that were not directly observed by those who wrote about them.
>
> The idea that because they were right in their doctrine of immortality and salvation they must also be right on all other subjects is simply the fallacy of people who have an incomplete understanding of why the Bible was given to us at all.[29]

When asked in an interview decades later if the "Big Bang" theory was inspired by his religious faith, Lemaître's reply complements the preceding response:

> As far as I can see, such a theory remains entirely outside any metaphysical or religious question. It leaves the materialist free to deny any transcendental Being. He may keep, for the bottom of space-time, the same attitude of mind he has been able to adopt for events occurring in non-singular places in space-time. For the believer, it removes any attempt at familiarity with God.... It is consonant with the wording of Isaias speaking of the "Hidden God," hidden even in the beginning of creation.... Science has not to surrender in face of the Universe and when Pascal tries to infer the existence of God from the supposed infinitude of Nature, we may think that he is looking in the wrong direction.[30]

It is worthy of note that the ideas expressed by Lemaître in physics and cosmology complement well—indeed, overlap with—those of Teilhard in geology and biology. However, while the evolutionary biologist-mystic was silenced for expressing his thought, the theoretical cosmologist was lauded and quoted. The difference might well be that Catholic Church officials sought to use Lemaître's ideas to justify a literal interpretation of Genesis 1, and that he was well-recognized internationally as a scientist who was simultaneously a priest: Church condemnation would have raised an international

---

29. Cited in Farrell, *Day Without Yesterday*, 203.
30. Ibid., 207–8.

furor, while Church recognition was seen as a way to justify and promote Church doctrine.

In the twentieth century, then, two priest-scientist contemporaries experienced contradictory receptions from the faith community and the scientific community. Teilhard de Chardin had been accepted by scientists and rejected by the Catholic Church because of his scientific work; Georges Lemaître had been accepted by the Church because of his scientific work and rejected by numerous scientists because of his religious faith. Both priest-scientists ultimately influenced—and continue to influence—the fields of religion and science which they pondered in depth and about which they spoke and wrote.

Thomas Berry, CP

In the latter half of the twentieth century and into the beginning of the twenty-first century, the foremost proponent and expositor of Teilhard's ideas was the Passionist priest-"geologian" Thomas Berry (1914–2009). Berry was not only an advocate of Teilhard's thought (he served for decades as the President of the American Teilhard Society; in lectures and essays he promoted Teilhardian thinking); he made his own significant contributions to religious cosmology and science-religion discussions and reconciliation.

Thomas Berry was born in Greensboro, North Carolina, the third of thirteen children. He was named after his father, and baptized William Nathan. When he was eleven he had a profound experience in nature that stayed with him throughout his life and continually oriented him toward a sense of communion with the cosmos. Crossing a creek at the bottom of an incline on a new home site to which his family had moved, Berry stood at the edge of a meadow and was absorbed in the sights and sounds of the moment:

> The field was covered with white lilies rising above the thick grass. A magic moment, this experience gave to my life something that seems to explain my thinking at a more profound level than almost any other experience I can remember. It was not only the lilies. It was the singing of the crickets and the woodlands in the distance and the clouds in a clear sky. It was not something conscious that happened just then. I went on about my life as any young person might do. . . .
>
> Yet as the years pass this moment returns to me, and whenever I think about my basic life attitude and the whole trend of my mind and the causes to which I have given my efforts, I seem to come back to this moment and the impact it has had on my feeling for what is real and worthwhile in life.

> This early experience, it seems, has become normative for me throughout the entire range of my thinking....
>
> Religion, too, it seems to me, takes its origin here in the deep mystery of this setting. ... [I]n this little meadow the magnificence of life as celebration is manifested in a manner as profound and as impressive as any other place I have known in these past many years....
>
> We might think of a viable future for the planet as renewed presence to some numinous presence manifested in the wonderworld about us. This was perhaps something I vaguely experienced in that first view of the lilies blooming in the meadow across the creek.[31]

In 1934 Berry joined a monastery of the Passionist Order in New York City, and chose Thomas as his religious name;[32] he was ordained a priest in 1942. He earned his doctorate at the Catholic University of America, Washington, DC; his dissertation, published in 1951, focused on Giambattista Vico's philosophy of history. Vico's emphasis on poetic wisdom and creative imagination in Western intellectual history would be an enduring influence on Berry, but not a restrictive one: he learned Chinese, Japanese, and Hindi, and studied Asian religions, culture, and history. After studying in China (1948–9) he and Theodore de Bary co-founded the Asian Thought and Religion Seminar at Columbia University, New York. Berry taught Asian Religions at Seton Hall University (1956–60), St. John's University (1960–66), and at Fordham University (1966–79), where he established the History of Religions program. While at Fordham he founded the Riverdale Center of Religious Research on the Hudson River, which he directed for more than twenty years (1970–95). During this period he wrote numerous books and articles on Buddhism, Confucianism, Taoism, and Hinduism. He died in Greensboro, North Carolina in 2009, and is buried in Green Mountain Monastery, Greensboro, Vermont.[33]

Thomas Berry advocated for a cosmic consciousness. In *The Dream of the Earth* (1988), he declared that since all that exists is related, the complex cosmos is a "universe" in which "everything is intimately present to

---

31. T. Berry, *Great Work*, 12–14, 20.

32. It was customary in most male and female Catholic religious orders and congregations that novices would change their name, indicating a change of life, when taking their first religious vows. Usually, they selected an admired saint's name; Berry chose Thomas in honor of Aquinas.

33. As of this writing, there is little detailed biographical information about Thomas Berry. The information here presented is from the most comprehensive site, that of the Thomas Berry Organization, which has articles written by Mary Evelyn Tucker and John Grim; see http://www.thomasberry.org/Biography/tucker-bio.html.

everything else.... Nothing is completely itself without everything else. This relatedness is both spatial and temporal....The universe is a communion and a community."[34] To confront the ecological crises caused by human exploitation, Berry stated that "a Great Work is needed," a "process whereby consciousness of an integrated human-Earth relationship is restored, and humans live in harmony with each other, with all life, and with Earth."[35] This Great Work, he states in a book by the same name, is a movement that gives "shape and meaning to life."[36] It results from, and in turn gives further meaning to, human perceptions of place in the cosmos: "every reality of the universe is intimately present to every other reality of the universe and finds its fulfillment in this mutual presence,"[37] and "nothing in the universe could be itself apart from every other being in the universe, nor could any moment of the universe story exist apart from all the other moments in the story."[38] For Berry, "the universe is the primary sacred reality," and people "become sacred by our participation in this more sublime dimension of the world about us."[39] People should realize that "the spiritual and the physical are two dimensions of the single reality that is the universe itself."[40] In words reminiscent of the teaching of Maximus regarding *Logos-logoi*, Berry states that "each being in its subjective depths carries the numinous mystery whence the universe emerges into being," and adds that "the universe itself is the primary sacred journey."[41]

Thomas Berry's *The Great Work: Our Way into the Future* explores human consciousness in a cosmic context, and human responsibility for the place(s) that humankind inhabits. It complements the Earth Charter: both works focus on respect for Earth, biota, and human community, and provide principles and sub-principles to promote and realize these. The essential ideas of both writings can be extended into the area of cosmic consciousness and concern, and guide ecological and biotic engagement with the places and lives that humankind will encounter in its exploratory travels in its home universe and, in current theory, in other universes or dimensions thereof.

Berry's most direct foray into the science-religion arena came when he coauthored, with theoretical mathematical physicist Brian Swimme, *The*

---

34. T. Berry, *Dream of the Earth*, 91.
35. "Afterword," in Hart, *Sacramental Commons*, 236–37.
36. T. Berry, *Great Work*, 1. A more extensive elaboration of Thomas Berry's work is found in Hart, *Environmental Theology*.
37. T. Berry, *Dream of the Earth*, 106.
38. T. Berry, *Great Work*, 33.
39. Ibid., 24.
40. Ibid., 49.
41. T. Berry, "Reinventing the Human at the Species Level," in *Christian Future*, 121.

Universe Story (1992). Having noted both the disparity among the creation stories of diverse religious traditions and their allegorical character (even if the latter was not accepted as such by adherents of these traditions), Berry and Swimme propose a common creation narrative that would flow from current scientific thought, with its firm foundation in material reality and scientific theory. In a later essay, "Women Religious," he stated that the universe story "is the story that the universe tells of itself. It is the story told by every being in the universe, by the stars in the heavens, by the mountains and rivers of Earth, by every wind that blows, by every snowflake that falls, by every leaf in the forest. To know this story of the universe as our sacred story is to have an adequate foundation for the tasks before us. This story tells us who we are and how we came to be here and what our lives are all about."[42]

Berry understood well the significant and necessary role that science should play in religious consciousness:

> Never before have any people carried out such an intensive meditation on the universe and on the planet Earth as has been carried out in these past few centuries in our Western scientific venture. Indeed, there is a mystical quality in the scientific venture itself. This dedication, this sacred quest for understanding and participation in the mystery of things, is what has brought us into a new revelatory experience. While there is no need for us to be professional scientists, there is an absolute need for us to know the basic story of the universe and of the planet Earth as these are now available to us by science.[43]

Berry recognized, too, the shortcomings of scientific thinking. In words complementary to those of Seyyed Hossein Nasr, he stated: "Scientific insights led to an alienation from the former sense of the cosmos as the locus for the meeting of the divine and the human. This became especially clear as science went on to discover that the universe was a self-emergent and self-organizing process. At this time the sacred character of the universe was overshadowed and even lost."[44] In words reminiscent of Maximus and Teilhard, Berry writes that "Each being in its subjective depths carries the numinous mystery whence the universe emerges into being."[45]

Berry was concerned, too, about the environmental crisis. In "Christianity and Ecology," he declared that "the survival of the planet Earth in its integral reality is, it seems to me, the basic issue that confronts us here. . . .

42. T. Berry, "Women Religious," in *Christian Future*, 81.

43. T. Berry, "The Gaia Hypothesis: Its Religious Implications," in *Sacred Universe*, 116.

44. T. Berry, "The Universe as Cosmic Liturgy," in *Christian Future*, 114.

45. Ibid., "Appendix," 121.

There is no way in which the human project can succeed if the Earth project fails."[46]

Thomas Berry, unlike Teilhard, did not consider at length, in his published works on cosmic origins and humans' cosmic place, the possible existence of ETI. However, in considering humans' resident cosmic consciousness, their role as a part of the cosmos that reflects upon itself, and their responsibilities for Earth and in the cosmos, he left open the extent of intelligent life in cosmogenesis. He observed that through the story of the universe humans "learn something about how the primordial mystery of the universe brought the planet Earth into being as the most blessed of all the planets *we know of*,"[47] acknowledging that intelligent life might have emerged elsewhere in the cosmos, and leaving open the question of whether or not there are other intelligent beings, like humans, through whom the universe becomes self-reflecting.

Teilhard, Lemaître, Berry, and other pioneering scientists and religious thinkers and leaders have become increasingly important not only for their provision of insights for humans' inquiry into the human place and prospects in the cosmic commons, but also for advocating and integrating complementary views from religion and science, directly or indirectly.

In the twenty-first century, conflicts between science and religion persist despite, in some situations, affirmation of scientific discoveries by religious leaders. In many respects, though not without internal conflicts and external struggle, the Catholic Church has become a leading religious proponent of science and scientific theories and discoveries—a far cry from its action against Galileo in the seventeenth century, and against Pierre Teilhard de Chardin in the twentieth century.

Spirit in Space

In past eras, religious teachings and religions' leaders advocated thinking about and praying to divine Being(s) as "Other," as transcendent and even distant (dwelling in a heavenly abode located "above" Earth). The divinity/ies might on occasion be solicitous of the well-being of a particular "holy" individual, or a "chosen" person or people, and offer them particular guidance or assistance. "God" or "Goddess" or "gods" were usually understood to be more solicitous of human beings than of other material beings. In a Western perspective, all were linked in a "Great Chain of Being," effectively a cosmic pyramid atop which humans were situated, and other biota were placed

---

46. T. Berry, "Christianity and Ecology," in *Christian Future*, 59.

47. T. Berry, "Religion in the Twenty-First Century," in *Sacred Universe*, 87 (my italics).

below to serve or provide for them; God was, of course, above and beyond all of this.[48]

Sometimes religions' members viewed human conflicts, including wars, as surrogate or simultaneous expressions of deities' battles in a heavenly realm. In the Christian tradition, especially after the Medieval Synthesis in which Thomas Aquinas "Christianized" Aristotle by incorporating his Greek philosophical thought and categories into Christian theological terms, God became viewed not only as a transcendent Other but, in that light, as a Being who was perfect and happy in "himself." Such a God would be personally unmoved and unaffected by biotic suffering and biota's experience of tragedy, and by the sadness that might follow and flow from harm to humans and other biota—all while God was understood to have characteristics of love, mercy, and justice that would stimulate occasional divine guidance and strength to overcome social evil and natural catastrophes. In this view God, although impassive, desired that a gradual dynamic process would enable the divine characteristics mentioned to be ever more operative in human consciousness, compassion, and conduct. In the twentieth century, however, some Christians, Jews, and other religious believers came to understand that divine Being was not impassive, but immersed in and experiencing, and transforming in some manner in a spiritual dimension of reality, the sufferings of humankind and all biota as their evolutionary existence continued. They recognized that God who is Love could not be impassive amid the real material and spiritual realities of the suffering of those whom God loves.

Materiality and Metamateriality

The interrelationship of religion and science has been disrupted to some extent and for some time because of the reaction of the Catholic Church in previous centuries, and of Protestant fundamentalist churches in the twentieth and twenty-first centuries, to scientific knowledge and theories. This rupture in relationships, whose earlier source was the mechanistic view of the universe that began to emerge in the Enlightenment and the Renaissance, has been addressed by a few notable figures from the fields of science and theology beginning early in the twentieth century. While respecting both areas of inquiry and knowledge as distinct fields, they noted, too, their

48. Religions other than Christianity have understood the divine to be related to all biota and even to Earth, but such an understanding, except for the thought and actions of insightful figures such as Francis of Assisi, did not begin to enter Christian thought until the twentieth century, when the concepts of a "web of life" and a "biotic community," and an understanding of God's transcendence-immanence and concern for creation as a whole—Earth and all life—began to emerge, to be disseminated, and to provide links to other religions' teachings.

complementarity. This became evident particularly as both addressed Earth's increasing ecological devastation and deterioration, and social injustices that worsened as a consequence.

What has emerged in recent years, especially among thoughtful scientists, theologians, environmentalists, and ethicists, is an understanding that the two fields have distinct areas of interest, research, and competence. Science is concerned with materiality, the matter-energy reality of Earth and the cosmos whose aspects are observable, quantifiable, and replicable, theories about which must be materially falsifiable; religion is concerned particularly with metamateriality, the spiritual dimensions and implications of humans' and other biota's existence, while seeking to ground this concern in, or at least relate it to, material existence. Neither field's representatives are competent to assert that ideas, teaching, and data from their own area of specialization and speculation are superior to what results from the work in and ideas of the other area. Science, by its nature, cannot make metaphysical claims, that is, profess knowledge about metamaterial reality, or even deny that such a dimension of reality exists. Religion, according to its nature, should not promulgate or retain doctrines that contradict scientific knowledge, or seek to impose a religious ideology on scientific understandings. The relationship of materiality and metamateriality in human thought and "on the ground" in concrete, material existence within the integral being of the cosmos can provide a figurative intellectual common ground that would enable fruitful exchanges of ideas and mutual collaboration in projects to effect socioecological well-being. In the words of Harvard biologist Edward O. Wilson, people from science and from religion can "meet on the near side of metaphysics" to attain their shared goal of protecting and conserving the creation in which all live and work. The three thinkers who have been the special focus in these pages represent well the possibilities for bringing materiality and metamateriality into thoughtful and respectful conversation.

Spirit and Science in Solidarity

In considering the speculations and investigations of Teilhard, LeMaître, and Berry religion and science have the opportunity to develop further their own complementary efforts to understand the meaning of discoveries already made and the potential meaning of theorized or unanticipated discoveries that might yet be made. This approach, rather than competitive endeavors and even invective to discredit the other or to claim philosophical or material superiority because of their respective bases of thought and theory, will provide a context for greater collaboration between, and mutual enrichment of, both science and religion.

Transformed understandings of religious inspiration, interpretation, and myths, related to scientific understandings and integrated with a shared vision of a common future, could be embodied in evolving common principles and projects. Common conduct in the commonweal could continually catalyze development of concrete social structures and ecological relationships that would craft from what is now an initially envisioned and expressed distant *utopia*, a future *topia* in which our human—and other intelligent beings', should they exist—descendants would live in an age yet to come.

6

Earth Commons and ExoEarth Commons

Contextual Considerations

The human exploration and colonization of planets and celestial bodies in our own and distant solar systems eventually will provide people with new homelands in habitable settings. In the probably faraway future on such faraway places, humanity will have an opportunity to integrate concretely the transformed consciousness and conduct needed to avoid making the social and ecological errors whose harmful impacts have permeated humans' and other biota's life on Earth, and degraded Earth's environment. Were this to occur, justice would be inherent in human communities, and environmental well-being would be an important and integral aspect of humans' relatedness to other biota and to extraterrestrial milieus. In distinct *praxes*[1] in these settings, the dialogic interaction between consciousness and conduct, between ethical theory in the abstract and ethical practice that is engaged with diverse contexts, humankind (Earth-evolved and -originating intelligent beings) and, should Contact occur, extraterrestrials (exoEarth-evolved and -originating intelligent beings) will adopt from and adapt to each other, separately or collaboratively, after their encounter in unique circumstances in newly engaged cosmic places.

Earth and ExoEarth in Dialogic Relation

When people travel to distant planets, initially there probably will be only two places in which humans will live: Earth and exoEarth, wherever that might be. Perhaps Mars will be the first planet colonized. The focus here will

---

1. *Praxes* are social settings that provide diverse places of dialogic interaction between socioecological ethical *theory* and socioecological ethical *practice*; often they are milieus of conflict between social groups.

be on the relationship between the people and planet in each locale in itself, and in their connection and interaction. ETI will be noted periodically, but the initial discussion will be about people and places. A particular consideration will be how humans have improved their origin planet and how they have taken what has been done or not done at home into space, and to what extent it will be replicated elsewhere. Ethics-ecology-economics-ecclesia will be subjects considered.

Intelligent beings' awareness of their continuing presence within and as part of integral being will enable them to analyze a new (to them) place's geophysical characteristics and biotic forms, and to have a certain comfort level even when confronted by events and biota that would have been considered extraordinary in previous places which they explored or where they settled. They will become aware of newly found social interactions in some places as they and resident intelligent biota discover each other, and newly found ecological dynamics: among biota, and between biota and the home and habitat provided by their planet (or other celestial body).

Adaptation in cosmic contexts of an Earth-formulated, -developed, and -implemented socioecological consciousness and conduct that embodies *socioecological ethics* would provide at least an initial impetus for humans to do better elsewhere than they have on Earth. They would see the "new world" as it is and as it might become, for better or for worse, when they as intelligent life arrive to colonize, and as they strive to carefully conserve pristine or inhabitant-conserved places while using them respectfully to meet human needs. They would seek to compassionately and collaboratively consociate with the biotic community with whom they will be interrelated, interdependent, and integrated.

In the persons of Earth's explorers when they become settlers on a planet elsewhere in the cosmos, new 'Adams' and 'Eves' will arrive to dwell in an analogous 'Garden of Eden.' The 'garden' planet would not be a primordial paradise, but a place of new beginnings where humans could relate respectfully to their planetary environment—including both its abiotic context and its existing biota at whatever their level of development and consciousness. In a biblical sense, for the religious believer, humans might be in this instance 'images of God' as God is presented in biblical texts: a divine transcendent-immanent being who regards all creation as "very good," walks solicitously among Earth's biota (Gen 1–2), and cares for all creatures (Gen 6–9; Job 38–41), and whose relationship with all creatures is unending, as symbolized by the rainbow, in a God-creation covenant (Gen 9). The well-being of each place is ensured when its respective biotic members live in an integrated, interdependent relationship with each other and fulfill roles with relative ecological responsibilities that correspond to and are derived from each species' evolving capabilities. The Discovery Doctrine has to be erased from human theory and practice for this to occur. Human settlers would not want to plant in their new "Eden" the seeds for social and ecological catastrophe. Continuation of Discovery in the cosmos would lead to Earth redux, *ad infinitum*.

Current Curiosity

Global interest in the implications of and the potential for benefit or harm from extraterrestrial explorations and encounters is extant in the twenty-first century. Human curiosity has been piqued by a variety of Earth and exoEarth events, begun dramatically with the USSR's Sputnik satellite, and continued by technological achievements in near and distant space by spacecraft launched by the USSR, US, and China, some of whose conceptual development, construction, and technical personnel have come from other nations in collaborative efforts—the most dramatic and visible being recently the gradual construction and operation of the International Space Station.

The 2012 landing of the US Curiosity land rover on Mars and the video data it has transmitted catalyzed a heightened interest in finding evidence of life there, and potentially knowledge of the cosmic origins or development of life generally. Entrepreneurs hoped that geological data might indicate the presence of surface or subsurface natural goods, which might be suitable for industrial development, commercial enterprises, and human well-being. Rover Curiosity has stimulated global curiosity about possible future exoEarth exploration and settlement, perhaps including, beyond a Mars colony, a lunar base and then a voyage to Europa (one of Jupiter's moons). A lunar base itself would provide a launching area for space vehicles to voyage throughout the solar system. Should a similarly hospitable place be found further out in the solar system, a launch site might be constructed there for travel to ever more distant stellar settings. The International Space Station might well be a portent of things to come, if nations can continue to collaborate with each other in technology, personnel for spacecraft, and mutually agreed upon and equitable distribution of natural goods discovered, with a particular concern for countries and peoples that lack, individually, communally, or nationally, economic resources sufficient to provide for their material well-being.

Contact Considerations

Interdisciplinary intellectual explorations by scientists and ethicists before and during Contact will result in innovative ideas and suggestions for appropriate human conduct in terrestrial-extraterrestrial engagements as scientific efforts expand and encounters occur. *A priori* reflection should prove invaluable for the general public, the scientific community, technologists, government officials, military personnel, members of religious traditions, and business entrepreneurs. It could be a useful foundation upon which to build the potential for positive encounters.

Discussions prior to Contact would provide, too, ideas to help people on Earth accept or at least tolerate a new sense of displacement if they become upset religiously or psychologically because of evidence of other habitable worlds and life. Humans on the ground in exoEarth places, separated from their homeland, would be similarly aided to adapt to their new physical and psychological milieu. Both Earth and exoEarth populations would be stimulated to develop a new sense of place as Contact events and colonization occurred.

Integral Being

Integral being is the totality of all being. In it all being and beings have originated—it is the 'stuff' from which all being has been and is being comprised, formed over eons and intertwined into an intricate web of interrelated being. In the cosmos in which humankind lives (the sole cosmos with which humans are familiar), which might be one of many cosmoses (scientists theorize the possible existence of a *multiverse*, a "universe of universes" of which only one is known to human science), existence began at some "point" just before the singularity, the initial explosion in space (LeMaître's "day without yesterday") which catalyzed the existence and expansion of the cosmos, and has continued through eons thereafter. Integral being births all being, which emanates from and in it. For the theist, integral being is the dynamic cosmos as a whole, existing in, engaged with, and sacred because of, divine presence; it originates in a creating transcendent-immanent Spirit who permeates all that is. For the secular humanist, integral being is solely the complex cosmos in itself, with all its diversity of being, whose origin before it burst into its present, dynamic existence is unknown.

Integral *being* is in reality integral *being-becoming*: it retains its essential properties and potential, and "experiments" with their characteristics and with its possibilities given the perennial or sometimes apparently ephemeral properties in play at particular moments. Integral being continually brings about new combinations and new forms of existence, perhaps much in the manner of DNA in organisms' evolution as they respond to altered environments and new ecological relationships. The macrocosmos and microcosmos witness abiotic and biotic birth, death, and rebirth. Stars are born in cosmic settings, including in areas called "stellar cradles"; they live for the time that is possible to them and then die, often after an expansion followed by a fiery explosion; new stars are born and enter in turn into their respective cycles. Species emerge evolutionarily in biotically habitable planetary settings, find their niche, adapt as they are able to new environments into which they foray or are forced, develop new species, subspecies, and individuals, all of whom are born, live for a time, and die, some of which are reborn through reproduction or mutation, and then become extinct when conditions necessary for their existence disappear.

For the *theist*, the concept and reality of *integral being* describes the dynamic, complexifying cosmos existing in, engaged with, and sacred because of, divine Presence. All being has originated in Being, a creating transcendent-immanent Spirit, since before the singular event that Belgian priest-scientist Georges Lemaître in his "hypothesis of the primeval atom" called a "day without yesterday." Lemaître's conjecture is known today in popular parlance as the "Big Bang" theory, to which scientists' subsequent astronomical observations and discoveries have corresponded and corroborated. Lemaître had predicted, too, that discoveries would be made and knowledge of cosmic dynamics would increase over time as scientific instruments and methods became more refined and accurate.

In the long "day" prior to cosmic creation, Spirit had envisioned and effected the essential characteristics and parameters which have continued to be operative and manifest for billions of years. Divine creativity continues today (and will continue tomorrow, through Earth time and through cosmic time), not in specific divine acts which place new creatures into living environments, but by a process divine action initiated at the primary formative moment of creation, and imbued and inhered with creative freedom to develop—unconsciously, unknowingly, and experimentally—abiotic dynamics and biotic evolution.

For the *humanist* (atheist; agnostic; [non-Spirit] spiritual), integral being is the dynamic, complex cosmos by and in itself, with all its diversity of being. The pre-origin of the universe—that brief or lengthy moment for which it existed before a primordial singular explosion initiated its ever-accelerating inflation into an ever-developing, ever-complexifying entity—is unknown. Theories continue to be advanced about how that state originated, and from what forms and forces it might have self-formulated its primordial existence.

Integral being, in both theist and humanist perspectives, requires intelligent beings' ethical consideration, and their commitment to responsible relationships. In integral being, the human being (the human species) and human be-ing (human existence) have ontological meaning and meaningfulness in the universe. Humans need not lament their appearance, evolution, and existence as a chance event with no significance other than what is self-assigned to evade ennui, hopelessness, or even despair in the face of apparent meaninglessness. Complementarily, humans need not consider themselves as the cosmic center and a hierarchically superior being created by Spirit to have that status, in order to be prompted to find, recognize, and comprehend a sense of individual and species place, placement, and self- and species-fulfillment, and corresponding responsibilities. Humans and other forms of intelligent life should experience humility-in-awe as they discern and ponder the extent, intricacy, and interactions of the cosmos, and realize that it did not originate or develop (whether by autopoietic unfolding or by continuing Spirit-initiated creation), solely for one species in one segment of its vast and complex expanse. The cosmos cannot be considered in existence for, nor solely populated by, a single intelligent species evolving on a single planet. However, humankind and other intelligent beings are responsible to care for the particular places in the cosmos where they become an interactive presence relating to other biota and to abiotic place.

Humankind might gradually come to recognize, acknowledge, and accept its unfolding and increasingly complex significance in cosmic space and spaces, all related within integral being.

The realities of integral being, in both theist and secular humanist perspectives, require ethical consideration as humankind seeks to understand the cosmic unfolding, planetary dynamics, and biotic evolution that impact life's initial development, length of tenure, extent of well-being, and demise, including through both natural and intelligent life-caused extinctions; and, intelligent beings' responsibilities toward their own community, other biota, and their common abiotic habitat during their time of planetary habitation and interrelated, interdependent interaction. "Integral being" is a theoretical concept and a cosmic reality that unites theistic, atheistic, and agnostic worldviews, and scientific, social scientific, and humanities modes of thinking, research, and action.

Integral being originates all being through the interaction of cosmos (order) and chaos (disorderly events, or events apparently dis-orderly in terms of currently known or theorized orderly dynamic systems and beings). Integral being has integral natural rights by virtue of being the dynamic origin-being, which requires respect and protection so that its complex processes might continue to weave cosmic being-becoming. All that exists–biotic and abiotic being, and cosmic energies–has intrinsic natural value and natural rights, since all existents are interrelated components of integral being; all intrinsic natural value and natural rights are universal, derived from the integral natural value and natural rights of integral being. It is apparent, therefore, that the concepts of *natural value* and *natural rights* must be acknowledged, accepted, and respected in the cosmos as a whole and in all its interrelated, interdependent parts.

It might well be the case, in integral being, that even the most minimal movement, the apparently least or non-consequential event, has far-reaching impacts. If, as has been theorized, a butterfly flapping its wings in a Brazilian rain forest can impact the climate elsewhere on Earth, so too, on a grand scale integral being in response to even a very minor event absorbs it entirely, or absorbs it and leaves it intact and passes it on, or absorbs it and alters or transforms it and passes it on. The world of quantum physics has described unusual, inexplicable relationships and phenomena such as what happens between two electrons that had previous contact and became separated again. Later in (Earth) time, if the spin of one is reversed, the other reverses its spin simultaneously—even without additional contact and at cosmic distances.

Human beings are Earth's most complex reflective consciousness, and together with such evolved intelligent beings as might exist elsewhere in the universe share a reflective *cosmic* consciousness. This consciousness enables intelligent life to come to realize that the cosmos as a whole and in its intertwined existents–elements, energies, entities, events, and entropy–manifests and mediates integral being. Cosmic being is interrelated, interdependent, and interactive in integral being.

When humans have a cosmic consciousness and recognize that they are part of a larger, integral being, they extend their understanding of "natural rights." The latter human construct has been reserved traditionally, anthropocentrically, and even androcentrically solely for human beings. In Western philosophical and political milieus in particular, the term "natural rights" ordinarily refers, erroneously, only to *human* natural rights. Natural rights must include rights for all Nature, for all biota and their abiotic contexts, in the Earth commons and in the cosmic commons. "Natural" must refer, too, to more than just what is appropriate for "Nature" on Earth. Humans' acceptance of natural rights for all Nature and for integral being would prevent humans' extension, to cosmos planets and celestial bodies, of the types of destruction of planetary natural systems that "civilized" humans have perpetrated on Earth's bioregions, in their homeland and in newly "discovered" areas.[2]

Respect for integral being requires that exoEarth perceptions, priorities, and practices–during space exploration, encounters with other worlds' living beings, and scientists', government leaders', and commercial entrepreneurs' discovery of other worlds' places and goods that might benefit humanity–should be influenced by consciousness of and considerations about cosmic

---

2. An in-depth discussion of natural rights is found in Hart, *Sacramental Commons*, 117–38.

interrelationships, interdependence, and integration as a whole (universal interactions) and in part (intelligent life interactions). Respect for integral being would mean, too, that humans would be open to awareness of how the encounter with extraterrestrial life (ETL), and Contact with extraterrestrial intelligence (ETI), should either or both of the latter occur, might aid in understanding the significance and place of evolving life in a dynamic universe. Respect for integral being would mean that scientists and government authorities, among others, would be concerned about pathogenic transfers to or from Earth, and adjust their research and field projects accordingly. Finally, respect for integral being would mean that humans carefully consider how their relation to newly found beneficial organisms or extraterrestrial goods might or even should influence research and field work on Earth and elsewhere. In this regard, too, cosmonauts and their funding governments and commercial or industrial sponsors should be concerned and careful as they consider how they might directly or indirectly, intentionally or inadvertently, adversely affect life or life's possibilities (including through habitat alteration) during space exploration and colonization, and thereby impact the present or potential reality and well-being of integral being.

Socioecological Ethics

Socioecology[3] is the study of the interdependent and integrated relationships within particular human communities in their time and place, in the present and intergenerationally; in engagement with distinct regional or globally distanced human communities in the present and intergenerationally; and with diverse biotic communities in an Earth context that provides their common ground and shared abiotic habitat and home. Socioecology considers the social complexities and dynamics of human community within both the evolutionary dynamics of the biotic community, and the geodynamics of the Earth environment shared by all biota.

---

3. I conceived and developed the terms and concepts "socioecology," "socioecological," and "socioecological ethics" several years ago to integrate my lifelong interest and involvement, as an activist and then as an activist teacher-scholar, in issues of social justice and eco-justice. I have used *socioecological ethics* over the years, in my university courses and in public lectures that I have presented in diverse venues. Several of my doctoral students at Boston University have incorporated the theory and practice of socioecological ethics into their respective doctoral dissertations, with appropriate attribution, and into their writing, public lectures and sermons, professional work in academic settings, and contextual community engagement. The elaboration of both concepts in *Cosmic Commons* provides for one and all a written source to stimulate consideration of the concept and its use to promote the well-being of planet (including biotic and abiotic constituents thereof) and people.

The term *socioecological* concentrates on and focuses consideration of holistic contexts; it incorporates the interrelationship and interaction of social justice in, between, and among human communities as they interact with each other in diverse social and environmental settings, and strive continually to develop ecological relationships among themselves, between humans and other biota while ensuring ecological justice for biota, and between humans and their Earth context: all while promoting environmental well-being for planet Earth.

As an adjective, "socioecological" categorizes and focuses the noun it modifies. It indicates an intimate relationship between human communities and their locally or globally shared environmental context, and the ecological connections among human communities, between humans and other biota, and between the latter and their shared Earth milieu.

Social interactions occur not in isolation but in relation. In these relationships, people concerned about human well-being, biotic well-being, and environmental well-being should engage others (humans; biota; Earth) with awareness and implementation in context of *socioecological ethics*: striving for social justice in human communities while simultaneously solicitous of the integrity of creation. Human just conduct seeks to ensure human betterment and continuing advances toward a more holistic human community, and ecological well-being for all biota and their Earth context. Ecology (interrelation) is practiced in specific social and planetary environments (material contexts).

Socioecological ethics provides a dialogically integrated and dynamic theory and practice (theory$_1 \leftrightarrow$practice$_1 \rightarrow$theory$_2 \leftrightarrow$practice$_2 \rightarrow$theory$_n \leftrightarrow$practice$_n$) that would be socially and environmentally engaged in context to promote social and ecological justice through humans' responsible practice of and projects for societal and planetary well-being. It provides a relational, theoretical, and rhetorical foundation for humans' right conduct that interrelates and integrates social justice for human communities, ecological justice for all biota, and environmental well-being for Earth. It provides a way to ensure the commons good, responsible use of commons goods, regard for the common good, and redistribution of common goods to meet the needs of all segments of society.

Socioecological ethics provides an entrée, in concept and in context, into dialogic relationships between and among distinct and diverse cultures, to promote human common good and creation common good (more generally, including biotic—of which humankind is a part—common good and Earth common good, conjointly). Socioecological ethics stimulates both care for creation and care for community. It is a process dedicated to fostering, in context, relational consciousness, relational conduct, and relational community.

Socioecological ethics dialogically unifies theory and practice; it is socially and environmentally engaged to promote social and ecological justice through humans' practice of societal and planetary responsibility. It utilizes a dynamic method of analyzing the current context in all its complexity, as far as is possible; responding to context while bearing in mind, and with a commitment to, the concepts, constructs, and practices that previously have promoted fundamental social justice and ecological well-being. In this process it considers and reformulates in context, as necessary, previously held beliefs, values, and understandings, and develops innovative material projects that embody them. Humans' survival as a species requires that humans evolve beyond the cultural, economic, religious, and political imperialism that received a specific stimulus in the formulation and activation of the Discovery Doctrine, and subsequently became incorporated not only in federal and state laws, usually implicitly, but in international jurisprudence as well.

Praxis Ethics

Praxis is the locus and focus of dialogic interaction, mutual influence, and integration of theory and practice, text and context, and principles and projects, socioecological ethical *theory* and socioecological ethical *practice* (both of which are informed by and evolve in context to adjust to then-current socioecological conditions). "Praxis" is not simply "practice," but the contextual engagement of both principles and projects for social, ecological, or socioecological justice and well-being. Theory and practice, text and context, and principles and projects do not operate solely in relation to each other. Rather, they interact dialogically beyond their primary linkage: theory becomes expressed in text; theory becomes embodied in principles; theory becomes an experiment in context; theory is reworked during projects; practice is described in text; practice is oriented to, and derived from, context; practice stimulates formulation of new principles, or amendment or rejection of existing principles; practice takes varied forms in distinctive projects in context; context influences principles; etc.

Praxis ethics[4] is a dynamic, communal (community-developed and community-oriented), context-to-content formulation of consociated ethical

---

4. The concept of *praxis ethics*, as consciousness and conduct in context, emerged from my decades of social involvement to promote human and civil rights, economic justice, and ecological responsibility. It originated with me several years ago, as I considered a way by which people of disparate backgrounds and perspectives might work together to effect progressive socioecological change in the United States and internationally. I have presented it at Annual Meetings of the Society of Christian Ethics, and have circulated a

principles and ethical practices developed from, adapted to, and integrated with and within diverse social and ecological settings, and oriented toward social and ecological transformation.

Praxis ethics is a fluid formulation that, like water which retains its essential elements but incorporates different minerals from the areas through which it flows, retains a substantive core while incorporating new insights and practices. Praxis ethics is a dynamic, relational flow of perspectives and principles that emerges in context and in projects on common ground.

Praxis ethics, since it situates and develops ethical principles and conduct in social settings, rather than derive them solely from academic speculation, recognizes the evolving nature of ethics. It is distinct from both applied ethics and contextual ethics; effectively, it integrates both by conserving core ethical principles but including also secondary, negotiable principles that might be modified in dynamic contexts. It proposes an ongoing, historical, dialogic relationship between theory and practice. It is "dialogic" rather than "dialectical," and, in fact, prefers "dialogic" to "dialectical." In praxis, where theory and practice interact, a *collaborative* process might emerge in some contexts; there need not be the *conflictive* process of thesis↔antithesis→synthesis that is characteristic of "dialectic."

Praxis ethics, then, considers the past in the present for the present and the future. It considers the past while analyzing and reflecting on the present, while working for the future: in order to promote the well-being of the present and the future—for human communities, interrelated and evolving biota, and Earth. It is characterized by concern, compassion, and creativity. It is intergenerational in nature (in the Earth milieu, with its succession of descendants) and reflective in nature (in that which defines the human: a progression of consciousness)—not only in its considerations of how to better the future, but also in its continuity through socioecologically dedicated intelligent beings in the present for the future.

Praxis ethics has a single core principle as its principal guide: love. For theists, it advocates "Love the Spirit and love every neighbor"; for secular humanists, it declares: "Embrace the transcendent and love every neighbor." The extended "neighbor" includes people, biota including ETL and ETI if and where Contacted, and Earth and other planetary bodies. Love of God and neighbor is expressed in the cosmic commons within the developing, being-becoming relational Spirit-biota community of integral being.

Praxis Ethics expresses a passion for justice and progressive change, and hope for the future well-being of humanity and other intelligent life;

---

book manuscript with the same title to several publishers. Its original focus was on the human common good in relation to the biotic common good and the planetary common good; here it is extended to considerations of the cosmic common good.

Earth and extraterrestrial places; and the biotic community of the cosmic commons. It bridges chasms between conflicting groups as it invites people from distinct ethnic, economic, and religious or humanist backgrounds to participate together in social change efforts according to their abilities, commitments, and potential entry level. It invites people to participate in social and ecological transformation, and to have a process to do so effectively in collaborative communities—by seeking community *common ground* on communities' shared Earth *common ground*: on Earth and on extra-Earth celestial bodies.

Socioecological Ethics and Praxis Ethics

Socioecological praxis is a place of dialogic engagement of theory and practice, of current principles and of contexts that challenge those principles, focused on social justice for human communities, ecological justice for all biota, and Earth's environmental well-being. It might have a broad, generic consideration that includes all of the factors mentioned in the previous paragraph, or a focused concentration on one of them. *Praxis ethics* in itself might focus solely on social oppression or ecological oppression. *Socioecological ethics*, by contrast, includes reflection and action stimulated by social and ecological issues integrated within a particular place and historical time period. Every type of praxis ethics is ethics-in-context (which is *not* "contextual ethics," derived in and from context) based on a current experience and formulated on site: praxis ethics carries to context previously pondered principles, but, unlike normative or deontological ethics that will be "implemented" in context, it is open to *adopt from context* what is already present and appears compatible to social and ecological progress in this unique situation, and *adapt to context* those of its prior principles as seem reasonably negotiable, and are considered neither fundamental in all contexts, nor firmly related to core, nonnegotiable understandings, beliefs, values, and practices.

The relational engagement of socioecological ethics as understood and as embodied in diverse planetary contexts is enabled by mutual consideration and contextualization of praxis ethics. Socioecological ethics is complemented by, informs, and provides suggested but dynamic principles for a particular type and setting of praxis ethics.

Socioecological Praxis Ethics

When the concepts of socioecological ethics and praxis ethics are related and linked, the outcome is *socioecological praxis ethics*. That is, the understandings which emerge in the dialogic context of praxis ethics are guided, extended, and deepened by considerations of social justice and ecological justice, rather than by such alternatives as philosophical considerations and philosophical ethics, or theological considerations and ethics, etc.

Praxis ethics is not *applied* ethics; praxis ethics is *derived* ethics. It is not formulated abstractly, but developed in organic dialogue with social realities. It is derived not just from the social setting, but from both that setting and from what has been brought to the setting, including the ideas, experiences, values, and principles of participants. Praxis ethics, then, is not contextual ethics; *praxis ethics* is *ethics in context*.

As ethics-in-context praxis ethics views and analyzes the present in light of the past (How did we get *here*?) and anticipatorily ponders or projects, from the present place in time, the future (How will we get from here to *there*?) in regard to the issue: eliminating a condition of political, racial, economic, ecological or other oppression for this group in this place, and replacing it with a just situation in a just society. It projects what just conduct is required (and what principles might be prioritized and invoked to propose the conduct) to affect this context in a beneficial way for the people(s) harmed by their current situation in this place, anticipating resolving the condition or issue in a way that future social impacts will be beneficial and promote ongoing well-being, or at least neutral and eliminate current adverse conduct in this place for these people and their social and ecological setting. It is deontological-contextual-teleological ethics: previously accepted principles are brought into context as consulting proposals that will assist formulation of just actions that will provide or at least set the stage for an improved socioecological milieu. In this place at this historical moment in this community actual places and principles from the past will be recalled and discussed, and initial ideal places in the future will be formulated. In the present *topia* which is a *dystopia* for these people and places in this time is envisioned to be transformed: *utopia* becomes both a vision and guide in the present place, and the projected place for which the community hopes and toward which it aspires. Here and now, historically developed deontological ethics is consulted in this community, socioecological ethics is being developed for this community-in-context in *topia*, and teleological ethics provides guidance for practices that will progress justly toward realizing the envisioned *utopia*: (deontological ethics→contextual ethics←teleological ethics)→socioecological praxis ethics.

World Charter for Nature and Socioecological Praxis Ethics

Consideration of one statement in the United Nations World Charter for Nature (1982) will facilitate understanding the theory and practice of socioecological praxis ethics. Recall that the WCN was developed internationally and accepted overwhelmingly in the UN. The WCN states in II.7: "In the planning and implementation of social and economic development activities, due account shall be taken of the fact that the conservation of nature is an integral part of those activities." The passage addresses social development and economic development and states that conservation of nature is integral when either type of development is activated in formerly pristine contexts (thereby altering the character of natural, undeveloped local contexts either *in situ* if left in place but diverted to other uses than what is presently the case or if extracted to be used elsewhere; or, if human constructs are placed atop previously untouched sites).

In exoEarth contexts, humankind would strive to effect and ensure the continuation of socioecological well-being upon arrival and during the time thereafter while people live or work there. In accord with the WCN, in the heavens as on Earth, humans as individuals and as communities would assess planetary contexts. They would determine the place(s) where they would have least impact in the short term or long term, depending on their purpose and length of stay. They would decide what types of minimum impact, site-integrated structures they would build and use for industrial, commercial, or residential purposes. Every effort should be made to avoid or to mitigate quickly adverse impacts on nature.

An ongoing respect and solicitude for the natural environment will characterize people's lives and work in their new home or place of livelihood. Buildings will be constructed in such a manner and with such materials as are respectful of the terrain occupied. At home and commercial and industrial workplaces efforts will be made to generate as little waste as possible, and such waste as is generated will be disposed of in a way that does not endanger local abiotic places, resident biota, and the ecosystems in which they are related and integrated. Human communities should be egalitarian, solicitous always for each and all of their members, as communities seeking the common good—which will be evident when common human well-being is realized in the Earth commons and in exoEarth commonses.

Should the colony interact in this way—among its members, with native fauna and flora, and with abiota—it would conserve nature and ensure the human common good. People will not do in the heavens what they did on Earth: replicate the types of adverse impacts Stephen Hawking cites in his rationale for humankind to emigrate into space. The exoEarth commons good and common good would be assured intergenerationally if humans continued to have a respectful contextual consciousness, evidenced by responsible human conduct.

Terrestrial and Extraterrestrial Species Prioritization

Conflicting claims by intelligent biota to territory and to natural goods appropriation and use might well cause difficulties in outer space exploration and development as they have among humans on Earth, although perhaps not inevitably. Competition for land and natural goods will result, whether nonhuman biota be intelligent or not, or benign or bellicose by nature. Biota prioritize the survival of their own species and its individual members. Competition need not become conflict and combat to obtain and keep goods, however. Interaction might lead to reassessment of actual needs, and proposals for possible substitutes for goods deemed at first absolutely essential and irreplaceable. It might lead, too, to collaborative development of new products or new uses for old products, both of which would help to provide for the essential needs of either or both species.

Prior research on and development of alternative natural and human-produced goods to meet humankind's needs would be advantageous both now and in the future to benefit people and planet: humankind and Earth, and colonizers and colonized planets. Innovations are available now, sometimes through grassroots organizations and community-oriented corporations, to provide new sources of energy (such that, for example, environmentally devastating and health endangering dirty coal, and geologically destructive fracking to acquire natural gas, would not even be considered for energy generation on Earth or elsewhere because coal and gas would no longer be needed, or would be needed minimally); new or revived and renewed nutritious food sources; and construction materials and methods that complement abiotic contexts and utilize renewable rather than reductive natural goods. These would include (if available on colonized places or regions thereof) use of plant species to provide biofuel or medicine; volcanic ash for mortar and regionally available calcium carbonate clay for bricks; and bamboo to replace iron rebar in concrete support systems such as building foundations, or to provide urban sidewalks.

Inter-Species Exchanges

Cosmic Contact might have beneficial cross-species influences. Terrestrials and extraterrestrials might observe in the other's ideas, principles, and social arrangements some transferable aspects that would enhance the well-being of their respective communities and planets. In such a case, the dialogic relationship between human present and future contexts, as described previously, would take the form of a dialogic relationship between terrestrial and extraterrestrial intelligent beings, for mutual social and ecological progress in thought and action, and distinct or collaborative benefits.

New People in a New Society

The hope and expectation of Earth advocates who envision a society in which the well-being of the commons provides a stable setting for justice and well-being for all humankind is that when it is realized a profound social reorganization to promote the common good will have been instituted. A new kind of people (oriented toward and acting to benefit ordinary citizens and the populace as a whole) will have come into place—and places. New people will interact in a new society. In order for this to materialize—literally and figuratively—significant change is needed in concepts of community and in economic theory and practices, in terms not only of effecting a greater sharing of natural and produced goods, but also of enabling an equitable allocation of lands.

In such a society, sharing of places and goods to meet essential community needs would displace private sequestering in order to satisfy primarily individual or corporate wants. Remembering the impacts of the individualism expressed in the cultural and economic extremism of Adam Smith and the Discovery Doctrine, as contrasted with the communal ideas, ideals, and practices of indigenous peoples and early Christianity; reviewing the negative impacts of the current, corporate-controlled capitalist ideology and globalization that benefit a select few and harm the vast majority; recalling that when a community benefits so too do all its members benefit in some way, whereas when solely self-selected individuals benefit then the community suffers in whole and in part; and restoring consciousness of and commitment to the neighbor love that is (or is supposed to be) characteristic of followers of Jesus and of the teachings of Moses, Mohammed, and Buddha, among others, and leads to compassion for the "least ones" in any and all cultures: all "peoples of good will" on Earth (and intelligent "beings of goodwill" from other places whose evolutionary development surpassed long ago individual acquisitiveness and whatever exercise of a Discovery Doctrine they might have practiced) would work together to eradicate gross social disparities and egregious ecological impacts of current ideologies and practices.

Private property imposed in all contexts is neither socially nor ecologically beneficial. In particular, current terrestrial individuals' or corporations' ownership of extensive holdings in land, and private rather than public ownership of natural goods such as petroleum and water, accompanied by selfishness passed off as rightful acquisition, continue to afflict the human community. New forms of social arrangements would be made to ensure the survival and well-being of the majority of people and peoples, not just of a powerful and often greedy minority.

In order for this to be feasible, human explorers and colonizers should not pre-determine the kinds of social and economic arrangements that should characterize new human locales in space, or human acceptance of extraterrestrial concepts and concrete practices. Rather, several possibilities should be developed *a priori*, with prioritization given, as suggested by UN instruments, to benefit those most in need. It would be most unfortunate if private property ideologies were to provide a foundation for the continuation in the heavens as on Earth of the human suffering and planetary devastation that personal and corporate selfishness and greed have inflicted on Earth—particularly in the twentieth century and into the twenty-first. Humans need to be new people creating a new society, where the needs of each are provided, and the abilities of all are collaboratively integrated and employed to construct and conserve a new society.

Space would indeed be a "common province of humankind"—and speaking universally, "a provider of common goods for all intelligent beings"—when its material goods are justly distributed and its social arrangements are justly developed and implemented. Humans might view themselves as trustees, and regard the one for whom they hold things in trust and in deference to whom they use them responsibly, as a beneficent Spirit or cosmic Nature, the integral being in which all is interrelated. By contrast, the militarization of space, and an arrogant attempt by one species to appropriate exclusively for itself what should be "common goods," will promote unending conflict, social injustice, and enduring inequity of distribution of natural goods—in the heavens in the future as on Earth in the present.

Encounters between terrestrial intelligent life (TI) and extra-terrestrial intelligent life (ETI) seem inevitable, given the age, extension, continuing evolution, and ongoing complexification of the cosmos. Intelligent beings have most probably been gradually emerging and evolving throughout the universe. An emergent event might recur at every cosmic instant, and therefore be in various stages of development throughout the universe.

The relative eons of existence, diverse contextual experiences, and distinct cultures characteristic of TI and ETI civilizations at Contact, present possibilities of conflict or cooperation. TI and ETI might well bring to contextual Contact divergent as well as convergent understandings of justice and just individual and group conduct toward the encountered Other. Conceivably, both TI and ETI would both be "aliens" in some historical moments and places, when Contact occurs on a celestial body far from their respective home planets, their distinct places of origin. Optimally, both TI and ETI would have developed prior to Contact, or be open to upon Contact, such just principles and benevolent practices as would have been needed and retained for their own social and intellectual (and spiritual) evolution, and eventual development into spacefaring intelligent beings.

Peace at home, then, would provide a possibility for peace in space. Similarly, unfortunately, continuing conflict at home or, the latter having been resolved, a still existing conquest consciousness would translate into conflict with species considered culturally inferior, less technologically advanced, or less capable of defense against aggressors.

Already there are some Earth proponents for both a particular State's national rights over new territory and goods and for their privatization accompanied by private ownership rights. The latter would mean that transplanetary or transstellar corporations would have control of them and the ability to profit almost exclusively from them, much as transnational corporations have been doing on Earth. A modern form of Enclosure Acts and enclosure practices would result from implementation of this ideology in concrete circumstances. Discovery Doctrine would live on, exported into space, to the detriment of states and social groups excluded from extraterrestrial benefits, and potentially would provoke both global and interstellar conflicts. On exoEarth planets this most likely would lead to genocide or near-genocide of indigenous peoples, seizure of their territories and natural goods, and unrestrained exploitation and exportation of the natural goods upon which indigenous populations have depended to meet their sustenance needs. Life would imitate art; Contact would mirror *Avatar*.

In contrast to parties' resort to conflict as two or several battle to acquire goods they need, a carefully considered collaborative and even congenial response in context could be based on a principle that natural goods are to be considered the "common goods of intelligent being"—first and foremost the goods of those whose origin planet is under discussion. If there is mutual recognition of this principle, a foundation is established for collaborative research and evaluation to accommodate equitably the needs of each species. Once again, the assumption and hope underlying such a possibility of accommodation, which would mean to each corresponding to every one's needs, would be that intelligent beings throughout the cosmos, if they have the technology to traverse interstellar distances, would have evolved past war as a means to obtain territory and goods, or would have determined rationally that war is not ultimately beneficial, but rather harmful, to bellicose parties.

Whether on the Earth commons or in the vast cosmic commons, socioecological ethics would suggest evolutionary contextual principles and proposals that might become, through Contact and in diverse space contexts, adapted to other milieus, adapt from other milieus, and adapt other milieus to its ideals and responsible practices, while retaining core values.

It is necessary for planetary, interplanetary, and universal well-being that humans develop a new sense of their place in, and integrality within, Earth and cosmos contexts. When people have a cosmic consciousness and a practical foundation in terrestrial-extraterrestrial socioecological praxis ethics, their perspectives will be informed, enhanced, and stimulated as they explore cosmic realities, extend common responsibilities, and establish parameters for realistic concrete projects. The result would be responsible scientific research and outreach, respectful terrestrial-extraterrestrial engagement, and religious, spiritual, and philosophical reconciliation with the implications and impacts of Contact.

Ultimately and universally, cosmic collaborative efforts, no less than Earth collaborative efforts, would result in enhanced societies for all involved, and new social visions toward which they might aspire independently and conjointly. These would enrich each intelligent being separately in their own communities and contexts, and interdependently and relationally in new communities of communities across cosmic contexts.

PART III
TERRA INCOGNITA

7

Cosmosocioecological Praxis Ethics

Topia, Dystopia, and Utopia in Earth and Cosmos Contexts

The vast cosmos invites human exploration and colonization. The barely visible (to the naked eye) and almost invisible (through telescopes) distant stars promise the presence of previously unknown types of planets, and suggest the possibility that if intelligent life has evolved on Earth over the billions of years of Earth's existence, it might also have evolved on other worlds in a similar, shorter, or longer span of time. The age of the cosmos, and increasing discoveries of habitable planets viewed through the Hubble, Kepler, and other telescopes, suggest that life—not necessarily as humans know it—might await Contact with humankind or any other intelligent species. It might, too, already have voyaged into space in search of cosmic community and companionship with others who have reached a complementary or at least compatible stage of evolution.

In cosmic places, if intelligence has emerged in biological evolution and been accompanied gradually by cultural evolution, terrestrial intelligence with its forms of social organization, and extraterrestrial intelligence with its social institutions, would have originated and evolved in biologically, socioecologically, and geologically distinct places. In their respective processes of evolution, they likely would have developed not only diverse forms of social organization, but also distinct types of relationships with other biota and with their abiotic common context.

In habitable places throughout the cosmic commons, no less than is the case for terrestrial biota on the Earth commons, exoEarth biota live in integrated and interdependent relationships. In their *topia*, their place, intelligent biota everywhere—human and nonhuman—likely strive naturally, in comparative ways, to live together as individual social beings in community, for the mutual social benefit and material well-being of one and all. Less complex biota might have a similar tendency to organize cohesively in order to safeguard species survival and security or because of an innate gene-stimulated sociality.

In their manner of social organization, TI and ETI might have widely divergent types of *topia*, their historical, material, and culturally evolving space and place, on the same planet or on distinct planets. In order to gradually make their topia a better place, they might have an ever-evolving *utopia*, a vision for the actual presence of a socioecologically functional and mutually beneficial community, in which the well-being of all members is ensured equitably, to the greatest extent possible. Or, their place might be, contrastingly and unfortunately, a *dystopia*, a socioecologically dysfunctional and stratified community in which the well-being of a few is exclusively and inequitably ensured, and the social and material oppression of the many is endemic and enforced.

Differences in the topias present in distinct worlds of the cosmos might positively or negatively impact terrestrial and extraterrestrial intelligent beings on Contact. Positively, when TI or ETI are open to recognize the better socioecological arrangements of the other, and strive to replicate these on their origin planet or other worlds in which they reside; negatively, when this recognition does not occur, and TI or ETI strive to maintain their dystopian ideology and social structures in all or selected communities, possibly accompanied by ecological devastation; they might even to try to convince the other that they should establish dystopias on their own world(s).

Topia

A *topia* is a historically constructed community social reality. It is a complex of social groups, institutions, cultural expressions, practices, theories, and ideologies. It includes the competing ideas and activities of the individuals and groups that confront each other in the power struggles inherent in every social milieu, whether overtly or covertly so. A topia is therefore a particular place in time in which emerge, or continue to compete, claims for ascendancy by distinct social groups who have contrasting ideas, diverse cohesive organizations, and socially and intellectually creative individuals. A topia is a moment (which could be lengthy) and a milieu (which could be extensive) in the dynamic historical processes of intelligent beings. It changes with time and varies with different constituencies of the general population. It is a particular place at the present moment of its historical process. In it past personal and group topias converge. In it utopias and future topias are envisioned, formulated at times, and occasionally struggled for and achieved.

In diverse worlds of the cosmos, topia is a place or multiple places on a planet or celestial body characterized by some form of social cohesion. The cohesion might be based on culture, economic class, religion, or politics, integrated into a social whole, and either voluntarily accepted or coercively imposed. Each topia has some other form of formative and normative identity based on an autocratically or consensually implemented ideological discourse and institutional structure. It might have originated in a particular individual's or group's autocratic or democratic decision to establish a society that benefits primarily themselves or their own group, however defined; or, by contrast, a broader, perhaps global constituency to benefit all in the maximum way possible. Its laws, policies, and internal police activities might be based on a predominating individualistic and selfish consciousness or a prevailing communal and social consciousness. It might have been developed with the worst intentions—to benefit the few at the expense of the many—or the best intentions—to benefit all, without exception, to the greatest extent possible. If the prevailing ideology promulgates and promotes (usually coercively) provision of benefits for the few, but has a willingness to adjust, as necessary, to ensure that all have the sustenance they need, it will adapt to new contexts to ensure that this does happen. If the prevailing ideology's policy is to ensure a dominating individual's or group's continued supremacy, then coercive political and police force will continue to be used to ensure that policy, even in the face of—in fact, especially when confronted by—rising discontent among the economically, politically, ethnically, culturally, religiously, and otherwise socially deprived powerless people.

The components of a topia include structured institutions (Earth topias' diverse practices are in parentheses) such as governments (executive, legislative, judicial, and military components), businesses (industrial, agricultural, commercial, clerical, etc.), economic systems (capitalist, socialist, communist, cooperativist, etc.), educational systems (autocratic—students as objects—or participatory—students as subjects), political parties (or autocratic individuals or families), and religions (denominations, sects, major faith-oriented and often dogmatically controlled—especially by specified clergy and specific scholars—systems organized according to shared beliefs such as Buddhism, Christianity, Confucianism, Taoism, Islam, Hinduism, Judaism). Each structured institution is organized in a way that functions as a control over its own members, as an intermediary between the latter and members of other groups and, when and where possible, as a source of power over other institutions and groups.

When an understanding of a topia is sought as far as is possible through objective analyses of historical persons, movements, occurrences and relations, the imaginative and formative thought that expresses that understanding is a *theory* about what exists, or an intellectual projection of what might come to exist—a utopia that would be a beneficial alternative to the current topia, which envisions and expresses the ideals of the utopians and the tactics they need to overcome their social (which is usually simultaneously their socioecological) oppression. Every topia contains within it the conditions creative of a different topia, but those within every topia who benefit from the status quo fight against the development and implementation of a potential alternative topia. Consequently, topia and utopia confront each other in context as each topia gives birth—however unwittingly and unwillingly—to dispossessed or disadvantaged people's utopian visions of an alternate, just topia.

Dystopia

In human historical-cultural time, in populous places on a planet one group might, through various means and even ruthlessly (lacking the compassion evidenced by Ruth in the biblical book of the same name, who was solicitous for others), come to dominate all others—particularly economically and politically, but also religiously, racially, culturally, sexually (in terms of gender and orientation), and ethnically. Political repression, economic impoverishment, and even slavery might exist in societies that are permeated by or possess particular characteristics of such a place, a *dystopia*.

A society judged to be (by those who suffer in it and by external observers) a *dystopia* is characterized by political, economic, psychological, ethnic, cultural, racial, and, often, religious (except when embodying a form of theocracy that is voluntarily accepted by a local or global populace) oppression. It is controlled by an individual or a minority elite whose intent is to benefit themselves rather than the commonweal, who will take whatever coercive measures are deemed necessary to seize and sustain absolute societal power and consequent financial benefit. It secures its domination through brutal force via a national police force, secret police, and/or military. *Dys-topia* means "bad place" or "worst possible place." A dystopia is a destructive departure from a topia, and a devastating deviation from a utopia.

Terrestrially, dystopias have existed in the historical past, and might exist in the historical future—an open-ended period that is already unfolding as a consequence of our consciousness and conduct in the present. A dystopia contradicts the best of human hopes and aspirations. It is permeated by injustice and controlled by a form of governance that benefits a powerful minority of citizens; it has no ideological, political, or economic intent to develop social constructs that embody ideals of a society characterized by or at least oriented toward and striving for justice for all and promoting the well-being of the commonweal and the common ground that all share.

Twentieth-century Earth examples of such "anti-utopias" include Nazi Germany; right-wing dictatorships in Latin America, usually coercively and brutally controlled by the military under the direction of the elitist and wealthy upper economic class; the Stalin regime in the USSR; and Iran under Shah Reza Pahlavi. At the time of these dystopias' greatest power, Latin American states and Iran were controlled by the United States for its own political and economic interests and benefit, while Eastern Europe was controlled by the USSR for its own political and economic interests and benefit. Both "superpowers" exercised control to provide particularly for the affluent lifestyle of their own economic-political elite—ironically, in the USSR, the communist party elite, among whom the ideals and vision of Marx were subverted. In its origin, the USSR had been promised to be a workers' utopian dream, but it became a nightmare for many as the Party, not the proletariat, controlled society. Some socialist critics called the USSR structure "state capitalism" rather than "communism" because the Communist Party, complementing what capitalists did in the United States and Europe, controlled and was the principal beneficiary of the economic system.

Before the imperialist reign of the two superpowers, Hitler's Nazi Germany seared the conscience and scarred the soul of people worldwide because of its dystopia's brutality and Holocaust, by which some six million Jews, and hundreds of thousands of Gypsies, leftists, and the physically or mentally impaired, were murdered en masse by Nazi authorities and their sympathizers, subordinates, and collaborators.

In the history of the Americas, as discussed earlier, sociologists and historians debate the number of native peoples—Indians in the United States and indios in Latin America—who were killed in what Indian scholars Ward Churchill and David E. Stannard call an "American holocaust";[1] current estimates vary between sixty million and eighty-five million native peoples. This was—and for many native peoples, continues to be—an Americas genocidal dystopia.

---

1. Stannard, *American Holocaust*; Churchill, *Genocide*. Evidence of *dystopia* is seen in the high rates of poverty, alcoholism, powerlessness, joblessness, and depression documented on reservations and in concentrations of Indians in urban areas. It is symptomatic of people plagued with a dominant culture's ethnic and cultural disrespect, external political control, suppression of traditional spirituality, control of economic development, treaty violations, and human rights abuses

Dystopia Dreams

In a dystopic milieu, those who are oppressed or suppressed envision a utopia, a better place the like of which they have not experienced previously, but whose possible future existence for them is present elsewhere in their world, present among other intelligent beings elsewhere in the cosmos, present in the historical memory of their own people or other people on Earth or on an exoEarth world in the cosmos, or an aspiration for the future toward which others strive and which others have come closer to achieve. Any of these manifestations of a just social order could provide a model able to be adapted in whole or in part by oppressed people in their own topia, at first as a vision of what might—should—be, and then in actuality when the vision is realized (real-ized: made a reality).

Dystopia Dream Dangers

A danger from dreams about and action toward a utopia from within a topia permeated by injustice is that a politically, economically, and culturally worse *dystopia* might result from efforts to effect change, whether in Earth or exoEarth societies. If those struggling for justice are not successful in their quest, and repressive governments are successful in their attempt to eradicate any possibility of further efforts to effect a transformation of the dystopia to a just society in a healthy planet, then their next social situation will become even more dystopic than their present one.

The struggle to transform topia so that it becomes a commonweal, an envisioned, just future, might be a success and result in an approximation of a utopia embodying the best aspirations and dreams of the people; or, it might be a serious failure and result in a worse dystopia, a society even more rampant with coercive structures embodying the worst fears and nightmares of those who had hoped to create a more just society.

In the world of fiction, dystopias are described in George Orwell's *1984* (written as Hitler's Germany was dying) and Aldous Huxley's *Brave New World*, published five years after the USSR launched Sputnik 1, the first artificial satellite to orbit Earth.

Dystopia in the Heavens as on Earth?

Cosmically, a dystopia would mean that the worst scenario presented in diverse ways in science fiction literature—that invading conquerors would kill intelligent life on, and steal the natural goods of, invaded planets wherever they travel—would become a reality—in terms of both Earth's invaders

(Martians as extraterrestrials who devastate Earth in *War of the Worlds*, for example, and its successor stories in books and films) and Earth invaders (for example, Earthlings as extraterrestrials who devastate Pandora in *Avatar*). Dystopias might, unfortunately, still exist on Earth when ETI arrives; dystopias might also be present in extraterrestrial contexts and cultures in the heavens when humans arrive. If benevolent intelligent ETs arrive, they might seek to educate humankind away from dystopic locales and ideologies (or, if they have a Discovery Doctrine they might try to displace it with their own superior civilization). If malevolent intelligent ETs arrive, they would have an additional rationale, if they needed one, to act toward humans as some humans were acting toward others. For their part humans, as space voyaging extraterrestrials ("extra" from Earth and "extra" in the perspective of intelligent inhabitants of exoEarth places) arriving in distant places, might act in a similar manner: they might be malevolent or benevolent aliens descending on and making Contact with native inhabitants of humans' newly found ("Discovered"?) planets and celestial bodies.

In exoEarth contexts just as in Earth contexts, every historical context creates the conditions for its own transformation. In some milieus, social change can be effected gradually and *internally* in an evolutionary transition from topia to utopia. In other settings, internal change would not be possible because those who hold power—political and economic, often enforced by a pliant military—reject and coercively suppress efforts at evolutionary change. Oppressed intelligent beings in the latter case might conclude that, regrettably, an immediate revolutionary response is required at some historical moment to change the oppressive status quo. When the conditions converge—intelligent beings' aspirations for a better society have become widespread sufficiently, and the passion for change becomes almost overwhelming—a catalytic time arrives in which to attempt social transformation. Intelligent beings' success in their attempt to achieve change to a more just society depends on the extent to which they are willing and able both to analyze their context accurately to ascertain the moment when change appears possible, and to engage in a sustained struggle, if necessary, for its realization. The quest for justice in a social setting in which intelligent beings experience injustice enables them, as they analyze their situation before they engage in either evolutionary or revolutionary action and during their efforts to effect transition or transformation, to envision and formulate their utopia—a commonweal in which injustices currently experienced will no longer be present, and the well-being of all will be ensured to the greatest extent possible.

In other contexts, where the conditions for social transformation exist but oppressing governments are excessively and extremely violent in silencing opposition to their dictatorship, oppressed groups (and even an entire nation on Earth, or its exoEarth geopolitical entity) might need *external* assistance. On Earth a case in point was the oppressive apartheid regime in South Africa, where strong and unified international economic sanctions and political pressure assisted the majority native Africans to dismantle apartheid imposed by the minority Euroafrican settlers. If humankind, or humans conjointly with ETI with whom they had had prior Contact and had established collaborative relationships, were to arrive on a dystopic planet or dystopic regions thereof their proposal to change the existing dystopia might be welcomed by native inhabitants. The difficult decision in such a scenario would be whether or not to interfere in the situation to assist native residents to establish a new, just topia. An Earth parallel would be the United Nations Peacekeeping Force (recipient of the 1988 Nobel Peace Prize[2]), which might be sent, by decision of UN nations through the UN Security Council (whose equivalent in space might be a Cosmic Commons Council),[3] to stop genocide or severe oppression in a nation. A policy of absolute "noninterference" in planets' existing internal affairs, however unjust they might be, is apparently a sound ideal in *Star Trek* and other works of science fiction, or in the stipulations of international law, but should not be, literally, "universalized" to all places and times if a society or an entire planet is entrenched in a dystopia, or is heading toward inevitable self-extinction. The latter would carry with it, in all likelihood, extinction of numerous other biota, and perhaps devastation of an entire world to the point where it would become uninhabitable.

In interstellar matters and relationships, ETI might be observing Earth nations' conflicts and a current human tendency to resolve international disagreements or to maintain dictatorships (some of which have a veneer of "democracy") by violent suppression of any attempt at change and have concluded, based on such observation, that they should not make direct, public Contact with humans because of these factors. They might not, however, have an absolute "noninterference policy," if events at US and Russian nuclear missile sites are accurate.[4] Their policy and practice might be not to coerce change, but to suggest change by symbolic actions.

---

2. The Nobel Committee stated that the Peace Prize had been awarded because "the Peace-Keeping Forces through their efforts have made important contributions towards the realization of one of the fundamental tenets of the United Nations."

3. See chapter 11.

4. See chapter 9: possible ETI activity at Malmstrom USAF Base ICBM sites in Montana, and the same type of ETI action in Russia.

Utopia

Utopia, u-topia, means "no place," a place and type of social organization that could never exist, or that does not yet exist but might come to be, in some distant future—now only envisioned, but in which intelligent beings hope, for which they dream, and toward which they aspire.

Utopia could be understood to be an ongoing "no place," unrealistic and unrealizable, an idealized projection of a dreamed or fantasized place that can never come to be; it will always be, in concrete and material reality, no place that is socially and institutionally visible and viable. (It might still be socially helpful, however, if it continually draws people forward to effect at least some measure of positive, progressive alteration of their topia.) Opponents of innovative insights and creative constructs might pronounce, in this case, "that's utopian" or "that's an impossible dream" in order to disparage and dismiss even consideration of a suggested utopia, a just alternative to an unjust status quo—especially when they benefit from the status quo and want both to maintain it and to shatter the aspirations and dreams of those who suffer from and in it. Even among most of the general public, the prevailing understanding of utopia, the word and the concept, is that it is an unrealistic fantasy, however desirable a particular utopian vision might be for specific people or populations. In common parlance, "that's utopian" has become a way of dismissing innovative ideas.

Alternatively, utopia could be understood as a "not yet place," a place and societal organization that might exist someday as a result of ongoing efforts to effect its concrete, material, and institutional reality and continued existence. In this case advocates continue envisioning and dreaming about a new social state, perhaps even a commonweal in which all community members benefit and not just those who are part of their particular constituency; but the vision is not openly expressed, except among themselves, and the dreams and hopes continue.

A utopian vision expresses the aspirations of downtrodden or socially disadvantaged people of a particular topia to alter their unjust position in an unjust social order. Their response to their topia is both a "*denunciation* of the existing order" and an "*annunciation* of what is not yet, but will be; it is the forecast of a different order of things, a new society."[5] Without the denunciation, which identifies the types and causes of oppression, utopia would be only an unrealizable dream. Without the annunciation, it would be only the unleashing of the previously unused destructive force of the oppressed side of a power struggle whose adherents suffered from injustice but would have no theoretical basis or vision for an alternative topia. The revolutionaries

---

5. Gutiérrez, *Theology of Liberation*, 233.

might establish, if successful, their own coercive dystopia—only the rulers and beneficiaries of the unjust social structures would change; the structures would stay essentially intact to benefit those who formerly suffered under them. Conversely, the coercive destructive military and police forces of the previously oppressing side of the power struggle, those who benefitted from injustice, if successful would impose a more coercive dystopia than what had previously existed in order to preempt any possibility of a future attempt at revolution and even of gradual evolution into a partially revised topia.

Utopias can be *absolute* or *relative*, in theory and in practice, in text and context, in dreams and in formerly envisioned social realities that have been approximated or even fully realized. An *absolute utopia* is the culmination of all the hopes and visions of a utopia, as initially and subsequently expressed by visionaries, in all aspects of its *being*. A *relative utopia* is a temporally and spatially interim objective whose achievement will signal that the state of society as a whole or in one or several of its specific aspects provides, in the utopian vision's gradual *becoming*, a step toward attainment of collaborative social visionaries' goal of an absolute utopia.

An absolute utopia-become-topia is realistically understood by its proponents to be a distant goal, to be attained over time, even intergenerationally, through gradual social transformations that achieve relative utopias.

In *The Feast of Fools*, Harvey Cox integrates "fantasy" and "reality." He suggests that "fantasy" does not mean something negative or disconnected from what is considered "real"; it is "advanced imagining."[6] Fantasy "reveals man's capacity to go beyond the empirical world of here and now. . . . Out of it man's ability to invent and innovate grows. Fantasy is the richest source of human creativity." Cox continues by saying that in fantasy, people image God: "Like God, man in fantasy creates whole worlds *ex nihilo*, out of nothing."[7] Relating fantasy to the gospel narratives, Cox states:

> Man's openness to a really new future is dependent on his capacity for fantasy. Fantasy thrives among the dissatisfied. This suggests that insight into the future and willingness to move forward may require an element of alienation from our present society. Could this be why Jesus insisted that only the poor and the disinherited could really grasp the Kingdom of God?

Relating "fantasy" directly to "utopia," Cox observes that fantasy "is essential to the survival of our civilization, including its political institutions. . . . [I]t can inspire new civilizations and bring empires to their knees."[8]

6. Cox, *Feast of Fools*, 73.
7. Ibid., 69.
8. Ibid., 98.

Fantasy does this "through that particular form of fantasy we might call 'utopian thought,' social vision, or perhaps simply 'political fantasy.' . . . It envisions new forms of social existence and it operates without first asking whether they are 'possible.' . . . Utopian thinking . . . provides the images by which existing societies can be cracked open and re-created."[9]

It is important, given the preceding, just who decides (and who coercively enforces or revolutionarily effects) the category of the type of utopia that is proposed. In Cox's words: "The charge that some idea lacks 'realism' can become a spear in the arsenal of reactionaries, hurled testily at anyone whose vision might upset existing privilege and power."[10] Utopian visionaries reject that what they propose amid oppression or deprivation is a pipe dream, a forever "no place." They hold fast to their political fantasy, that the utopia they suggest is a realizable aspiration and social vision, a context-transcending and eventually, a concretely-surpassing "not yet place" that becomes a new place, a new topia, a new society—even, perhaps, a "new fantasy" conceptually, prior to actual historical realization.

Antonio Gramsci, Organic Intellectual

Italian grassroots social philosopher and Member of Parliament Antonio Gramsci (1891–1937), who was imprisoned by Mussolini because of his ideas, writings, and political involvement for social transformation (and who died as a result), describes individuals who develop a consciousness that surpasses prevailing ideologies and class biases as "organic intellectuals." For Gramsci, an "intellectual" is not just someone with a high level of formal education, but any perceptive person, including a laborer or a member of a minority ethnic group, who sees social injustices and their systemic and corporate causes and becomes a visionary who proposes alternatives to the status quo. The organic intellectual rises above the consciousness and conduct of others in their social group and the broader society in order to theorize about, commit themselves to, and work for the vision they have. They advocate for the well-being of their own group in its own right and for society as a whole, with the hope of bringing about a better society for one and all. In Gramsci's words, "'organic' intellectuals [are] the thinking and organising element of a particular fundamental social class."[11]

Organic intellectuals are distinguished less by their profession, which may be any job that is characteristic of their class, than by their function in "directing the ideas and aspirations of the class to which they organically

9. Ibid., 98–99.
10. Ibid., 100.
11. Gramsci, *Prison Notebooks*, 3.

belong."[12] The emergence of such visionaries is enabled because "every social group . . . creates together with itself, organically, one or more strata of intellectuals which give it homogeneity and an awareness of its own function not only in the economic but also in the social and political fields."[13]

Martin Buber, Utopian

In *Paths in Utopia*, religious and existential philosopher, university professor, and mystic Martin Buber (1878–1965) provides important insights about what "utopia" means and can mean in contemporary society. Writing initially in Hebrew in the spring of 1945, as World War II was ending and in the aftermath of the horror of the Holocaust, Buber offers hope for a better human future. The topia through which he had just passed was devastating for humanity, and he thought that human hope and creativity might bring about a more just future. In order to imagine and realize a better world, people had to reflect on where they were and had been, and envision and work toward what and where they would like to be. In his words, "To be a 'Utopian' in our age means: to be out of step with modern economic development."[14] His words complement (and his life exemplifies) Antonio Gramsci's view that utopian thought is initiated by the "organic intellectual" able to rise above his or her particular social circumstances. As discussed earlier, "utopian" and "utopia" are used derogatively against the ideas or projects of people who find the status quo unacceptable. In an effort to dash the hopes and destroy the visions of the oppressed, those who benefit from current social structures deride these words and dreams as unachievable fantasies, and therefore not worthy of attempt. Buber counters, "It is of no avail to call 'utopian' what we have not yet tested with our powers."[15] Thus, by definition u-topia means no-place, as critics claim, but it can also mean not-yet-place. "What is at work here is the longing for that *rightness* which, in religious or philosophical vision, is experienced as revelation or idea, and which of its very nature cannot be realized in the individual, but only in human community."[16] In other words, to link Gramsci and Buber, the "organic intellectual" cannot accomplish a better society by themselves: their vision must be seized upon as credible by a social group or groups who view it as expressing their own

---

12. Ibid.
13. Ibid., 5.
14. Buber, *Paths in Utopia*, 3. Martin Buber is most noted for his classic *I and Thou*; other writings include *Tales of the Hasidim*.
15. Buber, *Paths in Utopia*, 6.
16. Ibid., 7.

aspirations, and embrace it, commit themselves to it, and strive to achieve it despite its *a priori* rejection by dominant and dominating others.

Buber declares further, in a passage that merits more extensive citation:

> All suffering under a social order that is senseless prepares the soul for vision, and what the soul receives in this vision strengthens and deepens its insight into the perversity of what is perverted. The longing for realization of 'the seen' fashions the picture.
>
> The vision of rightness in Revelation is realized in the picture of a perfect time—as messianic eschatology; the vision of rightness in the Ideal is realized in the picture of a perfect space—as Utopia. The first necessarily goes beyond the social and borders on the creational, the cosmic; the second necessarily remains bounded by the circumference of society, even if the picture it presents sometimes implies an inner transformation of man. Eschatology means perfection of creation; Utopia the unfolding of the possibilities, latent in mankind's communal life, of a "right" order. Another difference is still more important. For eschatology the decisive act happens from above, even when the elemental or prophetic form of it gives man a significant and active share in the coming redemption; for Utopia everything is subordinated to conscious human will, indeed we can characterize it outright as a picture of society designed as though there were no other factors at work than conscious human will. But they are neither of them mere cloud castles: if they seek to stimulate or intensify in the reader or listener his critical relationship to the present, they also seek to show him perfection—in the light of the Absolute, but at the same time as something towards which an active path leads from the present. And what may seem impossible as a concept arouses, as an image, the whole might of faith, ordains purpose and plan. It does this because it is in league with powers latent in the depths of reality. Eschatology, in so far as it is prophetic, Utopia, in so far as it is philosophical, both have the character of realism.[17]

The future should not be based solely on the past, but on the possibilities of what intelligent life can do with what is newly imagined in the present, in light of both past and future. In the Introduction, it was suggested that for a successful thought experiment we humans should not postpone to a distant planet in a future time that which we realize we should be doing now, in regard to the well-being of both Earth and humanity. We should begin now,

17. Ibid., 7–8.

and use our experience to develop further here and now, and later there and then, what our ideals and vision suggest as better courses of action.

Buber, a religious socialist and a strong proponent of cooperatives, stated:

> I declare in favour of a rebirth of the commune. A rebirth—not a bringing back.... An organic commonwealth—and only such commonwealths can join together to form a shapely and articulated race of men—will never build itself up out of individuals but only out of small and ever smaller communities: a nation is a community to the degree that it is a community of communities.... By the new communes—they might equally well be called the new Co-operatives—I mean the subjects of a changed economy: the collectives into whose hands the control of the means of production is to pass.... Only a community of communities merits the title of Commonwealth.[18]

Buber saw cooperatives, individually as independent entities and collaboratively as a community of cooperatives, as the new type of commune:

> The era of advanced Capitalism has broken down the structure of society.... Capitalism wants to deal only with individuals; and the modern State aids and abets it by progressively dispossessing groups of their autonomy.... [T]he heart and soul of the Co-operative Movement is to be found in the trend of a society towards structural renewal, the re-acquisition, in new tectonic forms, of the internal social relationships, the establishment of a new *consociation consociationum*.... [This social structure] is based on one of the eternal human needs: ... the need of man to feel his own house as a room in some greater, all-embracing structure in which he is at home, to feel that the other inhabitants of it with whom he lives and works are all acknowledging and confirming his individual existence.[19]

The thought of Martin Buber and others whom we have discussed, when harmonized, interrelated, and integrated, and promoted though socioecological ethical theory and practice, could help to ensure peaceful and mutually beneficial interaction among Earth communities, between Earth and exoEarth human communities, and among terrestrial and extraterrestrial intelligent beings in diverse cosmos contexts.

A collaboratively envisioned utopia could make the current topia a better place. Individual peoples dispossessed of rights and dignity because of

18. Ibid., 136–37.
19. Ibid., 139–40.

some particular characteristic—race, economic status, religion, etc.—share with other dispossessed and oppressed groups the common ground of an adverse and harmful societal situation, an existence maintained by the topia's dominant and dominating group(s). When they realize their complementary contexts, their collaborative efforts in collective struggles for justice could alter their individual and group status and access to human rights, cooperative ownership of property, and shared natural goods. A shared utopian vision, even if it has distinct components provided by diverse groups, can draw all forward together with its integrated, holistic perspective. The collaborative effort enables each particular segment to become a more powerful force as they strive to realize a relative utopia that is at once a new, more just topia in which at least some of their goals are met as needs are satisfied—and gives birth to the next utopian vision and its relative utopias. At some future time, utopian thinkers hope, the envisioned absolute utopia will be realized materially in history.

Social Context↔Social Vision

Social context and social vision clash dialectically or dialogically, or both, for people experiencing injustice. Conflict and collaboration jointly or separately can stimulate transformation. People come to understand a particular expressed social vision to the extent in which they understand the context in which it emerges, from which it cannot be separated, and to which it responds. In the words of Karl Mannheim in *Ideology and Utopia*, it is necessary "to comprehend thought in the concrete setting of an historical-social situation out of which individually differentiated thought only very gradually emerges."[20]

Topia↔utopia is *dialectic* when an existing, oppressive topia is called into question by proponents of an envisioned utopia: even more so when those with the new social vision (or an old one applied to a new social setting) begin to complement constructive theory with concrete tactics "on the ground" to transform society.

Topia↔utopia is *dialogic* when those suffering in the existing topia are stimulated by social visionaries—who might be "organic intellectuals," no matter their level of formal education—to imagine, and then to create, a new social order, and visionaries—who might also be social activists—in turn listen to the oppressed group and converse with them in context. In the process, each incorporates the others' ideas and aspirations to shape both the social vision-in-process and concrete actions-in-progress to realize (realize) it. Then, the seeds are planted for the envisioned commonweal to be

20. Mannheim, *Ideology und Utopia*, 3.

established in a near or distant future. Efforts are initiated to realize them through an evolutionary process or by revolutionary practice.

From among the variety of utopian visions born in any topia, initially (and perhaps thereafter) the dominant utopia most influences development of the succeeding topia. (In fact, as noted above, it might become the new ideology and even a dystopia.) This historical process might be represented schematically:

$$\text{topia}_1 \leftrightarrow \text{utopia}_1 \rightarrow \text{topia}_2 \leftrightarrow \text{utopia}_2 \rightarrow \text{topia}_3 \leftrightarrow \text{utopia}_3 \rightarrow \text{topia}_4 \leftrightarrow \text{utopia}_4 \rightarrow \text{topia}_n \leftrightarrow \text{utopia}_n \ldots \Omega/\infty.[21]$$

The dialectic tension or dialogic relationship can be initiated at any moment throughout a short or lengthy national or regional history. It will emerge when a group, or visionary individuals within a group, consider their social status in depth, question its existence, imagine a different social context, reflect on whether or not social change to implement social justice is possible for their group, and quietly begin to discuss and then organize with others the type of actions needed to effect positive change.

Insightful organic intellectuals such as Antonio Gramsci emerge from historical times and societal places in which topia and utopia confront each other. Within every topia, until universal peace based on universal justice and love have become firmly and irrevocable established as the ongoing topia, individuals and groups strive to better that aspect of society for the group(s) with which they primarily and inherently identify themselves, or in which they are primarily or partially identified by others. (The social position of African Americans who identify themselves, or are described by others, as a 'black' male or female, for example, is decidedly different from the social position of Euroamericans who identify themselves, or are described by others, as a 'white' male or female.) In topia, movements at first exist societally subsurface, gradually develop, and then emerge, often led by the type of "organic intellectual" described by Gramsci. They remain subversive—stimulating change from below—not superversive—leading change from above—and thereby "radical": seeking change to the historical roots of society, struggling initially at the grassroots level.

---

21. The process can last indefinitely as each dispossessed group seeks, in turn or together with other groups, redress of its particular oppression and ever-improving refinement of their new status; or, it lasts until some point in the future when each group has achieved justice in relation to and in collaboration with every group. The subscript 'n' means any number; the ellipsis followed by omega and infinity indicates that the relative utopias continue to evolve in dialogic (and/or dialectic) relation with their respective topias until the absolute utopia is reached (omega) or, the process continues indefinitely as ever-new utopias are formulated (infinity symbol).

A historically gradual social progression toward justice of specific oppressed groups in a specific place in which the topia↔utopia dialectic and dialogic process has taken place over time may be seen in the altered and altering situation of representative economic, racial, ethnic, and/or gendered groups in the United States, over the course of more than two centuries.

Topia↔Utopia in US History

In the United States, prior to the signing of the Declaration of Independence, England's King George III ruled over and was the ultimate owner of all the lands in the colonies. Patriarchy dominated cultures in both Europe and the Americas. Slavery was an accepted institution. Slaves were forcibly seized in Africa and shipped to the Americas in horrible conditions aboard slave ships; many died en route. The colonies were subservient, politically and economically, to their British ruler and the British ruling class.

Enter the topia↔utopia dialectic.

The colonists rebelled against George III to attain independence from their previous homeland and its oppressive colonial political-economic system. Shortly after the American Revolution ended in 1783, a US Constitution was developed and implemented by white Euroamerican males who wanted to stabilize the new nation. They established as their new *political* system a republic they called a democracy, and as an *economic* system the right of Euroamerican men to own property. In the Constitution, and in its Bill of Rights which was promulgated shortly after the Constitution had been ratified, the principal—indeed the sole—beneficiaries are Euroamerican male property owners, who solidified their political and economic independence *from* England and other European countries, and *for* themselves. The first utopian vision in dialectic conflict with the topia from which it emerged—self-governance accompanied by individual property rights—became concretized (at least as far as men were concerned), incorporated into an ideologically expanded and new topia: topia$_1$ had become part of men's topia$_2$. However, much work had yet to be done in regard to realizing the utopias of other societal segments.

While the new US government and economic infrastructure benefitted men, women were excluded from power, from economic benefits that would have come with ownership of property, especially in land, and from political benefits: they neither could vote nor run for office. African slaves brought to the former colonies prior to independence remained slaves, without freedom of any type, a right to property, or political power, unless their slave owner voluntarily granted them freedom (which was a rare occurrence). Native peoples had no basic rights in the new society as envisioned and institutionalized by the "Founding Fathers" (an appropriate, ironic designation in several ways) who declared that Indians had no fundamental rights to property in land, to self-determined governance, or to mobility within the colonies. Since Indians had been "Discovered" by European explorers, and because the founding colonies asserted that by right of revolution and consequent independence Euroamerican states and citizens inherited all the provisions of the Discovery Doctrine, Indians had neither political nor property rights, or even fundamental freedoms.

The suffragette movement initiated coordinated efforts to secure for *women* the right to vote. Impetus was provided by the 1848 Seneca Falls Convention in New York State. Key efforts to achieve women's suffrage were made by dedicated leaders that included Elizabeth Cady Stanton, Susan B. Anthony, Jane Addams, and Lucretia Mott. The right to vote became US law after passage of Amendment XIX to the US Constitution by Congress in 1919 and its ratification by states in 1920. For Euroamerican women, $topia_1 \rightarrow topia_2$, and for Euroamerican men and women $topia_2 \rightarrow topia_3$.

The abolitionist movement advocated for liberation of *slaves* by federal law in order to end slavery in the United States as a whole. While individual states such as Massachusetts had outlawed slavery, they continued to extradite to the South runaway slaves who had escaped from slave states to "free" states. The runaways were considered still to be the property of the southern landowner from whose abuse they had escaped. Even in so-called free states, the civil right to "property" took precedence over the human right to freedom and over compassion for people: injustice was reinforced. During the Civil War, in 1863, President Abraham Lincoln issued the Emancipation Proclamation, which freed all slaves. This could not be enforced, of course, until the North had defeated the South militarily and economically, and the country's geographical and political unity was upheld under a single national government. In 1865, as the war was winding down, Amendment XIII to the US Constitution, strongly advocated by Lincoln, was promulgated by a reluctant, politically divided Congress. Five years later, Amendment XV gave African Americans and members of other ethnic minorities who were US citizens the right to vote. For African Americans, $topia_1 \rightarrow topia_3$.

Many unjust aspects of $topia_1$ continue today: economic inequality worsens as the gap between rich and poor widens, and as women still do not receive equal pay for equal work on a national basis, despite policies and laws intended to redress the imbalance; and African Americans, Latinos, Indians, and other ethnic minorities and gender distinctive persons continue to experience discrimination. As a result of a dramatic shift in US citizens' attitudes toward same sex orientation in the first and early second decade of the twenty-first century, discrimination on the basis of sexual orientation has diminished to some extent: in 2012 elections, several states' voters passed same sex marriage laws. In all areas of women's and minorities' civil and human rights (the latter especially in regard to Indians), much is still to be done to further redress the issue and, unfortunately, also to retain gains already made: some segments of US society seek to repeal laws and policies that have enabled social progress to be made.

ExoEarth Humans in Colony Contexts

When humankind colonizes planets and celestial bodies in space, the consciousness and conduct they take with them, and the extent to which these are continued in extraterrestrial contexts, will determine the kind of social organization with which a colony begins, and what it might continue to have after a decade or more of colonization.

If there is no resident intelligent life on a colonized place, several types of humankind-based social structures are possible: initially and over time. If socioecological ethics guides human conduct, the political foundation for a colony, designated $topia_1$, could be characterized as socioecologically sustainable: social justice and environmental well-being would permeate human consciousness, and human conduct would evidence a sense of social and environmental responsibility. From this initial orientation and practice, the possibilities for colonists' community and planetary relationships include:

$$topia_1 \leftrightarrow utopia_1 \rightarrow topia_2$$

Here, colonists adapt to their new place and, having envisioned ways in which to better the environment and their developing social arrangements in and with the environment, achieve what they have envisioned they would do on and for the colonized planet or celestial body; or,

$$topia_1 \rightarrow dystopia_1$$

In this case, colonists carried with them from Earth a destructive social consciousness and minimal if any environmental concern, or developed and implemented them as they became accustomed to their new place—to the detriment of people and place. The colonists would not have acknowledged or accepted, prior to or after departing from Earth, UN guidelines that state that discovered and settled places should be the common heritage of all humankind, nor would the colonists have embraced a communal vision and practice for all members of their Earth State(s) of origin.

If humans have not undergone a conversion of consciousness and conduct prior to leaving Earth, the possible outcomes would probably be limited to

$$topia_1 \rightarrow dystopia_1$$

or

$$dystopia_1 \rightarrow dystopia_2.$$

In these instances colonists have established private ownership or national ownership of territory and natural goods that the UN states had agreed would benefit all Earth peoples, with a special concern for less advantaged nations and less advantaged populations within any nation. Effectively, Discovery would have been extended to or resurrected in outer space, and resulted in inequitable social arrangements and irresponsible natural goods development and use.

It should be obvious that the UN vision, which should be also the vision of all people(s) intent on achieving and maintaining social justice and environmental well-being, is best exemplified by $topia_1 \leftrightarrow utopia_1 \rightarrow topia_2$, and its ongoing iterations:

$$topia_1 \leftrightarrow utopia_1 \rightarrow topia_2 \leftrightarrow utopia_2 \rightarrow topia_n \leftrightarrow utopia_n \ldots$$

This ongoing process would indicate continuing human efforts to promote socioecological well-being for people and planet.

Should there be resident intelligent life, or should ETI arrive subsequent to humans' colonization, conflicts are likely if dystopia, rather than a stable topia or a topia in continuing process toward utopia, characterizes human conduct operative on planets or celestial bodies.

Conflicting Topias and Utopias in Cosmos Contexts

If and when terrestrial intelligent beings and extraterrestrial (and extradimensional?) intelligent beings come into Contact in cosmic contexts—planets or celestial bodies—they might find when they meet that they have compatible existing or envisioned *topias* or *utopias*, or different and even conflictive topias and utopias. Reflecting on how these types of conflicts currently exist among humans solely on Earth, it seems perhaps inevitable that they might occur at Contact between intelligent species on Earth (should ETI encounter humans on humankind's home planet) or in exoEarth milieus. As each intelligent species strives to satisfy its sustenance needs (let alone if it seeks to satisfy superfluous wants), two or more species might concurrently find essential natural goods on a planet they coincidentally began to explore from widely separated planetary regions, each being unaware of the other(s) similarly engaged. If conflict is not to occur, they must find some accommodation of each to the other's needs.

In regard to the basic Contact impacts and implications that have been under discussion—ethical, ecological, economic, ecclesial—several types of such possible conflicts might emerge in context. Conflict would not result only because two malevolent intruders, or one malevolent and one benevolent intruder, came into Contact on a place which would serve the needs of each of them—whether for natural goods, for territory, or for both. In such cases, presumably both malevolent species in one case, and the malevolent aliens in the other, would be belligerent or even bellicose, and the benevolent aliens, in the event they are one of the conflicting species, would have to defend themselves and strive for some interspecies accommodation, if feasible. Conflict might also result because two benevolent species have a pressing need for the planet's natural goods, or a part thereof. In the latter case, some form of concessions, cooperation, and even collaboration would more likely result.

Economic-ecological-ethical-ecclesial conflicts of species' independently developed social structures in their respective topias and the different ideologies underlying their respective utopias might occur at Contact. A benevolent species in any of the conflicts just noted might have culturally evolved to have a compassionate and conscientious consciousness and conscience, which are evident in their social organization, institutional structures, and intellectual approaches to issues which arise.

In *economic* terms, this might mean that an intelligent species understands that the needs of all social strata (and perhaps the elimination of any economic strata) take precedence over the wants of any particular strata, and that to most ably provide for the needs of all a communal or cooperative economic system (without private property claims or ownership) is required. This would include recognition that commons goods are common goods for all life, and that the best way to ensure the common good—for an intelligent species itself and for all biota—is to conserve and safeguard the planet's geological, biological, and meteorological integrity. Harms to any of these planetary systems would stimulate adverse economic impacts (as, for example, in the current case of Earth's global heating impacts: drought-stricken farms and scorched forests, for example, impact respectively farmers' bottom line and foresters' jobs and income; rising seas imperil waterfront businesses).

In terms of *ecological* well-being, an intelligent species might have realized that healthy and viable ecological interaction and interrelationship necessitated that all biota in an integrated planetary ecosystem be respected and honored for their intrinsic value as they evolve while integrated and interdependent with other biota, and that the abiotic context of the planet be carefully conserved for the well-being of each and all. This would require that the intelligent species would exercise a "precautionary principle" when in doubt about the end result of a particular action so that it would not have catastrophic consequences that would extinct biota and cause irreparable meteorological or geologic harm.

In the field of *ethics*, an intelligent species might have come to develop respect for the intrinsic value of all life and its evolutionary becoming, and the intrinsic value of abiotic places that are common ground, habitat, and home to all life. The intelligent species would carefully consider the intrinsic value of biota or abiota relative to the intelligent species' own needs in all places and times. When an intelligent species considers determining that a species or its individual members have, at this time and in this place, a temporary, primarily instrumental value vis-à-vis the intelligent species because of the intelligent species' need, it would carefully consider the desired species' overall evolutionary and survival needs in that place and time and its intrinsic value in its ecosystem, and project its needs in that and future places and times to avoid endangering or extincting it; or if an abiotic place or places had a temporary, primarily instrumental value for the intelligent species, the intelligent species would ascertain the abiotic place's integration in its contiguous areas, and its contribution to ecosystem integrity and well-being and the broader common good (the latter understood to mean the good of all in the commons).

Finally, an intelligent species might accept, as a result of spiritual evolution, that some form of *ecclesial* or spiritual consciousness should guide its life in some way, and its spiritual understanding consequently had evolved to the extent that there are no institutional religious institutions: mystics, prophets, and spiritual leaders are chosen as guides by intelligent beings because of their engagement with spiritual realities (whether understood to be related to a universally acknowledged and accessible sacred Spirit Being-Becoming or Spirit Presence, or another form of transcending consciousness or reality) and their ability to counsel others on ways to develop a complementary relationship with spiritual realities, however understood.

Dialogic Reconciliation

Differences between terrestrials and extraterrestrials in any one or several of these economic, ecological, ethical, or ecclesial understandings would have to be reconciled or at least accommodated at Contact or thereafter. The perspectives and practices would have to be integrated, however loosely, to promote ongoing collaboration in resolving planet-wide issues of occupation and use of territory, and acquisition and use of natural goods.

What would be required upon Contact, to promote commitment to community among diverse intelligent beings, is the integration of prioritized complementary, congruent, and even contradictory ethical insights and practices. This would require candid communication between newly encountering and encountered cultures, and their openness to consider carefully in context each other's experiences, perceptions, and perspectives. The distinctive *socioecological ethics* theories and practices that each brought to the discussion from their respective places of origin would need to be received by the other in a setting of interspecies and intercultural respectful give-and-take, toward mutual formulation and development of a common ethical theory and practice in their new shared place.

Toward achieving this goal, intelligent beings initially would reflect on their respective *topias* as brought into the moment of Contact, and their now common *topia* at the time and in the place of Contact. Ideally, they would be able to communicate to each other what they each considered to be the best of what they had developed to date, and see what complementary ideas, ideals, principles, and practices they had formulated and realized in their pre-Contact topia that might be part of a collaboratively developed topia that would be their shared place, materially and ideologically. Effectively, the parties conversant would be discussing not only arrangements in their soon to be common topia, but also elements which, upon engaging with each other, they conjointly decided would be utopian goals—realized as relative utopias gradually attained in stages as they progressed toward an absolute utopia—which they would seek to achieve together.

A *cosmosocioecological ethics* would emerge from the exchange of ideas. Once mutually accepted and embraced on an intellectual level, it would become operative in the new Contact context as a dynamic *cosmosocioecological praxis ethics*.

Ultimately and universally, cosmic collaborative efforts, no less than Earth collaborative efforts, will result in enhanced topias for all involved, and now those engaged in dialogue would envision new utopias toward which they would commonly aspire and strive to realize. TI and ETI both could be enriched separately in their own communities and contexts, and interdependently and relationally in new communities of communities across contexts.

Cosmosocioecological ethics as a dynamic theory becomes *cosmosocioecological praxis ethics* in contexts. Its theory and practice might differ in distinct places of encounter in the vast cosmos, depending on the level of understandings and modes of conduct that characterized the respective intelligent beings. The ethical intellectual construct would become an ethical method to guide conduct in real-life settings involving interbeing Contact. It would facilitate human collaboration on Earth and in the heavens, and TI and ETI collaboration throughout the cosmos.

The independent development by TI and ETI of their own origin planet, culture-based *socioecological ethics* would be expressed at Contact and evolve in context as *cosmosocioecological praxis ethics* when they interacted interdependently.

Cosmoethics would result from collaborative TI and ETI interaction and their dynamic, evolutionary adaptation and integration of their diverse formulations of cosmosocioecological praxis ethics. Cosmoethics would come to expresses their previously respective modes of consciousness and conduct, now interrelated and unified. Cosmoethics would stimulate, and integrate in each Contact context, social justice and environmental well-being—to benefit the cosmos commonweal at that time and place, with potential integration into other cosmos times and places.

Cosmoethics would provide a firm (core, nonnegotiable primary principles) but fluid (negotiable secondary principles) foundation for a *Cosmic Charter* to guide inter-intelligent species collaboration and promote cosmic community.

Cosmic Commons Context: Cosmic Commons Good, Cosmic Commons Goods

In Earth contexts, planetary well-being is ensured and sustained when, ecologically speaking, the commons good[22] (the well-being of the Earth commons environment) in which human communities live and work, which they alter to meet their needs, or from which they take needed natural goods, is conserved and cared for. Similarly, in space: cosmic ecological well-being, the *cosmic commons good* (the well-being of cosmos commons environments) in sites used by intelligent beings, is ensured and enhanced in places of the cosmic commons whenever and wherever intelligent beings respectfully remove natural goods from them to provide for their needs without adversely impacting abiotic contexts from which they take needed goods (which might bring about the destruction of place and the deprivation from place of integrated abiotic geophysical contours, water, soil, air, and other natural goods); destructive practices are detrimental to the integrity of the cosmic commons in itself, and to the survival of biota dependent on the geophysical and ecosystemic integrity of that place.

In Earth contexts commons goods are abiotic natural goods present in the commons which are available for all biota to meet their needs; they are able to do so when equitably distributed and shared: where they are left in

---

22. The concepts of commons good, commons goods, common good, and common goods, in Earth contexts, are discussed in chapter 3 and in the Introduction. An extended, in-depth discussion may be found in *Sacramental Commons*, 140–52. The discussion there is extended here, however summarily, to a cosmos context.

place (for example, water which enables biota to satisfy their thirst); when they are extracted (copper for cooking utensils or kitchenware; bauxite to produce aluminum for more energy efficient means of aerial or ground transport); or when they are diverted (water channeled to agricultural irrigation systems to provide food for humankind, or to urban or rural communities to enable cooking, washing, and life sustenance). People responsibly take from commons goods what they need, rather than expropriate in excess of what they need in order to satisfy their wants, which would jeopardize the availability of natural goods for other biota. In cosmic contexts, *cosmic commons goods* are abiotic natural goods available throughout the universe to provide for intelligent beings' and other biota's needs. Here, too, if an intelligent species exploits and extracts these goods, the abiotic context will be adversely impacted and biotic needs will not be met—including eventually those of intelligent beings if they are excessively acquisitive.

In the cosmic commons context, all intelligent beings are invited, for the well-being of each and all, to conserve the cosmic commons good and carefully and conservatively take and use cosmic commons goods in order that the cosmos might be intergenerationally productive.

Cosmic Commons Community: Cosmic Common Good, Cosmic Common Goods

In Earth contexts the common good is the integrated, interrelated, and interdependent well-being of all biota in conserved abiotic contexts. In cosmos contexts, the *cosmic common good* is the well-being of all intelligent beings in their original and colonized cosmic contexts, wherever they voyage and with whomever they consociate for mutual benefit. To the extent possible, intelligent beings seek to ensure that the well-being is not jeopardized by their own or others' activities. The ways in which all biota and abiota are interrelated and interdependent in integral being are not known. For humankind, at least, such knowledge is a long way into the future unless people are educated by other intelligent beings who might live in and roam the cosmos. Even lacking such knowledge, humankind is called to be responsible for the abiotic contexts where they live or where they acquire needed goods, and for relating equanimously and interacting equitably with other biota.

In Earth contexts, common goods are goods available as natural goods, or as human-processed and produced goods made from Earth's natural goods, that should equitably be distributed and shared throughout the Earth community, in order that at least the needs of each and all are met, and their life and well-being enabled. Community well-being should be prioritized over individual wants in order that this might be possible. An egalitarian economic system is essential for this to happen. In cosmos contexts, *cosmic common goods* are available in all areas of the universe, able to provide sustenance and well-being for biota native to the places in which they are located, and for intelligent life traversing space seeking new places in which to settle or from which to acquire natural or manufactured goods—through trade with native inhabitants or through responsibly removing them. In a cosmic commons, native beings and immigrant beings are required by integral being relationships and dynamics to be solicitous of the needs of those who are indigenous to their place or arriving in their place. In this way the harmony of integral being is maintained, and the well-being of intelligent beings wherever they are and wherever they are from is assured to the greatest extent possible. Operative principles in intelligent beings' engagement, interaction, and exchange could be "do for others as you would have others do for you," and "the greatest good possible for all."

In considering the cosmic commons good, cosmic commons goods, cosmic common good, and cosmic common goods it is evident that no intelligent beings, of themselves or in association with others, can comprehend the entirety of the cosmos. Neither can they compel TI and ETI compliance with living in accord with the concepts and the ideals they present. Just as humans cannot be "stewards" of the Earth, since they are not put in charge of the entire planet nor could they control every abiotic aspect and biotic community of Earth, neither can the aggregate of intelligent beings be placed over or responsible for every planet or other habitable place and all present on it.

What the concepts do provide are ideals about the ways in which cosmic community might be effected, enhanced, and ensured throughout the cosmos in those places where, and times when, intelligent life inhabits or otherwise uses them to meet their needs and provide for their well-being—as an individual species or in conjunction and collaborative community with other intelligent beings.

The ways in which intelligent beings use cosmic commons goods and cosmic common goods will, in a manner likely imperceptible, impact integral being. Like the butterfly flapping its wings in the Amazon rainforest, even what appears to be minimal TI or ETI activity might, through unseen and unknown cosmic laws and connections, reverberate cosmically elsewhere.

The respective perspectives and practices embodied or expressed in TI and ETI cosmosocioecological ethics provide theories whose activation in context as cosmosocioecological praxis ethics will stimulate intelligent beings conjointly and collaboratively to integrate their praxis understandings into cosmoethics. They will be stimulated thereby responsibly to care for those places in the cosmos where they are present. In so doing, they will ensure, to the extent they are able, the cosmos commons good and the cosmos common good, and promote responsible cosmic activity that provides an equitable distribution of cosmos commons goods and cosmos common goods.

Discussion and amendment of specific statements in UN space treaties will enhance understanding of cosmosocioecological ethics and cosmoethics.

The Outer Space Treaty, the Agreement Governing the Activities of States, and Cosmosocioecological Praxis Ethics and Cosmoethics

The United Nations Treaty on Principles Governing the Activities of States in the Exploration and Use of Outer Space, Including the Moon and Other Celestial Bodies (1967), better known by its shorter title, the Outer Space Treaty (OST), was a landmark international instrument issued by the UN. It will be discussed in the next chapter. Article 1 will be analyzed here in order to consider how it would be contextalized in cosmosocioecological praxis ethics. In like manner, from the Agreement Governing the Activities of States on the Moon and Other Celestial Bodies (abbreviated to Agreement on Activities of States [AGAS]), which was issued by the UN the following year, Article 4 stipulations will be considered. The OST and AGAS principles will be considered conjointly in the discussion that follows.

The OST states in Article I that exploration and use of outer space "shall be carried out for the benefit and in the interests of all countries, irrespective of their degree of economic or scientific development, and shall be the province of all mankind," and that outer space "shall be free for exploration and use by all States without discrimination of any kind, on a basis of equality and in accordance with international law, and there shall be free access to all areas of celestial bodies." The AGAS declares in Article 4: "The exploration and use of the Moon shall be the province of all mankind and shall be carried out for the benefit and in the interests of all countries, irrespective of their degree of economic or scientific development. Due regard shall be paid to the interests of present and future generations as well as to the need to promote higher standards of living and conditions of economic and social progress and development in accordance with the Charter of the United Nations."

If terrestrial intelligent beings on a distant planet in space Contact extraterrestrial intelligent beings in that place unexpectedly, humans are supposed to be guided by the OST and the AGAS, which their respective nations would have signed (and on Contact human explorers would use the UN instruments in whatever iteration to which they have been amended, if any, since their initial promulgation).

In the Contact context, humans will have taken with them not only the OST and the AGAS, but also their personal histories and cultural perspectives, and possibly but unfortunately some residual nationalism, ethnocentrism, or human species-centrism. In this context, too, ETI might have equivalents to the OST and the AGAS to guide them, and their own personal, species, and cultural characteristics and "-isms."

In the perspective and method of cosmosocioecological praxis ethics, which seeks to promote and ensure cosmic context well-being and cosmic community well-being in themselves and to provide for the needs of intelligent beings, and in light of cosmoethics, whose goal is to integrate the respective cosmosocioecological perspectives and practices of TI and ETI, the revised and now integrated Articles would declare: "Exploration and use of space shall be the right of all intelligent beings, who should be able or enabled to benefit from the places and natural goods found in the cosmos, to the greatest extent possible, by equitable distribution prioritized according to need; all intelligent beings who are in greatest need shall be the primary beneficiaries of territories and goods according to their needs, without regard to their stage of social or economic development, and while incorporating intergenerational responsibility and being in accord with inter-intelligent being collaborative covenants."

Cosmosocioecological praxis ethics could relate TI to ETI for mutual communication in context regarding their respective sociohistorical, ideological, and ethical perspectives and their reasons for coming to this particular context of Contact (including their species needs and place of origin needs).

In a specific TI-ETI place and moment of Contact, information could be shared stating what the purpose was of their space voyage, what they had hoped to find on this planet that would enable them to be successful in their space quest, and what they had found that would enable them to meet their own needs or the needs of their community. If TI and ETI sought and found natural goods that were not needed by the other, they could allocate territory such that, and develop ways in which, they would not interfere with the other's endeavors. If they were seeking and have found the same goods, negotiation could pursue the lines suggested by the amended Articles: to determine whose need was greater, and how might it be met, and whether there would be a surplus that would be available for the other. Or, they might determine if there were other ways in which each or both could benefit from the natural goods present on the planet, and extract them responsibly.

Careful and considerate communication of the needs of each would enable both parties to collaborate, using the principles they brought to the discussion, in constructing carefully and creatively a shared vision—using teleological ethics principles or guidelines to assist their joint journey forward—of their common future in the Contact place. Striving to realize the vision over time could mean that both would be able or enabled to meet their respective needs.

A Cosmic Charter, an ongoing work-in-progress that includes primary, core principles and secondary, potentially negotiable principles, possibly by activating cosmosocioecological praxis ethics, would serve to guide humans' consciousness and conduct initially at Contact; thereafter it might help humankind and other intelligent species develop and promote a conjointly formulated cosmoethics, and thereby enable or enhance collaboratively responsible conduct as they explore for and extract natural goods, establish colonies, and extend their common territory. As noted previously, the Cosmic Charter would be a dynamic, evolving work-in-progress[23] as intelligent species engage in the mutual give-and-take needed to accommodate each other's values and needs to the greatest extent possible. A "conflict of civilizations" would be averted, then, preempted by a "collaboration of civilizations."

---

23. Proposals for a *Cosmic Charter* and a *Cosmic Charter Council* will be discussed more extensively in chapter 11.

8

ExoEarth and Extraterrestrial Transformation

Ecosociality in Cosmic Community

Human evolution—whether understood as entirely a material autopoietic process or as a material-spiritual ongoing development flowing from divine creativity—has resulted in biotic complexity and intelligent consciousness on Earth. Terrestrial intelligent consciousness is expressed primarily through humanity's thoughtful reflection on its Earth socioecological setting and its cosmic context. In so doing, humankind evaluates, enlightened by human conscience, existing and potential social disparities and ecological devastation. Humankind as a whole or in part might be prompted by its insights to adjust its conduct and respond accordingly, guided by a compassionate and community-related conscience. Through this process, people conscientiously, creatively, and contextually propose and promulgate projects that stimulate social and ecological well-being.

Terrestrial consciousness oriented toward terrestrial transformation provides an integrated foundation for humanity to consider how it might or should act in space. Will humans be converted away from unjust conduct toward members of their own species, other biota, and the Earth home they share in common before embarking on exploratory voyages in space, such that these voyages of discovery will not be voyages of Discovery? Will a newfound terrestrial consciousness lead to extraterrestrial conscientiousness or will it succumb to reviving prior human history, and conceive and conduct an exoEarth planets-adapted conquest agenda? Will humans seek to use human technological advantage, when and where that occurs, to subject extraterrestrial indigenous intelligent species, other species, and the world newly visited to culturally-based and -biased Earth planet ideologies? Will humanity replicate thereby the way in which Western 'civilization' ideologies were used on Earth to justify in the Americas and on other continents, in

European newcomers' minds, a terrestrial Discovery Doctrine that resulted in catastrophic control over all life and its planetary environment? What types of socioecological injustices might be generated then in space: intentionally or inadvertently, materially or culturally? What degree of openness to innovative socioecological and cosmosocioecological ethical theories and practices will humans have in new praxes?

United Nations Documents for Earth and Space

UN documents released to date express at least a minimal international consensus on first, resolving Earth's pressing ecological and social needs; and second, ensuring a just resolution of ecological and social issues that might become contentious in space. Terrestrial human rights documents were developed in the aftermath of the horrors of World War II, and on into the Cold War era and the national dictatorships period; ecological documents were developed as people became more aware of and concerned about ecological devastation around the world, and understood humans' role in its causation and continuation. The space documents' formulation was stimulated initially by citizens and governments particularly concerned about the US-USSR having simultaneously an "arms race" and a "space race"; it appeared that human international conduct and conflict on Earth might be extended into Earth's atmosphere and onto the moon and celestial bodies beyond.

When interrelated, the UN documents present informed consideration of and suggest policies for humans' integration of intentional and just national, international, interspecies, and interplanetary conduct on Earth and in the heavens. Absent an arms race and a space race, these international instruments might continue to be influential in the manner in which humankind embarks on collaborative ventures in space. Consequently, they would be influential on the ideological path humankind pursues in celestial praxes in relation to newly encountered intelligent species (which could generate mutual respect, shared responsibilities, and common projects), other biota, and their shared abiotic home. The envisioned universal socioecological integration, then, if activated in cosmic contexts where newly encountered aspects of integral being exist, would help promote social and ecological well-being in increasingly more diverse milieus.[1]

Earth-focused UN concerns were discussed in depth previously (chapter 3). Cosmos-oriented concerns are the focus of the current chapter. United Nations' documents will be elaborated and discussed in depth because they

---

1. A preliminary summary of part of the content of this chapter was published, when this book was still a work in progress, in the journal *Theology and Science* as "Cosmic Commons: Contact and Community."

represent the considered agreement of states around the world on principles, policies, and proposals that should be operative in present and future space exploration, and on possible settlements in space. They might be extrapolated to be applicable also in the event of Contact.[2]

Discussion of UN international space-related documents will be limited to selected significant documents. Two treaties will be discussed: the Treaty on Principles Governing the Activities of States in the Exploration and Use of Outer Space, Including the Moon and Other Celestial Bodies, and the Agreement Governing the Activities of States on the Moon and Other Celestial Bodies. Two principles-proposing documents will be considered: the Declaration of Legal Principles Governing the Activities of States in the Exploration and Use of Outer Space, and the Declaration on International Cooperation in the Exploration and Use of Outer Space for the Benefit and in the Interest of All States, Taking into Particular Account the Needs of Developing Countries. (In these and subsequent UN documents, provisions that apply to the "Moon" apply as well to all "celestial bodies.") The expectations of the signatory states, according to these documents and expressed in them to varying degrees, were that extraterrestrial exploration and use would enhance human well-being, promote international cooperation, and help develop among nations mutual understanding, good relations, material benefits (particularly to economically poorer countries), and peace.

Proposals, policies, and principles elaborated in these Earth-originated statements are relevant for incorporation beyond terrestrial socioecological ethics for Earth and exoEarth for humankind at home and abroad, into extraterrestrial cosmosocioecological ethics envisioned to be practiced throughout in the cosmos. In the future, themes, ideas, and principles that

---

2. The United Nations issued documents regarding outer space issues in two categories: (1) *treaties* agreed to by member states, only one of which bears the word *treaty* in its title; among the five treaties are two *agreements* and two *conventions*: Treaty on Principles Governing the Activities of States in the Exploration and Use of Outer Space, Including the Moon and Other Celestial Bodies (1967), Agreement on the Rescue of Astronauts, the Return of Astronauts and the Return of Objects Launched into Outer Space (1968), Convention on International Liability for Damage Caused by Space Objects (1972), Convention on Registration of Objects Launched into Outer Space (1975), and Agreement Governing the Activities of States on the Moon and Other Celestial Bodies (1979); and (2) agreements on *principles* endorsed by states in the the UN General Assembly, five in number, two of which are *declarations*: Declaration of Legal Principles Governing the Activities of States in the Exploration and Use of Outer Space (1963), Principles Governing the Use by States of Artificial Earth Satellites for International Direct Television Broadcasting (1982), Principles Relating to Remote Sensing of the Earth from Outer Space (1986), Principles Relevant to the Use of Nuclear Power Source in Outer Space (1992), and Declaration on International Cooperation in the Exploration and Use of Outer Space for the Benefit and in the Interest of All States, Taking into Particular Account the Needs of Developing Countries (1996).

originate on and from extraterrestrial places might emerge, whether through the creative, contextual thinking of Earth explorers and colonists who have traveled afar, or from the philosophies, perceptions, and practices that other intelligent beings have developed in their own praxis experiences: on their home planet, in space, or on celestial bodies.

The UN documents, then, despite their anthropocentric biases regarding humans in relation to other biota and to Earth, and their anthropocentric lack of consideration of the possibility that there might be other intelligent beings exploring and colonizing the cosmos, will serve as bases for pondering possibilities to enhance human beings' consciousness and conduct prior to and during space journeys and ventures. They will be helpful to develop consideration and exposition of humankind's extraterrestrial responsibilities in exploration and settlement and, linked to terrestrial responsibilities discussed previously, a related and integrated terrestrial-extraterrestrial consciousness, commitments, and conduct requisite for responsible space exploration and settlement. A thorough, in-depth discussion of these documents in their entirety is not possible here; particular segments pertinent to consideration of the impacts of contact with newly encountered worlds, and Contact with newly engaged intelligent species in the cosmic commons will be highlighted and explored.[3]

UN Extraterrestrial Visions and Principles

Human interest in the exploration and use of space was dramatically increased by the launching of the Sputnik satellite by the then-Soviet Union in 1957. The immediate and intense response of the US government was to promote an enhanced educational and scientific effort to enable the nation to match and then surpass the achievement of the USSR. The "space race" was on, generated by two "superpowers" seeking hegemony not only in their respective geographic areas but, beginning with the moon, in the cosmic expanse beyond both nations. The purpose in developing orbiting satellites and interplanetary expeditions was not to initiate an adventurous acquisition and then international dissemination of knowledge about cosmic structure and meaning, and information about celestial natural goods available for human use and benefit; nationalistic military, industrial, and commercial objectives,

3. The reader is encouraged to visit the United Nations Web site to download, print, or read online, and thoughtfully consider, the depth of perspective and the breadth of topics explored in the documents cited, which were developed collaboratively. The titles alone indicate nations' concern to supplement existing documents due to new technological developments or other issues not fully addressed previously. The seeds of peace and justice—on Earth and on other cosmic bodies—are carefully sown in these visionary, realistic (not idealistic) documents.

rather than international benefits for humanity as a whole, stimulated US and USSR space ventures.

International concern developed over the future of humanity when part of humankind, initially solely the two superpowers, demonstrated its potential for travel far beyond Earth. Peoples of the world had both periodic fears that the "Cold War" would erupt into a heated exchange–perhaps even a nuclear one–and ongoing fears that either one of the dominant nations might gain a nationalistic advantage through extraterrestrial development, and consequently exclusively control, solely for its own benefit, extraterrestrial territory and natural goods. These fear factors stimulated development of international instruments that were intended to govern and guide outer space endeavors. War and weapons of war, and acquisition and hoarding of planetary minerals and other goods, were viewed as contrary to the real and pressing needs of Earth's nations, their economies, and their citizens. Extension of such practices into space was rightly understood to be of even more pressing concern because their impacts near and far would be catastrophic for Earth and for human civilization.

Concerns among nations about the ongoing threat of a major, even nuclear war on Earth between the US and USSR, and about the possibility that terrestrial conflict would expand into near and distant space were expressed in the United Nations. This led eventually to the development and ratification of international instruments designed to prevent conflict in space, and to foster mutual, global benefit from fruits of space exploration—at least to the extent that their provisions would be observed by signatories to the documents. There is much repetition in successive UN documents. This is due in part to the distinct bodies promulgating them—delegates assembled in and voting as the General Assembly, or states through senior government officials whose top leaders would affix their signature on documents after their respective relevant political processes for treaties had been completed—and in part to a perceived need to reiterate and reinforce particularly needed provisions. The repetition will be evident in presentation and discussion of the diverse documents, in order that the reinforcement (and at times slight wording changes) can be noted, but some repetition will not be included in order to avoid excessive redundancy (as distinct from required redundancy needed to demonstrate nations' reinforcement of prior instruments' provisions).

Treaty on Principles Governing the Activities of States in the Exploration and Use of Outer Space, Including the Moon and Other Celestial Bodies[4]

The Outer Space Treaty (OST), as this Treaty on Principles came to be called, was the first major international instrument intended to guide member states toward collaborative rather than competitive exploration and colonization of space. Underlying this effort, as noted previously, was global concern that the US and USSR would use their technological advantage over other nations to secure for themselves benefits that United Nations members considered a common human province; and, that places in space might become a battleground for the two superpowers, in which case Earth, too, would become a place of extraordinary conflict, a planet imperiled and assaulted by weapons wielded in and from space, including from orbiting weapons-bearing satellites, and out to space in reply, and therefore subject to retaliatory response in turn, in unending succession until either or both sides and the nations around them had been reduced to abandoned cities and barren landscapes.

In the introductory section, the Outer Space Treaty signatories, the "States Parties" to the Treaty, begin on a very positive note: "great prospects" are opening for humankind because of human ventures into space. Humanity has a "common interest" to explore and use space for "peaceful purposes," and "all peoples" should benefit from these activities. The States Parties should seek international cooperation in legal and scientific areas; this will help to promote mutual understanding and stronger "friendly relations" among nations. In all of this the Treaty is expected to promote the purposes and principles of the original Charter of the United Nations.

The OST establishes its tone and purpose immediately after expressing its vision and intent. It declares that the exploration and use of outer space "shall be carried out for the benefit and in the interests of all countries, irrespective of their degree of economic or scientific development, and shall be the province of all mankind"; outer space "shall be free for exploration and use by all States without discrimination of any kind, on a basis of equality and in accordance with international law, and there shall be free access to all areas of celestial bodies" (Art. I). The Article states further that *all* nations should benefit from space exploration; all share a common interest in having humankind as a whole be the recipients of whatever scientific discoveries and economic benefits accrue from outer space travels and on-site exploration of new common ground. As nations other than the US and USSR, the two (competing) pioneers in such endeavors, acquire the capability to explore they should have an equal right to do so. The Article elaborates and reinforces the "peaceful purposes" and "all people" phrasing of the introductory statement.

4. United Nations, *Treaties and Principles*, 3–8.

In this Article and those to follow, the precedents established earlier in the Antarctic Treaty—nations' agreement on an international treaty regarding territory, sharing of natural goods, collaborative scientific research, nonmilitary use, and mutual rights to inspect others' operations—are clearly visible and operative. The stakes in outer space, of course, are much higher: a vaster "territory" is available to be seen (at first from afar), understood, and then experienced directly and analyzed in detail as potentially habitable planets and other celestial bodies are explored on site and eventually and extensively settled, and become enabled to provide natural goods beneficial to humanity; or, as such bodies are at least secured for a scientific research base to provide scientific information, and perhaps as a launch site for further exploration.

Carrying forward principal Article I stipulations, the OST states in the next few Articles that no nation may claim sovereignty over any part of outer space, whether by occupation, use, or other means (Art. II; this statement serves, perhaps unconsciously, as a safeguard, a barrier against a country's or cluster of countries' use of Discovery Doctrine tenets—acknowledged or not, stated or unstated—to keep extraterrestrial land and goods for themselves); all activities undertaken in space must be in accord with international laws, including the UN Charter, in order that peace, security, and cooperation among nations will result (Art. III; this presents another barrier to Discovery, in that the international community as a whole, not a particular politically separated and bordered [and gated] segment of that global community, will oversee exploration and use: nations and peoples that have been historically victims of Discovery would have been and continue to be especially sensitive to nationalistic sovereignty claims); nations will not put nuclear weapons or other mass destruction weapons in an Earth orbit, nor will they have military bases and installations, test weapons, or undertake military maneuvers on exoEarth places, except that military personnel are permissible if they are engaged solely in scientific work (Art. IV; a provision identical to one in the Antarctic Treaty); astronauts of all nations should be mutually hospitable, and assist anyone injured and away from their nation's base to return to that base; and, researchers should inform each other or the UN Secretary-General about any space phenomena discovered that might endanger the life or health of astronauts (Art. V); states are responsible for damages or harm caused by objects they launch into space (Art. VII); any object sent into space belongs to the country that launched it, no matter where it is as it traverses space or where it lands, whether on a celestial body or back on Earth (Art. VIII); "cooperation and mutual assistance" will guide the conduct of nations' explorers; "States Parties to the Treaty shall pursue studies of outer space, including the Moon and other celestial bodies, and conduct exploration of them so as to avoid their harmful contamination and also adverse changes in the environment of the Earth resulting from the introduction of extraterrestrial matter and, where necessary, shall adopt appropriate measures for this purpose"; nations should exercise research protocols to ensure that experiments done by one country do not negatively impact those of other countries (Art. IX; the Article expresses a strong ecological concern and specific ecological ethics principles intended to govern terrestrial citizens' conduct and thereby avoid extraterrestrial internal [on the moon and other celestial bodies] and external [back home on Earth] harmful impacts); states should allow other States to observe launched objects as they traverse space (Art. X); states should inform the UN Secretary-General, general public, and scientific community of their space activities (Art. XI); on celestial bodies, any nation might visit all structures and vehicles of any and all other nations (Art. XII); and, the Treaty's provisions apply to all activities on celestial bodies, whether conducted by one or several countries (Art. XIII).

The final four Articles discuss provisions for the signing, ratification, and Deposit of the Treaty, and the processes for, respectively, amending the Treaty, withdrawing from the Treaty, and accepting later signatories of the Treaty: all states shall be free to sign the Treaty, including states not among the original signatories.

The Outer Space Treaty, then, provided an in-depth and extensive consideration of international issues related to outer space—to the extent possible given the knowledge available and accessible in its historical context and moment.

Centuries earlier, when the US Constitution was being developed and after it had been signed (September 1787), those who had signed it and other citizens realized that the founding national document needed modification to include a more specific statement of states' and individuals' rights. Consequently, the Bill of Rights was developed and signed two years later (September 1789); after ratification by the new US states it was added to the US Constitution as its first ten Amendments. When subsequent generations realized that the Constitution should be altered further to address issues that had arisen in the time elapsed since the Bill of Rights had been incorporated, they added new Amendments to meet new needs.

Similarly, the OST (1967) signatory states realized that there were issues they had not considered earlier, in the Treaty's development, either because they had been unaware of them at the time, or because some of them arose only when further technological developments occurred. Thus, just one year after the Treaty was promulgated, another major document was issued, the Agreement (1968) that is described next. Later still, other statements would be issued, principally by the UN General Assembly comprised of delegates from all UN member states.

Agreement Governing the Activities of States on the Moon
and Other Celestial Bodies[5]

In its Introduction, the Agreement Governing Activities of States (hereafter: AGAS) cites achievements to date in space exploration, noting in particular moon-related activities. In light of the latter, the AGAS states that the moon, Earth's natural satellite, will be a significant factor in more distant space exploration: lunar sites will be useful for humankind's "further progress in the exploration and use of outer space." The UN expects that all states will share equitably in moon development and realize benefits from lunar natural goods, and seeks to avoid international conflict on the moon. Consequently, the member states decided to clarify and complement existing principles, presuppositions, and bases.

In language identical at times to that of the OST, the AGAS seeks overall to prevent irresponsible exploitation of the moon and other celestial bodies by one or a few nations or by private economic interests; militarization of the moon; and international conflict over lunar territory and goods.

In its Articles, the AGAS declares that provisions regarding the moon should "apply to other celestial bodies within the solar system," except for Earth (Art. 1); lunar activities shall be undertaken "in accordance with international law," especially as expressed in the UN Charter (Art. 2); hostile activities and military bases and weapons are prohibited (Art. 3).

Article 4, because it encapsulates and addresses myriad international issues related to space exploration, merits more extensive citation:

> The exploration and use of the Moon shall be the province of all mankind and shall be carried out for the benefit and in the interests of all countries, irrespective of their degree of economic or scientific development. Due regard shall be paid to the interests of present and future generations as well as to the need to promote higher standards of living and conditions of economic and social progress and development in accordance with the Charter of the United Nations.

In this one paragraph, several important socioecological concerns are raised, and socioecological ethics on-site principles and constraints are implemented to address them. First, all humanity should share in benefits derived from lunar territory and goods. This means that *humankind as a whole* should share in lunar goods found during exploration, whether by a single nation or several nations: the explorers who find them or uncover them, and citizens in their sponsoring state(s), should not be the sole beneficiaries of lunar goods

---

5. United Nations, *Treaties and Principles*, 27–35. The 1968 Agreement expanded and extended principles and policies elaborated in the 1967 OST.

found and acquired on or under the moon's surface. Second, and a corollary statement: the exploration and use should benefit *all nations*, no matter their level of economic well-being or scientific expertise. Third, intergenerational responsibility should be constructively considered: lunar goods should not be overused and squandered to benefit only people living in the present; goods should be carefully conserved so that, in the future, descendants of the present generation will benefit also from the moon's largesse. Fourth, the results of lunar activities should be used in a way that demonstrates that exploring nations embrace the vision and fulfill the requirements of the UN Charter: by enhancing the living standards, economic well-being, and social progress of all peoples. (It might be added that in discussions of socioecological crises and concerns on Earth, the UN-stipulated requirements for activities "in the heavens" might stimulate deepened and broadened discussion of and inspiration for realizing social justice and environmental conservation, and thereby socioecological well-being, on Earth—a significant aspect of the historical UN vision and *raison d'être*.)

The AGAS Articles declare that states should inform the UN about their current and intended activities (Art. 5), are free to undertake lunar scientific research and, in the pursuit thereof, to extract minerals and other matter; scientific information derived from lunar activities and scientific research should be shared (Art. 6).

Article 7, too, deserves extended citation. In exploring and using the moon,

> States Parties shall take measures to prevent the disruption of the existing balance of its environment, whether by introducing adverse changes in that environment, by its harmful contamination through the introduction of extra-environmental matter or otherwise. States Parties shall also take measures to avoid harmfully affecting the environment of the Earth through the introduction of extraterrestrial matter or otherwise. . . . [C]onsideration may be given to the designation of ["special scientific interest"] areas as international scientific preserves for which special protective arrangements are to be agreed upon in consultation with the competent bodies of the United Nations.

Article 7 makes a strong statement about ecological responsibility—on Earth and on celestial bodies. Governments should guard against ecological devastation that would be caused by environmental degradation or pathogenic transfer from Earth to the moon or from the moon to Earth. Since UN documents observe periodically that what is stated in regard to the moon should be applicable in regard to "other celestial bodies," the warning against environmental degradation and contamination applies to exploration and development elsewhere in space, not solely to lunar expeditions and subsequent lunar development. In the course of their work, scientists and the nations for which they work should consider setting aside areas of special scientific interest as protected international preserves that would retain, to the extent possible, the relatively pristine condition they were in when they were found and initially explored. These sites might also be valuable scientifically, by providing examples of areas in their original state with which areas developed by humankind could be compared and contrasted in the future; and they would remain as a reminder of what once was (and, perhaps, of what should be once again, in some circumstances and contexts), and a caution against proceeding on paths and practices of exploitation that would wreak on the moon the types of "adverse changes" and "harmful contamination" about which the Article expresses concern.

The Articles state that lunar explorations may take place on or under the moon's surface (Art. 8); lunar stations may be operated from afar or by personnel in place (Art. 9); states should safeguard the health and life of all lunar occupants, sent by their own and other nations, as per Article V of the OST (Art. 10).

Article 11 reiterates and reinforces earlier statements that global humanity, not individual nations or people, should benefit from the moon and its natural goods, which are humanity's "common heritage":

> The Moon is not subject to national appropriation by any claim of sovereignty, by means of use or occupation, or by any other means. . . . Neither the surface nor the subsurface of the Moon, nor any part thereof or natural resources in place, shall become property of any State, international intergovernmental or non-governmental organization, national organization or non-governmental entity or of any natural person. The placement of personnel, space vehicles, equipment, facilities, stations and installations on or below the surface of the Moon, including structures connected with its surface or subsurface, shall not create a right of ownership over the surface or the subsurface of the Moon or any areas thereof. . . . States Parties to this Agreement hereby undertake to establish an international regime, including appropriate procedures, to govern the exploitation of the natural resources of the Moon as such exploitation is about to become feasible. . . . [The purposes of an international regime are] (a) The orderly and safe development of the natural resources of the Moon; (b) The rational management of those resources; (c) The expansion of opportunities in the use of those resources; and (d) An equitable sharing by all States Parties in the benefits derived from those resources.

The sharing stipulated by (*d*) is to ensure that "special consideration" be given to developing countries' "needs and interests" and to the needs and interests of all countries which had some role in assisting lunar exploration. The Article requires further that states inform the UN, the public, and the scientific community about natural goods discovered.

Article 11 is very forthright in advocating that common benefits should be derived from the common ground of the moon, which is regarded as a common human province and heritage. It declares that neither the moon's surface or subsurface, nor any natural goods found in place in any lunar level, may be appropriated by the party that first comes upon them or first seeks to use them. In this Article in particular an undercurrent of concern about potential nationalistic claims is evident: that a nation or nations might seek to assert "ownership" of territory or goods based on the Discovery Doctrine, even if not explicitly stated. Note the elements of Discovery addressed: claims of sovereignty based on occupation or use; placing personnel or buildings (remember "forts" in earlier centuries) on or below the lunar surface (remember the planting of a national flag on, and placing inscribed lead plates below, Earth's soil as done by explorers both European—Christopher Columbus on behalf of Spain's monarchs, Ferdinand and Isabella—and Euroamerican—Meriwether Lewis and William Clark under US President Thomas Jefferson).

Discovery claims are not permitted to any national or international political entity, nongovernmental organization, or "natural person" (this would include people or corporations [which have "person" status legitimized by controversial laws in some Western countries] who have industrial or commercial interests). UN delegates, particularly those from, or those sympathetic to the needs of developing countries are concerned that powerful nations (or their citizen entrepreneurs and corporate executives acting through them) might seek to reserve lunar places or goods for themselves. Were any such Discovery claims to occur, it would mean not only that less politically and economically powerful countries would lose a share in the common human province and heritage, but that more powerful nations might resort to military action to hoard common ground and commons goods. This would lead to international conflict on the moon and on Earth (and, later, on other celestial bodies that are discovered if the nation first finding them claims to have Discovered them).

The OST and the AGAS express a shared concern about the preceding. Their less powerful signatories, based on historical precedents they experienced or observed on Earth, or about which they became aware otherwise, would be particularly concerned. As is often the case, those who have been subjected to an economically or culturally based or biased form of oppression such as the Discovery Doctrine as it was developed by and for European nations in the fifteenth to seventeenth centuries—and subsequently refined, especially in the United States, in the eighteenth century and thereafter—are most sensitive to and apprehensive about its implications and potential applications.

As in the Antarctic Treaty, the AGAS requires that States Parties are to have access to inspect each other's buildings, vehicles, equipment, and stations, after giving adequate advance notice (Art. 15); any nation party to the Agreement "may give notice of its withdrawal from the Agreement" to the Secretary-General, effective one year after notice is received (Art. 20).

An obvious and unfortunate weakness in the documents just cited is that the principal treaties, the OST (Article XVI) and the AGAS (Article 20), contain the latter provision. Any nation that is a Party to the Treaty may give notice and withdraw at will from the treaties and pursue its own perceived interests, thereby dissolving international cooperation and, potentially and probably, subverting peaceful exploration and use. Unfortunately, then, the outer space principles, which had been developed over time and on which international consensus had been reached after lengthy discussion, deliberation, and debate would become meaningless if one or more nations decided that their own supposed national interests should take priority over "international cooperation" in, or the "peaceful purposes" of, outer space exploration and use. Their statement of national interests might actually be giving voice to private interests, as has happened on Earth previously: for example, corporations' complaints about having to respect the sovereignty of nations other than their own, environmental protection requirements, labor requirements, and profit maximization limits (retired General Smedley Butler's comments about military campaigns to benefit transnational corporations are relevant in this regard).

Declaration of Legal Principles Governing the Activities of States in the Exploration and Use of Outer Space[6]

The Declaration (hereafter: DLP) had preceded the OST by four years. It provided a foundation for, and the General Assembly's first formulation of, the Treaty that UN member states would subsequently sign.

The introductory paragraphs of the DLP note that Earth's citizens are "inspired by the great prospects" presented to humanity because of humans' forays into space. The DLP states, too, that all humankind shares an interest in space exploration and the use of space for "peaceful purposes." Exploration

---

6. United Nations, *Treaties and Principles*, 39–40.

and use should better the condition of all humanity and benefit every nation, whatever their economic situation and level of scientific accomplishment. Signatory states in the General Assembly hope to prompt international cooperation in scientific endeavors, laws developed to guide space exploration and use, and the peaceful use of what space will provide. Cooperation will help mutual understanding and strengthen "friendly relations" among Earth's countries and cultures.

Nine guiding principles (many of which are mentioned or evident in, or inferred from, the OST) are proposed to accomplish the foregoing. First, all humanity should benefit from space exploration; second, all states should have free and equal access to space exploration, as provided by international law; third, "outer space and celestial bodies are not subject to national appropriation by claim of sovereignty, by means of use or occupation, or by any other means" (note the allusion to Discovery Doctrine); fourth, nations' space exploration and use should be conducted as required by international law, including the UN Charter, in order to promote peace, security, cooperation, and understanding among all nations; fifth, individual countries are responsible for all activities originating within their own borders, whether undertaken by official government bodies or other entities, and must ensure that all such activities comply with DLP principles; sixth, "cooperation and mutual assistance" should characterize states' activities and actions, with consideration for other states' interests and for potential adverse impacts on other states' experiments and activities; seventh, when an object is launched into space under a nation's registry, the ownership, jurisdiction, and control of the object and its component parts are retained by that nation, and any object or its parts must be returned to the state of origin; eighth, a nation which launches an object or commissions its launch is responsible for any damage to places or harm to people caused by the object; and ninth, astronauts are the "envoys of mankind in outer space," and should receive assistance when they need it from any state, not just their own, including on foreign soil and on the sea; astronauts in such a situation should be returned expeditiously and safely to the state in which their vehicle is registered.

Comparison of the DLP and the OST reveals how the former was foundational for the latter. It demonstrates, too, how these and other documents were gradually and tediously developed through international collaboration, and supplemented and strengthened in successive elaboration.

Declaration on International Cooperation in the Exploration and Use of Outer Space for the Benefit and in the Interest of All States, Taking into Particular Account the Needs of Developing Countries[7]

The General Assembly first noted the increasing level of international cooperation in efforts to ensure that outer space would be used for peaceful purposes, and then related that to other international cooperative ventures and to the necessity of promoting ongoing collaboration in outer space for the mutual benefit of all states engaged in space exploration. It issued the DIC to facilitate realizing the principle that "the exploration and use of outer space, including the Moon and other celestial bodies, shall be carried out for the benefit and in the interest of all countries, irrespective of their degree of economic or scientific development, and shall be the province of all mankind."

The DIC described several considerations that should be preeminent in international space exploration. First, international law, including as expressed in the UN Charter and the OST, should be followed in space exploration and used to ensure that space activities are undertaken for "peaceful purposes" and to benefit all states, whatever the extent of a nation's "economic, social or scientific and technological development." All territory and the goods therein should be the "province of all mankind"; space-venturing states' special concern should be to meet the pressing needs of developing nations. Second, states will freely decide how to explore and use space in an internationally cooperative, equitable, and "mutually acceptable" way; this includes ensuring that contracts for cooperative ventures are fair, reasonable, and comply with the rights and interests of all involved parties, including by safeguarding intellectual property. Third, all states should promote equitable and "mutually acceptable" international cooperation, conscious of benefiting "developing countries and countries with incipient space programmes" who work with other nations whose space capabilities are more developed. Fourth, international cooperation must utilize the best means available—governmental or nongovernmental, commercial or non-commercial, for example—to promote participation by nations at different development stages. Fifth, cooperating countries, conscious of developing nations' needs, should have goals such as joint development and application of space science and technology, promoting space capabilities in other states, and exchanging among states, in a mutually beneficial manner, space expertise and technology. Sixth, agencies providing development aid, developing countries, and developed countries alike should consider possible use of space applications to achieve their development goals. Seventh, the role of the Committee on the Peaceful Uses of Outer Space should be strengthened, including relative

7. Ibid., 55–56.

to exchanging information regarding international cooperation in space exploration and use. Eighth, all states should contribute to the UN Programme on Space Applications and to international cooperative space activities, according to the extent of their own capabilities and of their participation in such activities.

Assessment of United Nations Space Documents

The UN documents published to date present socioecological ideals, expressed in practical principles which, if followed, would ensure peaceful and collaborative international cooperation in space. They would help thereby to provide significant scientific knowledge and material goods benefits that would stimulate a globally equitable sharing of natural goods, and elevate the global human condition. The achievement of the type of universal social justice and ecological betterment that the documents envision would enhance the well-being of humanity and its Earth home. The potential for realization of the cosmic vision the documents present, which is proposed to characterize human consciousness and conduct in celestial settings, is ultimately and intimately intertwined with achieving an improved Earth socioecological context, founded on and expressing a transformed human consciousness operative in human conduct on Earth. A simultaneous, dialogic, Earth-celestial progressive development and implementation of complementary modes of thought and action would enhance the chances of achieving socioecological well-being wherever humans venture and live.

The documents are decidedly anthropocentric. This is understandable, perhaps, in an era when humans were very self-assured in assuming and asserting that they were the only intelligent life in the universe. Through the documents, UN governments claim that space and "celestial bodies" are a "human province" and are to be shared human property; and, they provide principles to govern ways in which the cosmos and its natural goods could be equitably shared among all Earth's peoples. Even the documents' concern for those countries that are not as technologically developed, politically powerful, and economically prosperous as are others carries an assumption that all territories and natural goods found in space will be divided and shared by human national entities; no other intelligent being is expected to be present when humans arrive—or even afterward, should other intelligent species be engaged in similar space exploration. The documents might well prove to be, despite their rejection of one nation's or one people's claims to territory and goods, an Earth-based and -biased Extraterrestrial Discovery Doctrine, writ large. Humankind and human culture might be presumed to be "superior" to other intelligent species and cultures, much as occurred when unequal technologies and populations clashed in the Americas and elsewhere around the globe as Europeans (and later Euroamericans) seized native peoples' territories and natural goods.

The AGAS has an even stronger anthropocentric bias (Article 11) than does the OST; it is written as though there were no other intelligent life or evolving life that might now or in the future have a right to acquire and use the natural goods present on celestial bodies. In its cosmic anthropocentrism, it assumes (in a perspective similar to that of seafaring colonial nations in the fifteenth to seventeenth centuries, and dominant powers ever since) that the moon and other celestial bodies, and the materials discovered on them, are resources for humankind, as represented by diverse national interests and an integrated international interest, with no consideration of the possible existence and needs of extraterrestrial intelligent beings or of evolving primitive forms of life that might exist now or in the future on those bodies or in a cosmic "neighborhood." In their shortsightedness, they assume unabashedly that the moon and other celestial bodies, and their natural goods, all belong to humankind. It should be noted, too, that the documents, while declaring that benefits of outer space ventures should accrue to all humankind on Earth, there is no thought given to eventual colonists' probable decisions to reserve territory and natural goods for their own benefit, much as these would have been left to native inhabitants encountered on arrival. The resolution of this issue might be through some future trade arrangements between humans on Earth and humans in the heavens.

The AGAS does seek an equitable distribution of human-designated goods. Article 11 declares that in terms of natural goods distributions, the "Moon and its natural resources are the common heritage of mankind." This "common heritage" means that the moon and other celestial bodies are "not subject to national appropriation by any claim of sovereignty." The Article strongly and unequivocally states that the moon and all other celestial bodies are not to be extensions of any of Earth's political or economic entities, and that no nation, or an economic elite or transnational corporation, can claim or use resources exclusively for its own purposes and benefit. Finally, the Article advocates an "equitable sharing in the benefits derived from those resources, whereby the interests and needs of the developing countries, as well as the efforts of those countries which have contributed either directly or indirectly to the exploration of the Moon, shall be given special consideration."

Social and environmental conditions on Earth provide evidence that humankind has not developed a responsible ecological-economic-ethical consciousness and practices on its home planet, and would hardly be capable, from that foundation, to transfer sound thinking and action elsewhere in the universe. The human consciousness and conduct carried into space cannot be the progeny of human conceptual constructs and "best practices" as currently constituted. Extraterrestrial exploration must not embody and seek to extend current Earth economic arrangements, which would provoke conflict with other intelligent species. A significant conversion of consciousness is required for a fruitful conversation with another species that also is looking to enhance its well-being with natural goods available in the cosmic commons.

In such a conceptual context as currently is the case, the UN extraterrestrial-related documents, despite their shortcomings, have the potential to promote cooperation among nations—and, even among intelligent species—in space exploration. Particularly beneficial ideas of the documents include their emphasis on the common interests and rights of humankind (which could be extended to include interests and rights shared with extraterrestrial intelligent species); their stipulations that scientific information should be shared (which could be effected through terrestrial-extraterrestrial intelligent species interaction); and their advocacy for special consideration of the needs of developing countries, particularly through the equitable distribution and use of natural goods discovered and developed as a result of exploration (which could be a guiding principle and policy for distributions intended to be made for 'people(s)' on multiple planets or other celestial bodies, where one culture has greater need for a specific territory or the material goods it provides).

The AGAS and the OST propose, on the whole, excellent initial insights and principles for ecologically responsible human conduct—on Earth and in outer space voyages and ventures. The rights of "humankind," for example, could be amended to accept the rights of "all intelligent beings," as just noted; and, statements regarding safeguarding the needs of developing countries might be rephrased to include not only their needs, but the needs of similar extraterrestrial beings, and additionally of evolving biota and evolving intelligent life. In these considerations, the collaborative development of extraterrestrial consciousness and conduct might serve as a bridge to draw together the complementary perspectives of religion and science, much as ecology has done in Earth settings.

Looking and Going Forward

Core proposals providing guidelines for humankind's responsible space exploration recur explicitly in or permeate implicitly the UN documents just reviewed, analyzed, and discussed. The provisions, principles, and recommendations expressed in UN instruments are relevant not only to guide nations' space conduct, but also as a foundation for theories and activities to be developed speculatively and contextually as principles and practices of extraterrestrial cosmosocioecological ethics. In the future, this will be especially the case when and where experienced explorers encounter situations or conditions not previously conceived or conceptualized. In praxis, the dialogic interaction of operative theory and principles with on-the-ground contexts can lead to innovative ideas and practices that respond appropriately to new social and ecological settings.

Although not expressed or developed in this way, the UN's efforts could have positive impacts on and implications for Contact, should that occur, and for a more "universal," rather than merely global, just distribution of control over newly found territories and natural goods, and interaction with resident biota—intelligent, evolving toward intelligence, or seemingly incapable of having or developing intelligence. Current human consciousness and conduct on Earth, and some human attitudes apparent in UN documents, would provoke conflict with extraterrestrial intelligent species, or at least erect barriers to congenial Contact and collaboration on Earth or elsewhere in the cosmos. By contrast, many of the UN-developed ideals, principles, and policies might well prompt proposals of possibilities for the types of alternative thinking and acting that would promote mutual understanding and efforts to interact respectfully in space, if those in Contact are open to responding to each other's needs.

Ideas based on UN statements would be beneficial bases for an eventual and constantly evolving Cosmic Charter that would incorporate elements of the UN Charter, the Earth Charter, and space-related UN documents, and thereby supplement and complement these international predecessors and the precedents they provide. Such a future Cosmic Charter could amend and extend relevant sections and ideas from earlier documents. "All States," for example, could become "All Intelligent Beings' Communities," and "Developing Countries" could be changed to "Developing Cultures." (The formulation of such a Cosmic Charter based on existing international instruments and on ideas from other sources will be explored in chapter 11.)

Military, industrial, and commercial objectives that stimulated initial US and USSR space ventures continue to prompt space exploration today, although in an apparently collaborative rather than conflictive way. Over the years, other nations have been incorporated into space efforts and have become involved in joint projects, such as the International Space Station, by providing scientific personnel, shuttle vehicles for cargo and personnel, construction materials, and scientific instrumentation or components needed for the operation of the Station and for scientific research and experimentation. Civilian contract employees and space vehicles have been employed in space projects if only, at this point, to shuttle cargo and personnel. This might save nations' financial resources for other purposes, but it might also lead to abuse of private corporations' power, and corruption of government personnel.

Current ventures embodying cooperation for mutual benefit, and commitment to a better common future, are hopeful signs for development of more extensive collaboration projects. Their integrated operation represents a slight but significant change in human consciousness and conduct. They might not only herald humans' theoretical and practical cosmic sensibilities, but also, should Contact with extraterrestrials occur, highlight similar existing or potential concepts and constructs that would catalyze cooperation.

Terraforming Extraterrestrial Planets and Places

The beginning of a new era in space has prompted people to speculate about human exoEarth activity beyond *exploration*; now the conversation includes *colonization*—of planets and planets' satellites, within the solar system and, eventually, in distant space. Humankind has journeyed far intellectually, spiritually, and theologically from resistance to and rejection of the theories and data of Copernicus, Kepler, and Galileo in the sixteenth century, little more than five centuries ago. Discussion of a lunar base has been going on for decades—including secretly, as in the US Army's once classified 1959 Project Horizon proposal, developed twelve years after the Roswell events, to build a lunar military base in order to defend Earth from potential ETI invasion and to enable the United States to make a Discovery-like claim to the moon.[8]

The idea of an actual colony on the apparently human-hostile environment of the planet Mars has begun to be discussed in more recent decades, and seriously, by NASA scientists, other scientists, politicians, and the general public. The latter might have become interested in and informed about a Mars colony potential initially through Ray Bradbury's popular novel *The Martian Chronicles*, in which human occupation and native Martians' opposition to Discovery (not named as such) are followed by terraforming to make Mars just like Earth.[9] More recently, the popular imagination was stimulated even more by another work of science fiction: in the classic Gene Roddenberry *Star Trek* films that were shown in movie theatres and later televised on movie channels. At the end of the movie *Star Trek II: The Wrath of Khan*, a battle between Captain Kirk and his crew and Khan and his criminal consorts over a top-secret "Genesis" device results in success for Kirk

---

8. Project Horizon is described in detail, including with engineering diagrams, in retired Colonel Philip Corso's *Day After Roswell*, discussed in chapter 8.

9. See chapter 8.

and company but also occasions the death of Mr. Spock, who is interred on a barren planet called Genesis. In the sequel *Star Trek III: The Search for Spock*, the *Enterprise* crew returns to the planet. Humans, who have developed the technology to effect terraforming, activate the Genesis device to see if it works to rejuvenate and transform the planet Genesis in order to make it humanly habitable and hospitable. Somewhat unexpectedly, Spock is resurrected in the new, lush habitat and rejoins his *Enterprise* crew, which is on the planet's surface surveying results of the device. (There is an obvious link, of course, to the creation stories in the first two chapters of the biblical *Genesis*.)

Terraform means, literally, to form the earth or Earth. As noted in previous chapters, *terra* can refer to planet Earth or to Earth's land or soil, earth. Terraforming means, then, to alter a particular part of a geographic place by forming its surface to be amenable to human use or, more recently as (untested) technology has begun to be developed, to attempt to alter an entire planet; in that case, terraform would mean to form as an Earth. In both *Martian Chronicles* and *Search for Spock*, humankind alters an entire planet. In these circumstances, if there already was any life in place, it should have been extincted or dramatically changed because of the intervention and intrusion of an alien process for its transformation—its previous customary habitat and the niches into which it had evolved would have been replaced by a new abiotic setting and biotic possibilities for emergence of new species, and evolution of new species and such old species as might have survived, from simple to complex. (In the film, terraforming apparently is able to rejuvenate what once existed on the planet, without genetic variation.)

The possibility that humankind will initiate and impose either or both types of change to benefit humanity, let alone the actual implementation of plans to do so, catalyzes consideration of what manner of human consciousness, and its mode of expression in conduct in new contexts, leads to and promotes terraforming. (It should be noted here that science fiction sometimes becomes science fact. Jules Verne's (1828–1905) novels *Twenty Thousand Leagues Under the Sea* (1869) and *From the Earth to the Moon* (1865), for example, were written well before the technology existed to make possible the imaginary undersea journeys of Captain Nemo—for a distance, not at a depth, of twenty thousand leagues—in his *Nautilus* submarine, or travel to the moon in a projectile launched from Earth.)

Concerning lunar bases, colonization anywhere in exoEarth places, and terraforming, in whole or in part, if the UN documents cited are to have any meaning at all, then all such considerations should be discussed and debated openly in UN forums and consensus reached prior to any efforts along the lines mentioned. The US Army's Horizon lunar base, proposed in 1959, if attempted today, would clearly be in violation of the space treaties and agreements, to which the United States was a signatory, that were developed and promulgated subsequent to Horizon plans. Proposals and projects, including terraforming, are supposed to benefit humankind as a whole and the body of nations as a whole, with special regard for deprived populations within member states, and for less technologically and economically developed, and therefore less powerful, nations. They should be undertaken, too, as collaborative projects: interethnic and international, much as were the interplanetary and interspecies efforts indicated by the membership and activities of the crew of the *Enterprise*, and of the federation of planets that they represented.

Terraforming might well be beneficial in barren environments. Before such massive experiments and efforts on other planets and celestial bodies are begun, however, humans must think about their possible impacts on abiotic terrain and biotic life and habitat, and scientific instruments should be developed and made available to assess biotic presence in whatever stage of evolution, and used responsibly and transparently to gather and analyze data. It should be borne in mind, too, that what were considered barren or vacant areas on Earth, from desert sands to volcanic fissures deep in the sea, have been found to host life unlike biota that humans had encountered and classified previously.

Potential Benefits of Terraforming

Terraforming can enable humankind to go where no one has ever gone before, and to live there on arrival if the place reached has the potential to be altered, when inhospitable for humans, to provide for human life and livelihood. This would be beneficial for the human species if its population has grown beyond Earth's carrying capacity; if Earth has been destroyed or extensively and harmfully altered by human-caused or -exacerbated disasters such as those projected by Stephen Hawking, which have made it uninhabitable by humankind in whole or in part; if Earth has been depleted of vital natural goods (including abiotic goods such as clean water, air, and soil without which humans cannot provide for even their most basic needs); or if unceasing conflict dominates Earth, and a peaceful remnant escapes to begin the human experiment anew in a place distant and perhaps hidden from the view from Earth, and even from possible efforts to find the remnant. All of these reasons could be elaborated in greater depth in order to illustrate how departure from Earth might be voluntary because of collaborative international efforts grounded in peaceful and just coexistence, or coerced by life-adverse circumstances created by international actual or pending conflicts that are unleashing or will unleash catastrophic harm.

Potential Harms of Terraforming

Not only would existing biota, if any, be affected by terraforming: new biota might emerge and mutate in a new environment when existing elements in the soil, air, and water are integrated with alien intruders into their locale, however benign the intentions might be of those who introduce them. Uncontrollable pathogens might develop and multiply, with no natural controls to keep them in check and safeguard evolving biota, and chemical agents might be introduced in the form of pesticides—insecticides and herbicides among them—which would replicate patterns of destructive human behavior, past and present, on Earth. The end result could well be more harmful than beneficial, not only to humans but to other life, intelligent or primitive, native to, or exported or emigrant to the space place. A planet or celestial body previously habitable by some organisms, perhaps undiscovered because they do not fit preconceived and even prejudicial notions of what constitutes "life," might suddenly become uninhabitable for native lives. A place rendered hospitable in the short term might not remain so in the long term because requirements for human existence might be depleted, or have deteriorated due to planetary factors not previously understood.

In the latter regard, consequences of terrestrial agricultural practices should inform terraforming considerations, especially the introduction of hybrid seeds and crops and of GMOs (Genetically Modified Organisms) into the human food chain. Even if such agricultural inputs have been lab-tested and field-tested, they might not be able to survive climate change, insects, intrusive vegetation, altered water quality, droughts, and floods, or any combinations of the latter that occur in cycles far longer in time, or occurring infrequently, than the time during which hybrids and GMOs have been developed and tested in agricultural experiment stations. By contrast, most traditional agricultural plants have existed and been evolving over millennia or more, and have the genetic capacity—individual genes and gene combinations—to survive, at least in part, significant interruptions in their lives, as individuals or species, when these dramatic and infrequent natural challenges arrive. Here, too, terraforming impacts should be explored in greater depth and breadth prior to a rush to implement the latest technology purported to be able to benefit humankind.

The science-based *precautionary principle* should guide decisions as to whether or not a body in space should be subjected to terraforming, in whole or in part. The precautionary principle states that a course of action should be undertaken only if consideration of all possible outcomes has been made, using available data, and there is not only a likely preponderance of benefits over harms, but also no foreseeable likelihood of a catastrophic impact that might result from one of the possible harms, no matter how unlikely its possible occurrence; and the proposed course of action should *not* be undertaken if there is a preponderance of evidence that a catastrophic consequence is highly likely to occur; if benefits and harms are balanced but an imbalance might occur and enable a catastrophic event to occur; or if the potential consequences cannot be foreseen with any certainty but might be harmful and even extremely so. Lessons should be learned (and be constraining factors) from past human ideologies and practices that have resulted in social devastation, biotic extinction, and environmental destruction upon Earth.

Terraforming, then, has the potential to transform, for human habitation and use, near and distant planetary, lunar, and other celestial bodies. Its implementation, however, will have major biotic and abiotic consequences on and in the places where this occurs. Foresight should attempt to project possible positive and negative impacts of the implementation of terraforming, and care should be taken, to the extent possible, to promote the former and eliminate or substantially reduce the latter. Exploration and colonization, and the pioneering excitement which precedes them, should be tempered by thoughtful consideration so that Discovery Doctrine, in whatever form, is not exercised; its consequences in space would be significantly more harmful than they have been on Earth. Human visions for a human future in space, when the visions and the projects they will stimulate are informed by cosmosocioecological ethics, might well effect, in concrete contexts, further stages of the *utopia* that invites humankind into relatively better biotic and abiotic *topias* in the future.

Instruments in Implementation

The transformation of social relationships and of habitat terrain in exoEarth contexts (and possible transformation of the Earth context, should a dialogic relationship be developed between these distinct and diverse places of human habitation), would employ varied types of instruments for its implementation. These might include *scientific instruments, social instruments* (ideological, political and economic), *UN Instruments*, and *UN military instruments* (as a last resort, utilized only if necessary).

The scientific instruments would enable adaptation to place as a principal consideration, rather than adaptation of place to meet human needs and wants. They should exemplify appropriate technology for the contexts in which they would be used, rather than aggressive technology. The social instruments would include an essentially communal consciousness, oriented toward the common good of humans, other biota, and context rather than private or nationalistic appropriation to benefit the few; consensual governance, perhaps embodying a true democracy, to benefit intelligent life as an integrated, interdependent, and interrelated community; and economic structures intended to ensure that the needs of all are met prior to any distribution that would satisfy the desires of a select few individuals or groups. The UN instruments would include those discussed earlier, and any subsequent documents, that have been developed by representatives of diverse ethnic, racial, cultural, class, and political groups, upon which a consensus has been reached; the "majority rules" should not be operative, lest the least powerful be deprived of what they need more than what others need or want. The UN military instruments (or their exoEarth complement) should be used only defensively and as a last resort to protect intelligent beings' imperiled lives, by international (and perhaps by intelligent beings in inter-species coalitions') forces under a unified command—peaceful measures must have been tried first. Military means should not be used offensively for Discovery or other imperial purposes to obtain territory or natural goods, or subjugate native residents already at home and in place.

Terrestrial-Extraterrestrial Transformation

If the UN documents on terrestrial transformation and extraterrestrial transformation were implemented collaboratively on Earth and in the heavens, a dialogic relationship between contexts would effect a profound and socio-ecologically positive difference between present Earth, permeated politically and judicially by the ideology of the Discovery Doctrine and threatened with social and ecological devastation, and future Earth and any other human habitation in the cosmos. The transformed human consciousness and conduct that would in part precede and in great measure partner with efforts to progress beyond the present Earth topia, and eliminate any threat that the current devastated and devastating Earth topia might be replicated in space, would be beneficial not only for human and planetary well-being. It would also facilitate mutual respect in communication with extraterrestrial intelligent beings, at home and abroad. The result could be collaborative efforts to provide for the needs of all intelligent species in Contact.

9

Cosmic Contact

ETI and TI Encounters: Competition and Collaboration

Philosophers, scientists, and theologians have pondered for millennia the possible existence of worlds similar to Earth. Their interest has been piqued periodically because of sightings of unusual bright lights traversing the sky, seemingly not meteors or other natural phenomena which they previously had observed occasionally, and with which they were familiar. Their observations, or stories they heard about others' experiences, prompted scholars and ordinary citizens to reflect on and wonder about whether or not intelligent life exists in the cosmos other than on Earth. Debates continue to the present day because there is not yet compelling evidence available for either side to make its case completely to the satisfaction of the other—especially as sightings continue, and are monitored by governments worldwide who disseminate distilled information regarding Unidentified Aerial Phenomena to the news media and thereby the general public.

As stated in the Introduction, in order for our thought experiment to work well as we consider the possible implications and impacts of terrestrial-extraterrestrial intelligent life Contact, this chapter is intended to present data and narratives from credible witnesses who state that they have engaged, from near or far, UFOs/UAP or ETI. The information will be presented here "as if" the events have occurred as described; that should stimulate serious consideration of ETI-TI encounters and what they might mean for Earth and for biotic communities inhabiting Earth, and for representative humans aloft and away on space voyages.

Two events, from widely separated sites in the United States, will be discussed in depth: Roswell, New Mexico, and the Hudson River Valley, New York. Two other incidents from other US locales will be summarily discussed: Phoenix, Arizona, and Malmstrom Air Force Base Minuteman

3 ICBM Wing, Montana. Other events from around the world—the United Kingdom, Belgium, and Brazil—will be noted in passing.[1]

A Plurality of Worlds: An Historical Overview

A rapid perusal of conjecture regarding ETI since the Greek philosopher Epicurus (d. ca. 270 BCE) to the present day provides a glimpse of the extent of speculation about both the possible plurality of worlds and, should such worlds exist, about the kind of inhabitants they would have and the extent of their intelligence. Epicurus responded positively on both issues: there is a plurality of worlds, and intelligent beings likely dwell on them. He declared in a letter written to Herodotus that "there are infinite worlds both like and unlike this world of ours" because atoms exist in space in infinite numbers. He stated further that "in all worlds there are living creatures and plants and other things we see in this world."[2] Almost two centuries later the Roman philosopher-poet Titus Lucretius Carus (ca. 99–55 BCE) stated that since "empty space extends without limit in every direction," and "seeds innumerable in number" course through the cosmos, "*it is in the highest degree unlikely that this earth and sky is the only one to have been created*" (ital. in original).[3] Plato (pro) and Aristotle (con) had contrary views on these issues and, centuries later, Thomas Aquinas (1225–74) supported Aristotle's perspective.

Christian Theological Conflicts

A theological problem arose subsequently in Catholic Christianity: absolute denial of a plurality of worlds (as expressed by Aquinas) would mean denial of God's omnipotence, since God has the power to create multiple worlds. Three years after the death of Aquinas (who had taught at the University of Paris) the Archbishop of Paris, Étienne Tempier, issued the *Condemnation of 1277*. The thirty-fourth of the 219 propositions condemned (excommunication was threatened for those who continued to affirm any of them) any assertion that "the first cause [God] could not make several worlds."[4] The anti-plurality of worlds view, held by Aquinas and others influenced by Aristotle, now was deemed worthy of excommunication by the Church. The official hierarchical position then evolved to state that while a plurality of

---

1. These and other events are discussed in greater depth in Hart, *Encountering ETI*.
2. Crowe, *Extraterrestrial Life Debate*, 4–5. Crowe's book is a treasure trove of primary sources from two millennia that provide insights on the topics of multiple worlds and their possible biotic inhabitants and civilizations.
3. Ibid., 6–7.
4. Ibid., 21.

worlds is possible (because God could use divine power to create them), that does not mean they actually exist (God probably did not choose to use divine power to create them).

Theological difficulties did not end with the Condemnation and its subsequent amendment, however. One reason that the Church declared that the existence of plural worlds was possible but not probable was that the actual or even theoretical existence of multiple worlds raised questions about the meaning and uniqueness of Christian dogmas that taught that a "universal," "original sin" had resulted from Adam's and Eve's "Fall" in a Garden of Eden on Earth, as described in the book of Genesis and understood literally; God's consequent Incarnation in Jesus; and "universal" redemption of humanity (and even of creation, in some theologians' views) via Jesus' sacrificial atonement for humans' "Original Sin" in his crucifixion and death.

The Church then and since did not recognize (or at least did not express) that some of its own dogmatic assertions limited divine omnipotence. Among these is the doctrine that God could only or would only become Incarnate or embodied on one small world within the extensive plurality of worlds in the expansive (and now known to be the expanding, or more accurately inflationary) cosmos, and that God required, as expiation for humans' sins, the human sacrifice of an innocent person rather than the capital punishment of someone who had sinned grievously. Aspects of the Catholic Church's official position on this subject would play a role a few centuries later in the Church's condemnation of Galileo Galilei.

William Vorilong (d. 1464) a Franciscan thinker, was apparently the first theologian to confront directly the doctrinal contradictions noted. He stated that "infinite worlds" exist in God's mind, and might be available to human perception by means of angelic or divine communication. In response to questions regarding "original sin" he said that if there are so many worlds, the inhabitants were not descendants of Adam nor would have "sinned as Adam sinned," but would have been placed on paradisal worlds by God after transport similar to that of Enoch and Elias. Finally, regarding whether Christ's death "on earth could redeem the inhabitants of another world, I answer that he is able to do this even if the worlds were infinite, but it would not be fitting for Him to go into another world that he must die again."[5]

Nicholaus of Cusa (1401–64), in *Of Learned Ignorance*, considered the issues in more depth and with greater openness to diverse possibilities about the nature and place of extraterrestrial beings than did his theological and ecclesial predecessors. He reasons that

---

5. Ibid., 27.

Life, as it exists here on earth in the form of men, animals and plants, is to be found, let us suppose, in a higher form in the solar and stellar regions.... [W]e will suppose that in every region there are inhabitants, differing in nature by rank and all owing their origin to God, who is the centre and circumference of all stellar regions.... Of the inhabitants then of worlds other than our own we can know still less [than we do about inhabitants of Earth], having no standards by which to appraise them.[6]

Nicholas's last point is worth keeping in mind today as theologians and fervent believers of several Christian faiths offhandedly and even cavalierly express the anthropomorphic assumption that all intelligent life must have sinned (since humans did) and require redemption in order to have eternal salvation; and, since Jesus' sacrifice is for all sins in the world and worlds, extraterrestrials on Contact should be instructed in a Christian faith and baptized for their own, eternal good.

The last of the earlier Christian thinkers to be considered here, on the topic of the plurality of worlds and the inhabitants who might dwell in them, is Giordano Bruno (1548–1600). Born just five years after the death of Nicholas Copernicus (1473–1543), Bruno became an early and ardent follower of Copernicus's astronomical hypothesis that Earth was a planet in a heliocentric system. Bruno complemented and supplemented that idea with strong assertions concerning the existence of a plurality of worlds, and speculations about the forms life might take on them. A monk and scholar, his writings include *The Ash Wednesday Supper*, *On the Infinite Universe and Worlds*, and *Of the Immense and Innumerable*. Bruno's public enthusiasm for and advocacy of his ideas, some of which directly or indirectly questioned traditional Christian dogmas, caused the Inquisition to become interested in him and his works.

The *universe* is, for Bruno, a "world of worlds" (not his term): each *world* has everything comparable to all that is visible to humankind on Earth, including stars and planets in their spheres in the heavens above; a world is not solely a planet and its surrounding atmosphere. Bruno stated that "There are certain determined definite centers, namely the suns, fiery bodies around which revolve all planets, earth, and waters, even as we see the seven wandering planets take their course around our sun."[7] In the universe these are "all those worlds which contain animals and inhabitants no less than can our own earth, since those worlds have no less virtue nor a nature different from

---

6. Cited in Crowe, *Extraterrestrial Life Debate*, 31.
7. Michaud, *Contact*, 60.

that of our earth."[8] Rather than be theologically threatened by this understanding, Bruno declares: "Thus is the excellence of God magnified and the greatness of his kingdom made manifest; he is glorified not in one, but in countless suns; not in a single earth, a single world, but in a thousand thousand, I say in an infinity of worlds."[9] Bruno was concerned about the reception of his ideas, and lamented that there were adversaries who "dispute not in order to find or even to seek Truth, but for victory." His concern proved to be justified. In 1600, he was burned at the stake for his "heresies," his ideas and teachings that were judged by the Catholic hierarchy to have deviated from officially accepted Catholic doctrines.

Rigid Religion and Reflective Refutations

Religions' dogmatic assertions, when they face questions not previously considered, are consistent with humankind's typical anthropocentric and anthropomorphic speculations and projections about the being, nature, and attributes of God. Many religious believers' ideas and imagination are limited by their strict adherence to existing dogmas, ways of thinking, methods of approaching current and projected profound ideas, and the theological boundary lines which believers are forbidden by Church command or self-censure to cross. Similarly, humanist scientists have professional and social barriers that prohibit them from professing or even pondering aloud assertions about the existence or nonexistence of divine Being. Such societal-, 'scientific'-, and self-limitation imposes intellectual limits, enforces closed-mindedness, and allows for only limited and limiting material "tools" to be available to verify or falsify claims of divine existence, distinctive dimensions of reality, and projections based on the foregoing that are limited by current knowledge and societal factors.

Some scientists make dogmatic statements parallel to religions' dogmatic assertions when they declare unreservedly, unreflectively, and unabashedly that neither divine Being nor extraterrestrial intelligent beings exist ("show me the proof; where's the evidence?"). In so doing, such scientists mirror the manner of Christianity and other religions whose dogmas are based on preconceived ideas derived from a "tradition" developed in cultural and historical contexts during prior centuries or even millennia, and are therefore somehow unalterable or not amendable even when new data are discovered. By contrast, the simple, direct understanding of the Creator Spirit, as expressed by the Muskogee people through spiritual elder and traditional

---

8. Crowe, *Extraterrestrial Life Debate*, 43.
9. Ibid., 44.

healer Phillip Deere, provides a corrective for humans' attempts to define and contain divinity in a human conceived- and -constructed box, even while affirming God's infinity, omniscience, and omnipotence: "The Creator is 'The Great Mystery'; the 'Mystery' can be anywhere and everywhere."[10]

Phillip Deere's spiritual understanding contrasts not only with religions' dogmatic definitions of at least the characteristics if not the nature of divine being, but also with Ludwig Feuerbach's assertion of the way in which God has been conceived (literally) and elaborated in human thought. Feuerbach theorized, in his critique of religion, that ultimately all religions' declarations about divinity and divine characteristics are merely projections in human imagination of human feelings, hopes, and characteristics. Said Feuerbach: "Man—this is the mystery of Religion—objectifies his being and then again makes himself an object to the objectivized image of himself thus converted into a subject."[11] People project what a divine being should be, based on their thinking, ideals, and behavior, and worship that being. To rephrase Feuerbach, but generally express his idea: in a reversal of Gen 1:26, humans of whatever religious faith are to some extent claiming (Feuerbach would omit "some extent"), ironically and effectively: "Let us make God in our image, according to our likeness."

When confronted by and thinking about the Christian dogma that God "saves" humankind by Jesus' sacrificial atonement at his crucifixion, Thomas Paine was prompted to reject Christianity entirely. Paine was the fiery rhetorician for the thirteen colonies' 1776 uprising and revolution against England, and author of *The Rights of Man*.[12] After the revolution, primarily while in France, Paine wrote *The Age of Reason*. In it, he narrates how as a boy of seven or eight he listened to a relative who read aloud a sermon about Jesus' redemptive death. Paine continues:

> After the sermon was ended, I went into the garden, and as I was going down the garden steps (for I perfectly recollect the spot) I revolted at the recollection of what I had heard, and thought to myself that it was making God Almighty act like a passionate man, that killed his son, when he could not revenge himself any other way; and as I was sure a man would be hanged that did such a thing, I could not see for what purpose they preached

---

10. Conversation with this writer at Phillip's campsite during the International Indian Treaty Conference on the Sisseton-Wahpeton Reservation in North Dakota-Minnesota in the summer of 1984. Phillip Deere was the spiritual leader of the American Indian Movement and of the International Indian Treaty Council.

11. Harvey, "Ludwig Andreas Feuerbach."

12. Paine's ideas, discussed in chapter 1, provided an economic reality check on, and counterpoint to, the claims and theories expressed by Adam Smith, his contemporary.

> such sermons. . . . God was too good to do such an action, and also too almighty to be under any necessity of doing it. . . . [T]he Christian story of God the Father putting his son to death, or employing people to do it, (for that is the plain language of the story), cannot be told by a parent to a child; and to tell him that it was done to make mankind happier and better, is making the story still worse; as if mankind could be improved by the example of murder.[13]

Paine elaborated a reaction to Christian theology that illustrates a concern the Catholic Church had had centuries before as it asserted that an omnipotent God could create a plurality of worlds. The Church realized then that the assertion could raise questions about traditional dogmas. Paine did precisely that. After acquiring knowledge of astronomy, he accepted assertions about a plurality of worlds. He found this understanding to contradict essential Christian doctrine:

> Though it is not a direct article of the Christian system, that this world that we inhabit is the whole of the habitable creation, yet it is so worked up therewith, from what is called the Mosaic account of the Creation, the story of Eve and the apple, and the counterpart of that story, the death of the Son of God, that to believe otherwise, that is to believe that God created a plurality of worlds, at least as numerous as what we called stars, renders the Christian system of faith at once little and ridiculous, and scatters it in the mind like feathers in the air. The two beliefs cannot be held together in the same mind; and he who thinks that he believes both, has thought but little of either.[14]

Paine continues his narrative, noting how full of life the world of Earth is as a whole; that even a tree or a blade of grass, because of the myriad creatures inhabiting them, is a world, too; that since no part of Earth is without living inhabitants, "why is it to be supposed that the immensity of space is a naked void, lying in eternal waste?": the universe has room for "millions of worlds," each of which is "millions of miles" from the other: the Creator made not one but innumerable worlds in space;[15] and that the "inhabitants of each of the worlds of which our system is composed, enjoy the same opportunities of knowledge as we do."[16]

---

13. Paine, *Age of Reason*, 223.
14. Ibid., 224.
15. Ibid., 226.
16. Ibid., 228.

Paine rejects the idea that the Almighty, upon whom so many worlds depend, would stop the care of all worlds except one, and come to Earth to die in it; or that God would have to travel from world to world, dying on each one, to provide for each world's need for redemption.[17] Paine's consideration of a plurality of worlds leads him to conclude, similar to Bruno: "Our idea, not only of the almightiness of the creator, but of his wisdom and his beneficence, becomes enlarged in proportion as we contemplate the extent and the structure of the universe."[18]

Religions' Acceptance of a Plurality of Worlds and of Intelligent Life

The prospect of Contact, or even the possibility that it might already have occurred, might seem to be something that would be viewed as a threat to established religions. Yet, from evangelical Protestants to the Vatican, this is not the case.

Evangelical Christians' Considerations

As the debate about exoEarth life raged in earnest in the nineteenth century, prompted by both astronomical discoveries—facilitated particularly by the development of ever-larger and clearer telescopes—and by theological speculation and argumentation, a previously obscure Scottish evangelical, Thomas Chalmers (1780–1847) entered the fray in a big way. Ascending to the pulpit of the Tron Church in Glasgow, he preached a sermon on astronomy and extraterrestrials that generated substantial publicity. This catalyzed an enormous increase in attendance at his services thereafter: people were eager to hear the sermons that followed. The overflow attendance was stimulated not only by his ideas but also by his eloquence: he was an excellent, exhortative preacher. Distribution and sales of the sermons, published in 1817 as *A Series of Discourses on the Christian Revelation Viewed in Connection with the Modern Astronomy*, were extremely successful: within a year, the book went through nine editions as it sold twenty thousand copies. The popularity of Chalmers' sermons, and sales of his published works, continued for years thereafter.

Chalmers wrote that "Freethinkers" assert that "Christianity is a religion which professes to be designed for the single benefit of our world," inferring thereby that God "cannot be the author of this religion because God would not "lavish on so insignificant a field" so much attention as is

---

17. Ibid., 229.
18. Ibid., 228.

described in both the Old and New Testaments. In response, Chalmers declares that "Christianity makes no such profession."[19] He went on to say that "we can lay no limit on the condescension of God or on the multiplicity of his regards even to the very humblest departments of creation.... [I]t is not for us, who see the evidences of divine wisdom and care spread in such exhaustless profusion around us, to say that the Deity would not lavish all the wealth of his wondrous attributes on the salvation even of our solitary species."[20] He continues by asserting that "Christianity has a far more extensive bearing on the other orders of creation, than the Infidel is disposed to allow."[21] Chalmers describes the vastness of space and of distances separating celestial bodies, and declares, "Why should we think that the great Architect of nature, supreme in wisdom as he is in power, would call these stately mansions into existence and leave them unoccupied?"[22] Considering this further in light of stellar bodies, Chalmers states, "Each of these stars may be the token of a system as vast and as splendid as the one which we inhabit. Worlds roll in these distant regions; and these worlds must be the mansions of life and of intelligence."[23]

Chalmers asserts that it is not presumptuous to state that moral considerations are not solely present on Earth, but exist in other, distant regions of space. They "are occupied with people" in whom "the charities of home and of neighbourhood flourish," the "praises of God are there lifted up and his goodness rejoiced in," "piety has its temples and its offerings," and "'the richness' of the divine attributes is there felt and admired by intelligent worshippers."[24] Sometimes Chalmers seems to infer that only Earth inhabitants sinned. When discussing redemption he mentions "the single world which had turned its own way," an act for which the Son atoned,[25] and he suggests elsewhere that in the universe, "one secure and rejoicing family" existed; only Earth was an "alienated world" which "strayed," and had the universe's "only captive members."[26] Regarding the universality (literally) of Christ's redemptive act, Chalmers states that according to reason, "the plan of redemption may have its influences and its bearing on those creatures of God who people other regions, and occupy other fields in the immensity of

---

19. Crowe, *Extraterrestrial Life Debate*, 242.
20. Ibid., 243.
21. Ibid., 244.
22. Ibid., 247–48.
23. Ibid., 251–52.
24. Ibid., 254.
25. Ibid., 257.
26. Ibid., 259.

his dominions;[27] that to argue, therefore, on this plan being instituted for the single benefit of . . . the species to which we belong, is a mere presumption of the Infidel himself."[28]

Chalmers's eloquence and popularity helped to plant the seeds for Christians both to accept that intelligent extraterrestrials inhabit other worlds, and simultaneously to remain secure in basic tenets of Christian faith.

In recent years, as speculation increased that extraterrestrials might well be encountered on Earth or in the heavens, some evangelical Christians who reflected in a small way elements of the perspective of Thomas Chalmers, have declared on occasion that since Jesus, the Son of God, died on this Earth world for all intelligent life, then ETI from all worlds should be converted to Christianity and baptized so that they, too, might have eternal salvation.

Vatican-Related Statements

The Vatican has considered periodically, primarily in the twentieth century and into the twenty-first, the possibility that not only life but intelligent life exists not solely on Earth, but elsewhere in the cosmos. Its primary research in this field has been undertaken by the Vatican Observatory, established in 1891 by Pope Leo XIII. Statements on the topic by scientists associated with the Observatory have been released to Vatican-related and independent news media. The Vatican Observatory owns and operates two of the world's best telescopes, one on the grounds of Castel Gondolfo, the pope's summer residence in Italy, and the other on the campus of the University of Arizona, Tucson where it has its own Steward Observatory. Its priest-scientists, most of whom since its inception more than a century ago have been Jesuit priest-scientists, have been recognized around the world for their scientific expertise in their respective fields.

The Vatican Observatory and the Pontifical Academy of Sciences sponsored the Vatican Conference on Cosmology in 2005. In the Address of Pope John Paul II to the participants,[29] the pope states that the natural sciences, especially cosmology, have made humankind "much more aware of *our true physical position within the universe*, within physical reality—in space and in time. We are struck by our smallness and apparent insignificance, and even more by our vulnerability in such a vast and seemingly hostile environment. Yet this universe of ours, this galaxy in which our sun is situated and this

27. Note the similarity to Teilhard on this point.
28. Ibid., 258.
29. John Paul II, "Address."

planet on which we live, is our home." Consciousness of our Earth home and its cosmic location "inspires us, taking us out of ourselves and forcing us to look far beyond the limits of our unaided vision." Human study of "the universe in all its immensity and rich variety serves . . . to emphasize our fragile condition and littleness." However, at the same time, "*We are made in the image and likeness of God*. Thus, we are capable of knowing and understanding more and more about the universe and all that it contains" and engage in this quest for knowledge "with questioning reverence and with awestruck imagination."[30] While human exploration of the material dimensions of the universe benefits from the scientific quest, some aspects of human life are not similarly capable of examination but "draw our attention to the realm of the Spirit." Art and poetry, and the human desire for justice, peace, and wholeness, lead people to observe "*an interiority in the universe and particularly in human life*" whose aspects require the use of "the arts, the humanities, philosophy and theology."[31] Technology used by astronomy and cosmology helps humankind "*to put ourselves and everything else into a larger perspective*, encouraging us to move beyond our own narrow and selfish concerns. Our view of ourselves, of God and of the universe is radically different from that of people in the Middle Ages," because people see themselves located in a complex, intricate, and delicate universe. Viewing themselves from the outside, including from the moon and elsewhere in the solar system, people realize that they must be "*more responsible for ourselves, our neighbours, our institutions, and our planet*."[32]

In his words, John Paul II anticipates to some extent the theme permeating this writing: we must embrace a change of consciousness and of conduct, on our Earth home and for and in the cosmos we explore, in order to correct what we have done socially and ecologically that has harmed the biotic community, including humankind, and its abiotic planetary context. His advocacy of the arts, philosophy, and theology as needed for holistic knowledge provide, too, considerations for who should be on board Stephen Hawking's space ark and other space vehicles, and help humans to settle elsewhere in order to start anew—with a new perspective and practice. The Catholic Church certainly has come to have a different perspective from that held by itself and others as the "people in the Middle Ages." The *Proceedings* of the conference, too, relate to the major themes under consideration here.[33]

José Gabriel Funes, SJ, director of the Vatican Observatory, attracted international attention in 2008 because of his affirmative comments when

30. Ibid., par. 2.
31. Ibid., par. 4.
32. Ibid., par. 5.
33. Stoeger, *Theory and Observational Limits in Cosmology*.

asked about the possible presence of intelligent life in the universe. In an interview titled "The Extraterrestrial Is My Brother" in *L'Osservatore Romano*, the Vatican's semi-official newspaper, Funes responded to the interviewer's questions about conflicts between science and religion, the origin of the cosmos, and the role of astronomy and astronomers, particular those of the latter who were members of religious faith traditions.[34] Funes' responses to other questions attracted widespread attention. They were focused on the possibility of whether or not life existed elsewhere in the cosmos and, if it did, if some biota had evolved to become intelligent beings. Since Genesis spoke solely about Earth and the animals, men, and women that inhabited Earth, the interviewer wondered if that excluded the possibility of other worlds or of living beings on them. Funes replied that "the universe is formed of hundreds of millions of galaxies, each one of which is composed of a hundred million stars. Most of these, or perhaps all, might have planets. How could one exclude that life has developed elsewhere too?" When asked whether such life could be similar to or more evolved that humans, Funes responded, "It's possible. . . . Certainly in such a great universe one cannot exclude that hypothesis." The interviewer wondered if that would cause a problem with the Catholic faith. Funes said it would not, and continued:

> Just as there exist a multiplicity of creatures on Earth, there can be other intelligent beings created by God. This does not contradict our faith, because we cannot put limits on the creative freedom of God. Speaking as Saint Francis did: If we consider Earth's creatures "brother" and "sister," could we not perhaps speak also about an "extraterrestrial brother"?[35]

In his thought, beautifully expressed through its association with the words of Saint Francis, Funes affirms a common family of intelligent beings, no matter their place of origin or their evolved material appearance. All are "brothers" and "sisters."

Funes addressed the question of whether or not other intelligent beings needed redemption and replied that if they did, Jesus had wrought salvation for all, but that "If other intelligent beings exist, it's not a given that they would need redemption."

---

34. Funes, "L'extraterrestre è mio fratello." All quotations were translated by the author from the original Italian.

35. Ibid. *Come esiste una molteplicità di creature sulla terra, così potrebbero esserci altri esseri, anche intelligenti, creati da Dio. Questo non contrasta con la nostra fede, perché non possiamo pore limiti alla libertà creatrice di Dio. Per dirla con san Francesco, se consideriamo le creature terrene come "fratello" e "sorella," perché non potremmo parlare anche di un "fratello extraterrestre"?*

José Gabriel Funes, an Argentine Jesuit priest, astrophysicist, and Director of the Vatican Observatory, has stated, then, that it is probable that there are intelligent beings elsewhere in the cosmos, given the cosmic expanse and limitless divine creativity. Consideration of Contact with extraterrestrial intelligent life, therefore, offers an opportunity for reflection on the complexity and diversity of creation—and of humans' role to relate to and interact responsibly with other biota in the interdependent and integrated Earth community and interdependent and integrated cosmic community, wherever their space explorations take them.

The Vatican Observatory and the Pontifical Academy of Sciences cohosted a weeklong conference about extraterrestrial intelligent life in November 2009. It was the first Vatican-sponsored conference on the newly emerging field of astrobiology; thirty scientists prominent in the field were invited participants. In a press conference at its conclusion Chris Impey, Director of the Vatican's Steward Observatory and Chair of the Department of Astronomy at the University of Arizona, Tucson, commented, "If the universe is abundant with life there is companionship in our future." If there is no extraterrestrial life, "people will realize that this planet is very special and so with that will come an extra obligation even if we didn't already feel it to care for this place and this special thing that happened here."[36]

When Catholic bishops from around the world were called to participate in 2012 in the XIII Ordinary General Assembly of the Synod of Bishops, the Special Guest invited to address them was Werner Arber. He is a distinguished, internationally known scientist, professor of microbiology in the Biozentrum University of Basel, co-recipient of the 1978 Nobel Prize in Physiology or Medicine, and the president of the Pontifical Academy of Sciences (a role held in the past by Georges Lemaître), a position for which he was selected by Pope Benedict XVI in 2011. In his lecture, Arber discussed current understandings about both cosmology and evolution. In a section titled "Cosmic evolution and biologic evolution as facts that reveal important laws of nature," Arber stated that

> the ongoing processes of evolution of the Universe and of life are now solidly established scientific facts that serve as essential elements of permanent creation. In recent centuries and increasingly in recent decades... scientific investigations have revealed that our universe is of a tremendously large size and contains, besides a very large number of solar systems, also still mysterious, so-called dark matter and dark energy. And this entire complex, in which our planet Earth is just a minute component,

---

36. Glatz, "Vatican-Sponsored Meeting." The story was covered by numerous media sources, including the AP, Fox, and *U.S. News & World Report*.

is known to be in a slowly progressing, steady evolution.... At this time, we assume that life must also exist on some extraterrestrial planets, but we are still waiting for scientific information for this assumption.[37]

Arber *accepts* as facts that Earth orbits within a solar system (a sharp Vatican departure from the Catholic Church's attacks on and condemnation of Galileo Galilei, for which Pope John Paul II apologized in 1992) and that biological evolution on Earth is a fact (a notable change of position since Charles Darwin published *On the Origin of Species*, and the Church had debated condemning him and his theory), and *assumes*, while awaiting scientific confirmation, that biological evolution has been occurring on other worlds in the cosmos and that this includes the evolution of intelligence.

On the latter issue, Arber (and, by extension, the Catholic Church and Benedict XVI, who had appointed him to head the Pontifical Academy of Sciences) indicates a third distinct area in which the theology and scientific ideas of the Catholic Church had evolved over centuries. The Church had been ambiguous since the time of Thomas Aquinas about the possibility, meaning, and theological impacts of a plurality of worlds, and had burned Giordano Bruno as a heretic for having forthrightly and firmly taught that there was indeed a plurality of worlds and that God might well have chosen to be Incarnate on them as God had chosen to be on Earth. Through Arber's words (not refuted or even questioned publicly by the Vatican), and through Funes' words in an interview with *L'Osservatore Romano* four years earlier (comments also not called into question by the Vatican), the Catholic Church signaled that it accepted at least a working assumption of exoEarth intelligent life (it might well have its own data on UFOs from its telescopes, gathered quietly and secretly over decades: J. Allen Hynek's secret survey of fellow astronomers to satisfy his curiosity about their observations of UFOs, taken when he worked on Project Blue Book, disclosed that 11 percent had seen UFOS, far above the general public percentage[38]). Possibly, prompted by increased scientific data, the Church has had theologians working behind the scenes to develop revised Catholic teachings and even doctrines that explore the implications of extraterrestrial intelligence for Christian believers on Earth, and to suggest possible means of religious or spiritual engagement with ETI after Contact on Earth or in the heavens.

---

37. "Intervention of the Special Guest, Prof. Werner Arber."
38. Hynek and his work are discussed in chapter 10.

UAP: National Governments' and Credible Witnesses' Narratives

Over millennia, as noted earlier, stories and songs have recounted numerous tales of UFO/UAP events, and possible TI-ETI encounters. While seven incidents from around Earth will be described, most of them briefly, the principal focus will be on two extensively described, analyzed, and discussed events in the United States:[39] narratives about a crash of an ETI craft on a ranch near Roswell, New Mexico, and numerous sightings of UFOs/space vehicles over the Hudson River Valley in New York State.

UAP activity in all of the locations was reported by credible witnesses. The US government has denied the veracity of all UFO/UAP reports, and used ridicule and pressure from federal agency representatives to suppress stories, coerce witnesses, or call into question the motives and character of witnesses. Despite these efforts, the elaboration of these events in newspaper accounts and books continues to mushroom. The factual accuracy of the narratives presented will not be debated here. Rather, the incidents will be used to stimulate further questions that would catalyze deepened engagement with UAP/ETI possibilities: What would be the implications and impacts of ETI Contact—whether it has occurred already or, if not, it occurs in the future? What might be possible ecological, economic, ethical, and ecclesial consequences of Contact?

A distinguishing feature of seemingly factual UFO accounts is that they present narratives from people who would be considered "credible witnesses" in a court of law. In *UFOs: Generals, Pilots, and Government Officials Go on the Record*, former NPR reporter Leslie Kean published a well-researched book that includes interviews with or statements by people who would be considered "credible witnesses."

In the Introduction, Kean discusses the COMETA Report, a ninety-page document released by the government of France; it was prepared by thirteen retired generals, scientists, and space experts, all working independently of the government. These consultants determined that 5 percent of incidents cannot be attributed to earthly sources: the narratives concerning them describe "completely unknown flying machines with exceptional performances that are guided by a natural or artificial intelligence." The COMETA Report's conclusion: "numerous manifestations observed by reliable witnesses could be the work of craft of extraterrestrial origin"; the most logical explanation for the "manifestations" is "the extraterrestrial hypothesis."[40]

Kean notes that about 90 to 95 percent of UFO sightings can be explained. She declares that "to approach UFOs rationally, we must maintain

---

39. In Hart, *Encountering ETI*, narratives concerning other sites will be analyzed.
40. Kean, *UFOs*, 1–2.

the agnostic position regarding their nature or origin, because we simply don't know the answers yet,"[41] and adds that "principled skepticism is the foundational premise of this book."[42] Kean's premise complements that of *Cosmic Commons*, and her objective presentation of material she assembled provides data useful for the thought experiment undertaken here. Neither a *pro* nor a *con* position will be taken on the issue of UFOs but, in order to stimulate in-depth and ongoing consideration, government documents, government and military officials' statements, and UAP citizen witness accounts will be presented as credible narratives.

Kean states that national governments around the world officially, albeit secretly, began studying the UFO phenomena beginning at least in 1950 (earlier in the United States): the United Kingdom, Chile, Peru, France, Brazil, Russia, Mexico, Uruguay, Ireland, Australia, Canada, Denmark, Sweden; many have released once secret files, beginning in 2004.[43] Her book contains numerous photographs and witnesses' drawings that illustrate the text, including several photos from diverse nations' military personnel.

International UFO Events

UFOs have been observed throughout the world. J. Allen Hynek discusses several events he investigated globally. These will be described in more detail in the next chapter, and in more depth in *Encountering ETI*. Only two international sites of encounters will be described below.

United Kingdom

In May, 2008 the British government, as a result of a Freedom of Information Act request, released extensive documents from the Ministry of Defence (MoD) about reports, and investigations of reports, regarding Unidentified Flying Objects (now also called Unidentified Aerial Phenomena/on, UAP) viewed flying over or landing on British soil. The British Broadcasting Corporation (BBC) examined these files and those released in subsequent years and provided accounts of their contents. The files discussed MoD efforts to track and document appearances of UFOs as reported by British citizens, military personnel, and at least two United States Air Force officers—one from his fighter plane, and another at an English base used by the USAF—from 1978 to 1987. The BBC perused the files and did two reports in 2008:

41. Ibid., 12.
42. Ibid., 13.
43. Ibid., 116–17.

one published (and at one time available) online, and the other an audio recording of a televised news broadcast.

The BBC news article published on the BBC Web site on May 13, 2008, is now unavailable. The subheadline was "Secret Files on UFO sightings Have Been Made Available for the First Time by the Ministry of Defence."[44] The article stated that the UFO files "include accounts of strange lights in the sky and unexplained objects being spotted by the public, armed forces and police officers." One file stated that "the United States Air Force filed a report about two USAF policemen who saw 'unusual lights outside the back gate at RAF Woodbridge'" in Suffolk in 1980. The latter event gained increasing notoriety as the "Rendlesham Forest Incident," and is discussed extensively in Leslie Kean's *UFOs* and in my *Encountering ETI*. The most startling content in *UFOs* includes Kean's transcription of a live recording of USAF Colonel Charles Halt's description of an alien craft approaching the RAF base where he was Deputy Commander, and photocopies of drawings and notes that Sergeant James Penniston, Security Police squadron member, made in his logbook at the time—including those regarding a landed space vehicle.

Belgium

In Belgium, Major General Wilfried De Brouwer was Head of Operations of the Belgian Air Staff during 1989–90, when extensive UFO sightings occurred. In his statement to Kean, he ruled out the possibility that Earth-based craft were seen. He declared that "I am approaching the UAP issue in a pragmatic way. I stick to the facts and avoid extrapolations to possible extraterrestrial activities." He added: "Nevertheless, I encourage scientific research which should be based on the objective analysis of a number of observations reported during the Belgian wave. Such research should not exclude the extraterrestrial option."[45] De Brouwer is unlike most of his military or government counterparts in the United States in two ways: he suggests that scientific research should be undertaken, focused on UAP events; he is open to scientific findings that indicate that the vehicles observed, probably having origins in space, were under extraterrestrial intelligent control.

Kean provided background data for the events De Brouwer considers. She states that "of the approximately 2,000 reported cases registered during the Belgian wave, 650 were investigated and more than 500 of them remain unexplained.... The findings were exceptional. More than 300 cases involved witnesses seeing a craft at less than 300 meters (1,000 feet), and over 200

---

44. BBC, "MOD Releases UFO Files." Searching the BBC site for this article by its title calls up no video, audio, or print file. Instead, the visitor is given a link to the audio file.

45. Kean, *UFOs*, 22.

sightings lasted longer than five minutes. Sometimes observers were right underneath the craft."[46]

United States Events

Since the nineteenth century, UAP have been sighted in every region of the United States by numerous credible witnesses, some of whose accounts were published in local or regional newspapers and, on rare occasions, broadcast on national media. The focus here will be on sites in four states: New Mexico, New York, Arizona, and Montana.

Roswell, New Mexico

Reports from New Mexico in 1947 about the crash landings in the Roswell-Corona area of two extraterrestrial, technologically advanced spacecraft, possibly controlled by extraterrestrial beings, one of which was piloted or at least occupied by ETI, are well known internationally. They are among the most startling UAP narratives available to date.

On the Foster Ranch located near the town of Corona, some seventy-five miles from Roswell, ranch foreman William "Mac" Brazel found debris that had resulted from a mid-June crash of one, or crashes of two, aerial phenomena. (Two conflicting stories collide at this point: Brazel saw remnants of either what he took to be two technologically advanced US military aerial apparatuses of some sort, or of two extraterrestrial space vehicles and the occupants of one, three of whom seemed to be deceased and one of whom was still alive, but injured.) Brazel contacted the nearest US military installation, the Roswell Army Air Force base—which, at the time, was the only military base in the world with atomic bombs and the bombers capable of dropping them, and included the crews and planes that had dropped the atom bombs on Hiroshima and Nagasaki. Base commander Colonel William Blanchard responded by sending his top-ranked officers who had security responsibilities for the base to investigate: Major Jesse Marcel Sr., base intelligence officer; and Captain Sheridan Cavitt, Counter Intelligence Corps (CIC) agent.

Marcel and Cavitt surveyed the scene, gathered fragments from the crash site, and returned separately to the base. Marcel put boxes of debris in the back seat and trunk of his family car, and stopped at his home that night on his way back to the base to show his wife, Viaud, and their son, Jesse Jr. what he had found. The family all handled the material, and Jesse Jr. discovered geometric shapes inscribed on an I-beam among the debris.

---

46. Ibid., 34.

Major Marcel continued on to the base, and presented to Colonel Blanchard the debris and his evaluation of it. Colonel Blanchard had the base public relations officer prepare and release a news story which stated that a "flying disc" had landed near Roswell. The press release startled media representatives and stunned members of the general public who saw the subsequently published original story.

As word began to spread about the "flying disc," the US government acted swiftly through the highest level military command in the region, represented by General Roger Ramey at the Fort Worth Army Air Field in Texas. Ramey ordered that the debris and Major Marcel be flown to him in Fort Worth. He quickly called a press conference about the "flying disc" event. At the press conference, Major Marcel was present and had been ordered to kneel on one knee next to the remnants of a weather balloon and hold its radar target. Ramey stated that the ranch crash debris came from a Top Secret Mogul Project weather balloon, held by Marcel and on the floor next to him. Major Marcel had been ordered not to speak then or at any time in the future about the Roswell matter; he followed orders, even though the combination of events made him appear to be less than fully competent in his professional capacity as the intelligence officer at the world's only atomic bomb base, where his complementary responsibilities included, among other things, investigating debris from crashed civil and military aircraft. Marcel continued in his air force career—he retired as a colonel—and continued, too, to be conscious of what he had seen, how his professional reputation had been harmed, and how he remained under a shadow for the rest of his career and in retirement as a result of Air Force and US government denial of what had transpired at Roswell and in other UFO events.

Over the years, the US government has tried to suppress discussion of Roswell in the media, especially through US Air Force press releases and press conferences and, ultimately through ridicule of people who researched and discussed seriously this and other narratives about UFOs in US airspace, extraterrestrials on US soil, and other incidents around the world. As questions arose about the supposedly crashed Mogul weather balloon, and then about stories describing alien bodies, the Air Force over several years issued five contradictory statements to the media about crash debris to try and erase media and public interest in Roswell. The USAF finally and arbitrarily declared that the matter had been settled—there were no UFOs other than natural phenomena that could be scientifically explained—and it would no longer discuss Roswell. Over the years, too, new data emerged as witnesses to Roswell-related events or their families or colleagues produced and publicized information not previously available publicly.

Colonel Jesse Marcel Jr., MD, son of Colonel Jesse Marcel Sr., is recognized today as a key witness to Roswell area events by those who accept the general narrative as historically accurate overall. Dr. Marcel is a retired Helena, Montana physician (an otolaryngologist, or ear, nose, and throat specialist); a veteran of the first US-Iraq war, in which he was a flight surgeon and a helicopter pilot who flew more than two hundred combat missions during a thirteen-month period in a 2004–2005 tour of duty;[47] and a retired colonel in the Montana Air National Guard. He continued to be very interested in Roswell and its aftermath (and apparently was reading *Encountering ETI*, to write an endorsement, when he died unexpectedly in 2013). He traveled extensively and gave numerous media interviews to correct false statements about the role and message of his father, then-Major Jesse Marcel Sr., who had investigated the Roswell event as the ranking security officer of the nearby atomic weapons facility, the Roswell Army Air Field (RAAF; at that time the separate branch of the military known now as the US Air Force had not yet been established).

Dr. Marcel pledged to his father, when the latter was on his deathbed, that he would correct the misinformation distributed by US government agencies and the United States Air Force about his father's professional capabilities and conduct. To that end, Marcel Jr. published *The Roswell Legacy: The Untold Story of the First Military Officer at the 1947 Crash Site*, written in collaboration with his wife, Linda Marcel; the book is dedicated to his father. Jesse Marcel's firsthand account of Roswell events is the principal source for discussion of Roswell in this chapter.

Marcel Jr. writes that "mine is a story of actually seeing and handling artifacts from the site."[48] He declares that the answer to the question, "Are we alone in the universe?" is "an emphatic no."[49] He includes in the book several photos not only of the principals in the Roswell event, but also of Air Force documents that verify his father's expertise in intelligence matters. Time and again, Marcel Jr. describes his father's proficiency in the areas needed to assess, as an intelligence matter, debris from crash sites of weather balloons or diverse types of aircraft; he establishes clearly that his father's training, the top secret assignments he was given, his successive promotions, numerous awards, and medals, and the highly positive evaluations and commendations that he received before and after the Roswell crash,[50] all indicate the regard in which he was held by his commanding officers; a major when the Roswell

---

47. Marcel and Marcel, *Roswell Legacy*, 23.
48. Ibid., 24.
49. Ibid., 25.
50. See especially the list in Marcel and Marcel, *Roswell Legacy*, 42–43.

crash occurred, he had been promoted to lieutenant colonel in the reserves by the time he retired from the military.

Major Marcel's responsibilities as an intelligence officer at the RAAF included investigating "aircraft accidents, or any problem that arose with security."[51] This was a highly important assignment, since the base had atomic weapons and was integrated within the Strategic Air Command (SAC).

Marcel Jr. describes in detail and for several pages[52] his recollections of the night in 1947 when his father came home, briefly, carrying several boxes of Roswell crash debris (three different types of material) that he was taking to the RAAF. The debris included a piece of an I-beam made of some kind of metal, and the boy Marcel noticed "something unusual about the inside surface"—it had a distinct color and, upon closer investigation, what appeared at first to be writing, but then was seen as a line of "geometric symbols . . . [which] seemed like part of its surface."[53] There were some thirty symbols which were "solid," and "not line drawings." Marcel Jr. states that his father later described to him something not in the boxes that he had brought into his house: "fine strands resembling fishing line," which might have been a form of fiber optics.[54] His father also told him, years later, that he had not seen any alien bodies, but had heard reports about them. After RAAF commander Blanchard issued the "flying disc" crash press release[55] and Ramey's press conference in Fort Worth, by July 10 all of the Roswell debris had been removed from the Foster Ranch and transported by a massive C-54 cargo plane to Wright-Patterson Air Force Base, Dayton, Ohio. Several witnesses, including the C-54 pilot, Captain Pappy Henderson, later reported seeing bodies carefully packed in one of the large crates that were placed on board the aircraft.[56]

Marcel closes *The Roswell Legacy* by expressing the hope that readers, as he was, will be "touched by the truth of otherworldly civilizations," and "look objectively, hopefully, and perhaps even lovingly upon the promise that future contact holds of humanity." Humans should "act responsibly when faced with evidence that the universe is big enough to be home to a great number of civilizations . . . to deny this would be to place human limits upon God's infinite capacity for creation."[57] (This twenty-first-century statement complements earlier ideas expressed by Giordano Bruno in the sixteenth century,

51. Ibid., 50.
52. Ibid., 51–60.
53. Ibid., 57.
54. Ibid., 59.
55. Ibid., 65.
56. Ibid., 73.
57. Ibid., 164.

Thomas Paine in the eighteenth, Thomas Chalmers in the nineteenth, and Teilhard de Chardin in the twentieth.)

In 2012, Jesse Marcel Jr. reflected back on his childhood experience, recalling his vivid impressions of handling the debris and his thoughts at the time. He reflected:

> In 1947 existence of planetary systems was only a theoretical possibility but recent discoveries have found that planetary systems are like grains of sand on a beach. When I saw the Roswell debris, I was pretty sure that there were other earthlike planets out there populated by intelligent civilizations, so I was gratified later that telescopes like the Kepler have proven that what I was thinking all along was true. There is no proof at present that would satisfy skeptics as to the reality of the fact that there are one or more civilizations that are aware of us and indeed have sent probes to our planet much like what we are doing in our solar system, but that will come with time. People have asked if the Roswell event altered my religious feelings, and I have to say that if anything it confirmed my belief in Intelligent Design and God. Someone once said that if we are all there is, there is a lot of wasted space out there.[58]

Through the years, once the Roswell crash came to the forefront in the public imagination, innumerable books and articles were published on that incident.[59] Only four of the most prominent, well-researched works written by credible writers will be mentioned here.

Retired Lieutenant Colonel Philip J. Corso authored *The Day After Roswell*, which focuses on decades of military research on the Roswell-Corona crashes of UFOs. What makes his narrative unique is that Corso is to date (as far as this author could find) the most senior military officer to acknowledge not only that there were bodies recovered from the Corona site, but that he had seen an alien body in temporary storage in Kansas soon after the Roswell crash—"a four-foot human-shaped figure with arms, bizarre-looking six-fingered hands—I didn't see a thumb—thin legs and feet, and an oversized incandescent lightbulb-shaped head. . . . [I]ts facial features . . . were arranged absolutely frontally"[60]—and, twenty years later, read a top secret report, in a hidden Pentagon file cabinet, that contained a copy of the Walter Reed Hospital pathologist's autopsy of an alien cadaver from

---

58. Jesse Marcel Jr., e-mail message to author, October 31, 2012.

59. Others of the most prominent, well-researched, and well-recognized books by credible writers, as noted earlier, include Friedman and Berliner, *Crash at Corona*; Carey and Schmitt, *Witness to Roswell*; and Friedman, *Top Secret/Majic*.

60. Corso, *Day After Roswell*, 34–35.

the crash site.[61] Corso documents government efforts to suppress and cover up all UFO reports, and to use ridicule against those who discuss them in order to discredit witnesses and dismiss incidents. The book includes UFO surveillance photos[62] in Army intelligence files that were used to advocate support of Army research and development efforts focused on the reverse engineering of Roswell debris; appendices contain copies of previously classified military documents that analyze reports about and implications of UFO sightings, and the military's proposal, complete with drawings, for a military base on the moon, "Project Horizon," to monitor and defend against extraterrestrial space craft[63] (all aliens and their craft, Corso says throughout his book, should be regarded as hostile, as threats to human civilization, because of their superior technology and accounts of alien abductions and cattle mutilations).

The "Roswell Incident" remains very much alive in public consciousness, stimulated especially by revelations from those who once had significant experiences with the event itself or with credible witnesses who were significant participants in or observers of the activities at the Foster Ranch, the Roswell Army Air Field, Wright-Patterson Air Field, or involved in or aware of military maneuvers and policies to suppress and silence UFO discussion.

Hudson River Valley, New York

The most extensive description about events in the Hudson River area is given in *Night Siege: The Hudson Valley UFO Sightings*, by J. Allen Hynek, PhD, an astrophysicist and astronomer who was for two decades the scientific consultant to the US Air Force when it was officially investigating UAP events; Philip J. Imbrogno, a science educator and author of books on UAP topics; and Bob Pratt, a journalist and author of books about UAP. In *UFOs*, Kean discusses briefly the Hudson River Valley UFO reports.[64]

Kean and the authors of *Night Siege* all observe that surprisingly, in contrast to other UFO events, there has been no shortage of credible witnesses willing to come forward and describe their experiences with UFO incidents in the Hudson River Valley. At a Brewster, New York, UFO conference in 1984 and in its aftermath, Hynek et al. report that "so many professional people—people who normally would not speak out in public about seeing a UFO because it would damage their reputations—were now going on record

---

61. Details from Corso are discussed in depth in *Encountering ETI*.
62. Corso, *Day After Roswell*, following 180.
63. Ibid., 305–62.
64. Kean, *UFOs*, 158–61.

as having witnessed whatever it was in the sky."[65] The authors—especially Hynek because of his previous role as Air Force civilian scientific consultant—are well aware of government efforts to suppress UAP information, and to use false data, ridicule, and denial to coerce members of the public into being fearful of social and professional consequences should they acknowledge any experience with these phenomena or, in some professions, merely mention an interest in the subject. People who are credible witnesses because of their profession or educational background are especially vulnerable to reprisals, and often practice self-censorship as a result. During the research and writing of *Night Siege*, hundreds of witnesses came forward to speak with the authors, witnesses buttressed in part because there were thousands of witnesses and newspaper stories reporting events; the witnesses had achieved a certain comfort level that enabled them to offer their testimony. Even so, an undercurrent of fear was present in some cases, primarily because of possible employment repercussions, to the extent that several witnesses related what they had seen but requested that their name and, for some, even their occupation, not be used in the book.

Night Siege describes Hudson River Valley UFO sightings chronologically; dates of the events described follow sequentially. Where media reports are available, these are cited to corroborate the testimony of witnesses. It was obvious that local newsrooms found stories credible; for some that was easier since some of their own staff had been observers. (National media, however, paid scant attention to this and other stories reported and published locally.) The book includes photos taken and videos recorded by witnesses, including by an officer of the Connecticut State Police, near I-84. Representative stories follow.[66]

On March 24, 1983, between 8:00 and 9:30 P.M., hundreds of people in the Hudson River Valley, most of them near Yorktown, New York, observed a low-flying, boomerang-shaped UFO with red, green, and blue lights. Although the Yorktown Police Department acknowledged receiving hundreds of calls, and although police investigated, afterward the police initially told people that they could not provide details because of the "sensitivity of the situation," and then said that people had seen planes in formation. Later one of the police officers called the authors who were investigating the phenomena and told them that he and others did not accept the official explanation because they knew that the object was not several planes in flight. He stated, "I saw the thing, and it hovered directly over my head for five minutes—and

---

65. Hynek et al., *Night Siege*, 150.

66. The Hudson River Valley incidents are described more extensively in Hart, *Encountering ETI*.

airplanes don't hover!"[67] The aircraft explanation—including an astounding statement that what witnesses had seen were ultralight aircraft flying in formation when there were very dangerous high winds—became the official FAA response to area UFO reports. There were a great number of occurrences for two days. Witnesses who went on record[68] included Hunt Middleton, a New York corporate executive; Ed Burns, an IBM program manager, who said that "if there is such a thing as a flying city, this was a flying city. . . . It was huge"; Bill Hele, chief meteorologist for the National Weather Corporation, who stated that "I have been around planes for the past twenty years—and . . . I realized that this thing did not have the appearance of any known object or anything similar to an airplane or group of airplanes that I'd ever seen"; John Piccone, Grumman employee who worked with numerous types of aircraft and spacecraft; Joe Trongone, mechanical engineer, who designs small aircraft and who was sure that the UFO was not an airplane; Albert Silbert, a physicist and high school science teacher, and his family, who did not disclose what they had seen until months later, fearing ridicule; and Robert Golden, an IBM executive.

Although government officials and government agency staffs continued to offer the light aircraft explanation, "a number of people who saw both the UFO and later the formation of planes said the planes looked nothing like the object."[69] The authors note that "The caliber of people who were willing to go on record as having seen this UFO was remarkable. Many of these people were professionals who had nothing to gain and much to lose by reporting they had seen a UFO—scientists, engineers, doctors, lawyers, pilots, police officers, and many other people who are stable, respected residents of the area."

On March 25, 1984, hundreds of people in and around Peekskill, New York, viewed a large, triangular- or boomerang-shaped UFO. On the following day, the daily newspaper for a several county region in New York, the *Reporter Dispatch*, carried this headline: "The UFOs Are Back and They're Right On Schedule." Once again, the Federal Aviation Administration assured residents that they had seen stunt pilots; this angered people who had called the police department, since they knew otherwise. One witness responded, "I am an educated person. Don't you think I can tell the difference between planes and something strange in the sky?"[70] David Boyd, a pilot who saw lights in the sky approximately eight hundred to one thousand feet overhead, thought at first that he might be seeing ultralight aircraft but was

67. Hynek et al., *Night Siege*, 17.
68. Ibid., 29ff.
69. Ibid. 51.
70. Ibid., 78.

doubtful. Then he saw the lights make a level turn (without banking). At home, he called the police and was told he had seen ultralights in formation: "I questioned that because I don't know many guys who fly ultralights in formation.... [Moreover] it was moving ... much slower than anything I have ever flown. Anything flying that slow would drop to the ground."[71]

On June 11, 1984, the headline in the *Reporter Dispatch* read: "UFO Buzzes New Castle." The news story elaborated:

> Once again mystery invaded the night skies of Westchester and Putnam Counties. The strange object was flying and it was unidentified.... Whatever it was, it disrupted the New Castle town board work session and caused dozens of other people to look up and wonder. New Castle Police Sergeant George Lowert burst into the board room at 9:35 P.M. to announce a UFO was cruising over the building. Within seconds, the entire board and a dozen spectators were scampering down stairs to catch a glimpse of the lights.[72]

The article reported further that a Westchester County Airport control tower operator saw an object on radar at 9:30 P.M. Later, in an interview with *Night Siege* authors, Lieutenant George Lowery, who had been the desk sergeant described in the *Reporter Dispatch* story, elaborated further on what he and other police officers had seen. Although the FAA issued its standard "airplane" line, Hynek et al. observe, "A police officer wouldn't interrupt a public

---

71. Ibid., 86–87. Complementarily, regarding the craft's sharp turn, witness Mark Galli (who saw UFOs in both 1983 and 1984, one year apart) said that "it made a sharp right-angle turn to the northeast in one quick motion." Coincidentally, I had a similar experience. I saw an object do the same maneuver in the Hudson River Valley in 1963, two decades prior to the sightings highlighted in *Night Siege*. While a college junior, sitting on the bank of the Hudson River near Poughkeepsie, New York, with two college classmates on a crystal clear night, I viewed the stellar splendor above. Unexpectedly, I saw the brightest, largest, and longest meteor that I had ever seen moving extremely fast south to north above the course of the river. Suddenly, without banking or slowing, it went perpendicular as it shot straight up into the sky at the same velocity, and vanished after being visible briefly on its new trajectory. I blinked, said nothing, but thought to myself, "I've studied physics; that's impossible to do: it should have shattered attempting that change of direction at that speed." After a moment, one of my friends said, "Did you see that bright light go suddenly at right angles to itself?" I said that I had, and one of us commented, "I guess we saw a UFO." We laughed, and I did not think further about the incident. A quarter-century later I was standing near the cash register of a bookstore in Helena, Montana, awaiting my son who was selecting a book from the children's section in the back. There was a stack of a certain new book on the counter; it concerned UFO sightings in the Hudson River Valley during the years 1962–65. I picked one up, skimmed through it, and said to myself, "I was there." To this day, I regret not buying a copy.

72. Cited in Hynek et al., *Night Siege*, 89–90.

meeting of the town board just to have the officials run outside to see some planes."[73]

On July 12, the largest number of sightings to date was reported. Witnesses included Ed Mulholland, engineer for the Perkin-Elmer Corporation, which "designed and built the optics and some of the electrical components for the NASA Hubble Space Telescope"; and Dr. Richard Long, university law professor.[74] On July 24 Bob Pozzuoli, a major electronics firm vice president, videotaped a UFO. An ABC technician reviewed the tape and said, "I've never seen anything like it before. Every time I watch it, I get goose bumps. It's weird." Next, Dr. John Baker, West Coast University, California, analyzed it, said it was not an aircraft, and withdrew from the analysis. In 1985 HBO wanted to use the tape for a television series, and sent it to Dr. Al Hibbs at the Jet Propulsion Laboratory, Pasadena, California. Hibbs used a highly sophisticated computer that "was used for imaging data from the Voyager and Viking space probes, which sent back pictures of Mars, Jupiter, Saturn, and Uranus. Hibbs concluded that the object in the Pozzuoli tape could not be identified," and this continues to be its status

On July 24, 1984, the same evening that Bob Pozzuoli had shot his video in Brewster, twenty miles away, a major "close encounter" occurred above the Indian Point nuclear reactor on the Hudson River. A New York State Power Authority police officer on security duty at the reactor complex called the authors to say that he and eleven other officers saw the UFO when on duty. A "giant UFO" had hovered three hundred feet above a reactor "for more than ten minutes." The guard said that the shift commander "was ready to order the guards to shoot the UFO down." When interviewed by the authors, with the reluctant permission of Indian Point administrators and under the watchful eyes of a security supervisor, "Carl" (a pseudonym) said that he and others, including Con Edison utility personnel, had seen a UFO on both June 14 and July 24. On June 14, the UFO appeared to have a boomerang shape. He estimated it was about three hundred feet in length, and moved at approximately ten miles per hour. He declared that "no small planes could stay in formation with the wind that night. The wind didn't faze these lights at all. When it hovered, it just stood there. I was in the service and I flew helicopters, and I know how hard it is to keep formation with small planes.... . [T]he lights were much too intense for a small aircraft." On July 24, a guard yelled, "Hey, here comes that UFO again!" after which two supervisors and three officers ran out to observe. It hovered only thirty feet above the reactor. A camera on a ninety-five foot pole was directed to videotape the UFO, which was "bigger than a football field," for some fifteen minutes. Usually,

73. Ibid.
74. Ibid., 104–15.

this camera and others constantly film objects near the reactor, but authorities said that no film exists of the event.[75]

Night Siege contains numerous other witness accounts (including the extreme speeds with which the UFOs flew even after beginning from a standing start) and government intrusions into the authors' efforts, which included a veiled death threat from a National Security Agency representative when he was not given a tape he wanted; and probable telephone taps.[76]

Arizona

In 1997, hundreds of Arizona residents in different locales reported observing a huge UAP low overhead, flying slowly and silently.[77] Concerned or alarmed Arizonans called the offices of Governor (1991–97) Fife Symington III. Governor Symington held a press conference the following day, and announced to the news media that Arizona police had apprehended the perpetrator of the event. State police brought into the room a costumed staff member dressed as a stereotypical alien replete with a large bald head and almond eyes: one of Symington's aides, handcuffed, was presented as the culprit behind the sightings. The press present for the meeting laughed. Symington dismissed the incident as a hoax, and, despite the stories of numerous people, the Arizona press did not follow up on the story. Probably no reporter dared to do so because of the ridicule, directed toward witnesses or curious members of the general public, that usually followed credible reports of UFO sightings.[78]

On the tenth anniversary of the UFO sightings, a documentary filmmaker decided to interview Arizona citizens and former governor Symington. When the governor entered the room where the camera was set up, before the questioning could begin, he stunned the interviewer by stating that a decade earlier he had seen the UFO. He elaborated in an on-camera interview, during which he noted his credentials as a former USAF pilot and officer, and stated that the craft was unlike any human-made object that he had ever seen. Fearing political repercussions at the time, he denied that a UFO had hovered over Phoenix.

75. Ibid., 159–68.
76. Ibid., 134–35.
77. Kean, *UFOs*, chapters 24 and 25.
78. In 2012, at a Boston University alumni/ae dinner, I mentioned this incident when conversing with a university professor, a BU graduate, who was teaching out of state and had asked about my current writing projects. Coincidentally and somewhat amazingly, he told me that the governor's aide whom Kean reports was present as an alien had been his college roommate, and had told him about the incident and his involvement in it as he played the role of an apprehended alien.

Malmstrom Air Force Base ICBM Wing, Montana

One of the most extraordinary UAP incidents Kean describes took place in Montana in 1967. Malmstrom Air Force Base controls the Montana "Wing" of the Minuteman Intercontinental Ballistic Missile (ICBM) system; a Wing has 150 missile silos in "flights" (a flight consists of ten missiles in their respective underground silos, about two to ten miles apart from each other; all are connected separately to a control center, in this case Malmstrom AFB). Kean writes that on March 16, 1967, near Malmstrom AFB, "nearly twenty nuclear missiles were suddenly shut down at Echo Flight while UFOs were in close proximity."[79] She further states that about one week later, on March 24, ten additional missile systems ultimately controlled by Malmstrom but under the immediate control of officers in control centers deep underground, were shut down when an oval object hovered above Oscar Flight; each of the ten missiles had an independent power system. Boeing engineers later suggested the probable cause as "some kind of electromagnetic pulse directly injected into the equipment. Whatever force was involved had to penetrate sixty feet underground to do its damage."[80]

The most extensive coverage of this and other Montana events is provided by Montana biologist and social thinker Joan Bird in *Montana UFOs and Extraterrestrials: Amazing Stories of Documented Sightings and Encounters*.[81] Bird did extensive research on Montana UFO sightings and the credible witnesses who reported them. At the beginning of chapter 2, "UFOs and the Minuteman Missiles," Dr. Bird quotes retired Air Force Captain Robert L. Salas, whom she interviewed in depth and with whom she corresponded extensively by email. Salas was stationed underground, observing the control panel which included each missile's individual control light and switch, when missiles in the "Oscar" missile flight bunker were shut down. Bird quotes Salas: "*I, without reservation, accuse the U.S. Department of the Air Force of blatant, pervasive and a continuing cover-up of the facts, deception, distortion, and lying to the public about the reality of the UFO phenomenon.*"[82] Salas's book[83] recounts events at Malmstrom; James Klotz, his coauthor, has a related Web site.[84]

79. Kean, *UFOs*, 139.

80. Ibid., 144–45.

81. Bird, *Montana UFOs*, chapter 2. Bird has a lengthy and in-depth discussion of the Malmstrom Air Force Base and related incidents in the US and USSR. (The latter, 1982 event, cited by Bird, was provided by a retired colonel from the former USSR in an ABC interview in 1994 in the United States, after the collapse of the Soviet Union.)

82. Ibid., 7.

83. Salas and Klotz, *Faded Giant*.

84. See http://www.cufon.org. The Web site includes declassified Air Force reports.

In a separate incident related to Malmstrom, officers at Echo Flight had reported an identical incident one week earlier: their ten missile controls were shut down one by one. Civilian corroboration of the Echo Flight incident was provided at the time by truck driver Kenneth C. Williams in a report he filed[85] with the National Investigations Committee on Aerial Phenomena (NICAP). Williams was unaware of the then-secret report from Malmstrom officers; neither were they knowledgeable about his statement at the time: his report was not disseminated publicly for decades; theirs was classified 'Secret' for decades.

Taking UAP Reports Seriously: Science and Transparency

Serious consideration should be given to credible witnesses' narratives about UAP sightings in or near Roswell, the Hudson River Valley, Phoenix, and Great Falls in the United States, as well as in England, Belgium, and Brazil. Such reports do not necessarily mean that what previously had been considered "science fiction" has become now "science fact." Rather, it means that an intentional, international UAP scientific research effort should be undertaken. This would require several concurrent steps: first, serious scientific research in the United States; second, federal and state governments' transparency in UFO-ETI matters, as would be evidenced by their release of all currently secret and secure files, and by open, responsible, honest, and transparent involvement in ongoing UAP reports; and third, intergovernmental and international scientists' collaboration, among researchers from all nations—including from the private sector, nongovernmental organizations, and government agencies—in current and future projects, and provision of publicly accessible documents emerging from its efforts-in-progress and the theories it generates and conclusions it draws from its research.

A new US government public persona and attitude, and different policies and actions, would have significant impacts on UFO/UAP research and the general public's attitude toward reports of UFO incidents, people who are credible witnesses to them, and the public's—and scientists' and pilots'—willingness to disclose and discuss their own experiences.

---

85. Bird located the Williams quote about and report from his extraordinary experience in Kaminski, *Lying Wonders*. Kaminski headed Boeing's investigative team that evaluated the Echo Flight shutdown to determine whether or not there had been a technical failure (Boeing had contractual oversight over missile systems maintenance).

On Mars as on Earth?

Earlier, Stephen Hawking's optimism regarding space exploration and settlement was discussed. Hawking declared that such ventures were urgently necessary. Humanity, in his view, was soon to destroy itself (and Earth) through environmental destruction, global war, or irresponsible human population generation. Therefore, it was imperative that a segment of humanity be safeguarded for species salvation: some selected people would be sent elsewhere in the solar system. Hawking specifically named a Mars colony as a possibility for representative humans' relocation. Realistic cautions should be set forth once again now, regarding Hawking's romanticism about the changed character, consciousness, and conduct of Earth's space explorers and settlers (as implied in his expectation that this human remnant would do better on a distant place, and not replicate on Mars what humans had done on Earth). It is exceedingly probable, given ordinary practice in such endeavors that, contrary to Hawking's scenario, human space expeditions would be comprised of people with specific scientific knowledge and military skills—whose objective would not likely be to establish an altered-consciousness colony of human settlers who would idealistically strive to be ecologically, economically, and ethnically responsible, particularly when engaging resident intelligent biota, if any. For colonization, skilled artisans would be added for construction of new "civilizations" for a new humanity in new contexts—physical and social constructs that might do violence to exoEarth places. During the quest to locate possible human-friendly planets or celestial bodies, if the current ideology of Discovery continues to permeate optical and radio (and other) telescopic exploration and subsequent selection of potential cosmic places for settlement, then Earth-developed human nationalistic, ethnic, or cultural conflicts will continue on Mars, and in the heavens more broadly. Human presence "in the heavens" would then be as destructive of planetary environments and indigenous inhabitants as it has been "on Earth."

As noted earlier, "Martians" have been vilified, even without known Contact, as malicious invaders of Earth. But the possibility that humans might be destructive invaders of Mars is very rarely a theme in fiction,[86] and not much discussed in nonfiction, either. We humans, as might be the case with other intelligent species, see ourselves as "good people" as we gloss over Discovery and a human history replete with intrahuman wars. If consciousness and conduct are not changed, then, prior to space ventures for colonization purposes, seeds of ecological destruction, biotic extinction, and human self-genocide will be sown on Mars and in other places—and likely

---

86. A significant exception is Ray Bradbury's classic *The Martian Chronicles*, which appeared originally as short stories in magazines and was published as a book in 1949.

be mirrored on Earth as nations battling over territory and natural goods on new worlds wage war on Earth to strengthen their position. Earth's foreseen endangered future will become in a sense, at least ideologically, the immediate present consciousness on Mars, and, as on Earth, human conduct will embody human consciousness—and both planets will be jeopardized as a result.

Religious Implications: Serious Speculation

At the end of the twentieth century and into the twenty-first, scientists and religion scholars began to reflect more extensively and deeply about the possibility of the reality of UAP and of extraterrestrial intelligent life. *Contact with Alien Civilizations: Our Hopes and Fears about Encountering Extraterrestrials*, by Michael Michaud, and *Many Worlds: The New Universe, Extraterrestrial Life and the Theological Implications*, edited by Steven Dick, are two of the most in-depth and in-breadth works. Each offers a diversity of topics elaborated objectively, for the most part; a selection follows.

Michaud offers a caution at the very beginning of his book. He balances the exuberance and overreach of some who advocate and argue for acceptance of extraterrestrial intelligent life: "If extraterrestrial life exists, most of it may be in simpler forms comparable to the one-celled organisms of Earth biology."[87] Regarding the opposing point of view on the existence of extraterrestrial intelligent life, he takes exception, obliquely, to the absolute claims of some skeptics who assert that there is no intelligent life other than on Earth in the entire cosmos and dismiss the possibility of intelligent life with the query, "Where's the evidence?" Michaud observes in response that "St. Augustine had condemned the idea of humankind's presence beyond Eurasia."[88] (Christian attitudes changed diametrically centuries later when they realized that Augustine's dogmatic statement on the issue was in error—explorers found previously unknown lands, and Church officials responded with the Europe- and Christianity-affirming Discovery Doctrine.)

In his summary of religious thinking over the centuries, Michaud notes that René Descartes in 1644 "suggested that an infinite number of creatures far superior to us may exist elsewhere.... Descartes claimed that our merits are not diminished by the fact that intelligent beings on other heavenly bodies have similar ones."[89] This is countered to some extent by William Whewell, Master of Trinity College in Cambridge, England, who in an 1853 treatise against the plurality of worlds sought to evaluate empirical evidence criti-

87. Michaud, *Contact*, 1.
88. Ibid., 15.
89. Ibid., 17.

cally. He suggested parameters of habitability within which life would have to develop; noted that "life" does not necessarily mean "intelligent life"; and stated that humans had not existed long on Earth. A Christian, he observed that if there were intelligent beings elsewhere in the universe, they would, Michaud summarizes, "have some kind of relationship with God. That would dilute any special relationship we humans came to have."[90] (This might be either an anthropocentric reason to decide out of hand that there is no other intelligent life, or a clarifying reason for advocating human humility should intelligent life be encountered.)

Michaud cites Isaac Asimov's assertion that a species is intelligent if it can develop a complex technology.[91] However, other possibilities to consider should be that an intelligent species might have *surpassed* complex technology—they evolved beyond it after they developed it—or *bypassed* complex technology—intelligent extraterrestrials never developed it, because they had no need for technology as humans understand it since they used their intellect and creativity in other directions and fields of inquiry. (This might well be, in fact, how they developed the types of space vehicles, energy generation, and mobility reported by credible witnesses.)

In terms of adaptability to different planetary contexts, a common assumption is that extraterrestrial and terrestrial space voyagers would carry special attire on board their craft that would enable them to explore worlds with differing elements or combinations of elements in their diverse atmospheres. This will likely be ordinarily true, but Michaud offers an intriguing possibility: that extraterrestrial life, intelligent or not, might evolve during migration and resettlement in order to adapt to new environments; ETI might live in different environments on their home planets than humans do on Earth, and might evolve further when they are voyaging in space vehicles, and yet further once on other worlds[92] (among which, it should be noted, Earth might be included).

Astronomer Paul C. W. Davies suggests, in both Michaud and Dick, ideas about the human presence in the cosmos. In Michaud, he asserts that consciousness is a natural development in evolution, a product of operative natural laws.[93] Davies does not speculate about what physical forms, social organizations, sciences, technological developments (if any), means of communication (including verbally expressed language or telepathic transfer), and economic arrangements ETI might have; possible genetic or other means needed or useful for their evolutionary development or adaptation;

90. Ibid., 19.
91. Ibid., 33.
92. Ibid., 61.
93. Ibid., 79.

and the types of mutations that might have occurred, are occurring, or can be enabled to occur given their genetic constitution and complexity as they engage with or relate to diverse contexts. All this data would provide useful—and perhaps necessary—knowledge for humans to have before, or to know soon after, Contact.

In Dick, Davies describes possible biological differentiation among biota evolving on different worlds in "Biological Determinism, Information Theory, and the Origin of Life." He suggests that life might have a diverse biochemistry in different locales: for example, it might have "the opposite chirality, i.e., its nucleic acids and amino acids would be the mirror image of ours," or "a completely different genetic code," or even "different sorts of molecules altogether." The universe might be "biofriendly," allowing for the evolution of distinct types of life.[94] This suggests another reason for the possibility that while "we are not alone" in terms of being intelligent life, the "we" elsewhere in the cosmos might differ dramatically from us biologically. (A possibility in addition to what Davies suggests might be that intelligence does not have a "life" at all, if by this is meant solely materiality, a biotic existence: there might be intelligent beings who have a different form of existence, unknown to humankind.) He suggests, too, that "if it turns out that life does emerge as an automatic and natural part of an ingeniously biofriendly universe, then atheism would seem less compelling and something like design more plausible."[95] (An alternative to a specific divine "design" is possible: that the creating Spirit gave the cosmos creative freedom to dynamically develop over time, and gave biota creative freedom to evolve and adapt over time. Rather than being "designed" by divine intelligence, the cosmos was provided with foundational laws to utilize in different combinations—which could explain a diverse biochemistry, should such be found to be the case cosmically.)

A provocative idea presented in Michaud by Nobel Biology Laureate Christian de Duve is that humans are not the final stage but a transient stage in *biotic* evolution,[96] which differs from Teilhard de Chardin's understanding, discussed in chapter five, that humans' current level of development is a transient stage in *human* evolution, not that humans would be displaced or replaced by another species during biotic evolution, as seems to be implied by de Duve. Along the same lines, Michaud comments regarding de Duve that "superior aliens might regard us as precursors of advanced species, as we look upon higher primates other than ourselves."[97] In Dick's book, de Duve states in "Lessons of Life" that "the majority of cosmologists believe

94. Davies, "Biological Determinism," in Dick, *Many Worlds*, 17.
95. Ibid., 27.
96. Michaud, *Contact*, 87.
97. Ibid., 86.

that there must be, in our galaxy alone, as well as in others, many celestial bodies with a history similar to that of planet Earth. If they are right, then the deterministic view leads to the conclusion that life is indeed widespread, a normal manifestation of matter in many sites of the universe, a cosmic imperative."[98] Where life exists, "In animal evolution, the direction toward increasing complex polyneural networks appears strongly favored by the fact that a more effective brain is advantageous under any circumstance. Thus, if life exists elsewhere in the universe, the likelihood that it may produce intelligent forms, some perhaps more advanced than the human form, is far from negligible." This threatens religious and humanist beliefs in a human-centered universe, he thinks.[99] He states further that humankind, as other species, continues to evolve and its future mode of existence is unknown: "It would be surprising if in the future development of life on Earth, vertical evolution toward greater complexity did not continue to take place, perhaps leading to beings endowed with considerably sharper means of apprehending reality than we possess."[100] This implies that the "greater complexity" and apprehension of reality might or might not be attained by the human species.

The latter ideas can assist people's efforts to understand and be open to forms of intelligent life in space that might not be "like us," who might even have evolved not only to a greater level of cosmic consciousness, but also have invented extraordinarily enhanced technology that provides the means to journey not only via interstellar, but also interdimensional, vehicles and techniques. This might provide indications of reasons for some of the characteristics observed in UAP such as immediate and extraordinary acceleration and change of direction, and appearance-instantaneous disappearance-instantaneous reappearance at some distance away from the observer—assuming, of course, that such observations are accurate.

Extraterrestrial intelligence—and perhaps terrestrial intelligence—might even have the potential to evolve to exist principally in spiritual dimensions of reality; some intelligent beings might have already reached that mode of being. This is not something considered by scientists, since it is outside their purview of investigation into and speculation about the nature of material reality. However, other scholars considering cosmic inhabitants' possible modes of existence need not be similarly limited by merely material considerations and the tools needed to assess them, but might responsibly consider other possibilities available to and presented by diverse fields and modes of inquiry.

98. de Duve, "Lessons," in Dick, *Many Worlds*, 6.
99. Ibid., 8.
100. Ibid., 11.

Michaud identifies a perspective that provides a stimulus to realistic speculation, rather than mere theoretical projection: humans define "civilization," even before setting out into space, on the basis of whether or not ETs use humanly detectable technologies.[101] As noted earlier, ETI might have bypassed or surpassed use of such technology. Commenting on culture, Michaud adds that "if we define culture as socially transmitted behavioral patterns, we must recognize that it has not been limited to humans"[102]—nor, it should be added, limited only to a particular human culture, for example, European or Euroamerican, as the sole or superior culture of humankind. Ironically, from the perspective of indigenous peoples, the preceding sentences might be understood to mean that Michaud rejected a "Discovery Doctrine redux," when he dismissed the human species' anthropocentric characterization of what kind of "civilization" should be exported and enforced "universally" in human space exploration—unless ETI has an equivalent civilization that will be accepted on Contact—but Michaud still retained a human description and limitation of what defines "culture," a description that he assumes would be used at Contact to assess ETI.

Similarly, Michaud quotes a NASA workshop statement on technological development as an indication of intelligence, and then provides probing insights related to it:

> Almost certainly once a species with the requisite intelligence, manipulative ability, and complex social organization has evolved, technological civilization will develop.... To go from a stone age culture to our present level of technological development required no biological evolution. All that was needed was the development of ideas, and their testing by trial and error.[103]

Michaud notes that "others argue that technological development is not inevitable; the motive must be present as well as the potential."[104] He elaborates by noting that in some cultures competition and weapons development are stimuli for technological advance. He seems to assume that technological development would be accompanied by military advances; should this not be the case a culture would be regarded as inferior. However, military "progress" and social competition should not be, in this author's view, used as criteria: a residue of Discovery remains when suggesting this and, moreover, such criteria would not be present much if at all in communal or community-oriented societies, which are more prone to peaceful interaction with likeminded so-

101. Michaud, *Contact*, 89.
102. Ibid., 91.
103. Ibid., 92.
104. Ibid.

cieties, where differences or disagreements can be resolved equitably without armed conflict.

Discovery Doctrine required that for a civilization to be Europe-like and therefore pass the first Discovery criterion to avoid invasion—possession of a superior culture—private property was required, whose dynamics included competition to own the best land. Certainly, too, the Europeans possessed more advanced weapons, but this military advantage likely was the result of ethnic and monarchic egotistic or pecuniary desires for land or natural goods available in another nation. Obviously, native communal cultures did not have competition or machine-developed and more advanced and harmful weaponry. Consequently, native civilizations and cultures were rejected preemptively by European ideology well before European arrival in what came to be known as "the Americas." This attitude, if present during space exploration, is both anthropomorphic—since it projects beforehand which terrestrial intelligence–developed characteristics of civilization are necessary and operative for extraterrestrial intelligence–developed civilizations to be recognized as such—and anthropocentric, since it implies human preeminence in cosmic cultural and civilizational evolution.

Michaud provides additional insights in a sidebar titled "Is Science a Universal?" He states that "many projections of how extraterrestrial civilizations will develop have been written by scientists extending trends far into the future. Yet, our own history casts doubt on straight-line projections. . . . Even when civilizations possess the technical skill, social or psychological forces may deter them from advancing further, though that advance may seem obvious to outsiders and successors. . . . Human interest in science may be rare, conceivably unique."[105] Two biases, anthropomorphism and scientism, are evident in the projections Michaud calls into question. The first, once again, is a human-centered view: that since human cultural evolution led to scientific and technological development, so too must all intelligent species and cultures have evolved and "advanced." The second is a solely scientific and social scientific speculation; historians and humanists generally seem not to have had much input into its formulation: science is first and foremost.

The citations from Michaud just presented were selected using the primary criterion of their relevance to a focus on the ideological and socio-ecological impacts of intercultural discovery and engagement—on Earth and in the heavens—bearing in mind how the European Discovery Doctrine has impacted indigenous civilizations in Earth history, and how a continuation of Discovery or its permutations might endanger if not extinct exoEarth intelligent beings' cultures. If an integrated human consciousness and conduct

105. Ibid., 93.

during extraterrestrial space voyages can do no better when encountering the 'other' than did a solely European-based Discovery Doctrine and its concomitant conduct after terrestrial ocean voyages, interaction with other intelligent species' representatives and cultures will prove to be harmful, not beneficial, to all parties in Contact. It will be, most likely, more harmful to humans when they meet ETI that is far more advanced technologically and, if it has a communal or community-based culture that is concerned with common good rather than individual desires, is more judgmental of an inferior human culture still locked in a more primitive individualistic perspective and conduct.

Michaud's work is rich with consideration of other issues, but their inclusion would require a significantly lengthier work. Insights from Steven Dick follow.

In *Many Worlds*, Dick reflects early on that "the new universe has implications for all areas of human thought and for all the world's cultures."[106] Discussion of terrestrial and extraterrestrial intelligent life, in the context of cosmic entirety, is not to be left to just one field of study but considered by all, in an interdependent, mutually informative manner.

Dick provides essays from diverse fields and distinct scholarly perspectives that offer insights on a variety of topics that overlap with or complement those in Michaud. Christopher McKay, SETI Institute astrobiologist, speculates that while in the vast cosmos life might have evolved differently and have diverse types of biochemistry, this is not the case on Earth. He observes, too, in "Astrobiology: The Search for Life Beyond the Earth," that "to understand the scope and diversity of life in the universe may well require that we search the cosmos just as understanding the diversity of life on Earth only came with exploration of the entire planet."[107] He states, too, that "the great diversity of life forms on Earth are really just morphological variations on a single fundamental biochemistry. Every life-form on Earth carries RNA and DNA that use just 5 nucleotide bases. The proteins that constitute the machinery of biochemistry are based on twenty left-handed amino acids. . . . Not only do all organisms on Earth share the same basic genetic code, but they all show clear evidence, in this code, of shared descent."[108] After all the scientific research and speculations concerning how life began in the first place, McKay states that "the origin of life remains a scientific mystery."[109]

In a similar vein, in "Life in Our Universe and Others: A Cosmological Perspective," Martin J. Rees states first that "planetary systems are (we

106. Dick, *Many Worlds*, xi.
107. McKay, "Astrobiology," in Dick, *Many Worlds*, 45.
108. Ibid., 47.
109. Ibid., 49.

believe) so common in our galaxy that Earthlike planets would be numbered in millions,"[110] and then adds, "We still do not know whether life's emergence is natural, or whether it involves a chain of accidents so improbable that nothing remotely like it has happened on another planet anywhere else in our galaxy."[111] Rees addresses, too, speculation about how common intelligent life might be, and what might characterize it: "We still do not know the odds against life getting started. Even when simple life exists, we do not know the chances that it evolves toward intelligence. . . . And even if intelligence exists elsewhere, it may be enjoying a purely contemplative life and doing nothing to reveal itself. Absence of evidence would not be evidence of absence."[112] He returns to the issue of origins of life by declaring that "Theorists may, someday, be able to write down fundamental equations governing physical reality. But no physicist will ever tell us what breathes fire into the equations and actualizes them in a real cosmos."[113]

Arthur Peacocke, Oxford University biochemist, theologian, and Anglican priest, provides an array of insights related to life on Earth and elsewhere in the heavens in "The Challenge and Stimulus of the Epic of Evolution to Theology." He declares that profound alterations of human knowledge about intelligent beings—including divine Being—are currently occurring: "We are now living through the most fundamental challenge of all to Christian Belief—the fundamental displacement of the basic understandings of nature and of humanity, and consequentially also of God."[114] Rather than being able to believe that we humans are an individual and special creation of God, data discovered demand that we acknowledge that "We are part of nature, part of an evolving cosmos, indeed, we are stardust become persons!"[115] This does not mean that the cosmos has begun and developed on its own: "For theists, the whole process is given its existence, with that potential capacity for life, by God,"[116] and the "genetic fallacy" must not continue of "explaining reductively" human and cultural development in terms of biological or cultural origins.[117] Especially relevant for considerations about the possibility of the existence of and material forms taken by evolved extraterrestrial intelligent life is Peacocke's statement that God is concerned for all life, not just humankind. Although these words were used by Peacocke in reference

110. Rees, "Life in Our Universe," in Dick, *Many Worlds*, 65.
111. Ibid., 66.
112. Ibid., 67.
113. Ibid., 73.
114. Peacocke, "Challenge and Stimulus," in Dick, *Many Worlds*, 91.
115. Ibid., 92.
116. Ibid., 94.
117. Ibid., 99.

to biota on Earth, they are aptly applied to biota throughout the cosmos: "we now have to escape from our anthropocentric myopia and affirm that God as Creator takes what we can only call joy and delight in the rich variety and individuality of other organisms *for their own sake*."[118]

Astronomer Jill Cornell Tarter in "SETI and the Religions of the Universe," and philosopher of science Ernan McMullin in "Life and Intelligence Far from Earth: Formulating Theological Issues," present contrasting views on whether or not ETI will have some kind of religious belief.

Tarter claims that long-lived extraterrestrials "either never had, or have outgrown, organized religion,"[119] and elaborates further that "we can imagine that elsewhere long-lived technologies may have been developed by intelligent creatures who never had the need to invent God(s) or religions, or who did so in their youth, but later replaced them with a more scientific world view."[120] Religion in Tarter's thinking is determined to be "unscientific" by those such as her who have no religion, either because they outgrew childhood religious education or never had been expected to hold religious beliefs of any kind. Anthropomorphically Tarter states in these proposals that since some scientists have become or always have been nonbelievers, so too must extraterrestrial species, whose intelligence has evolved far longer than that of humans, be universally atheists: they likewise "never had the *need* to *invent* God(s) or religions." The Spirit in her view is a Feuerbachian-like creation, human-developed and -projected, and worshipped because of a psychological or social material exigency. Such is her limited and limiting assessment for reasons why religion emerged on Earth. Tarter expresses a standard atheist scientist's belief about evolution of intelligent life: inevitably, life will become too intelligent to (have to) believe in either Sacred Being (however expressed in distinct cultures) or human projections of the Spirit as expressed in religions' dogmas. (A standard expression of this scientific belief is the saying that one cannot be both a scientist and a Christian; to have both identities, a scientist must "leave their brains at the church house door" before entering to worship.) Tarter's "organized" is a key word in her statement. Actually, contrary to her beliefs, based on and projected from her experience, ETI might well have evolved beyond institutional "religion" beliefs, practices, and rituals, and thereby been enabled to develop a higher "spiritual" understanding related to a transcendent-immanent Spirit. Rejection of religion need not result in abandonment of ideas about and experiences of a cosmic Being—in fact, it might lead to enhanced understanding of a Spirit unmediated by ideological boxes or confined within the metaphysical

---

118. Ibid., 95.

119. Tarter, "SETI," in Dick, *Many Worlds*, 145.

120. Ibid., 147.

doctrines or material buildings characteristic of most religions. (Throughout Earth today, people dissatisfied with organized religions and their clerical structures have moved away, intellectually, from demanded beliefs and the dedicated rituals that express them. Increasing numbers of people self-define as "spiritual" rather than as participants in a structured religion, and believers of its required dogmas.)

Ernan McMullin (who had been one of Georges Lemaître's physics students) observes that "were traces of life to be discovered elsewhere in our solar system today, it would favor the Augustinian idea that the 'seeds' of life were implanted in matter from its first appearance. Such seeds could presumably come to fruition anywhere where 'water and earth' provided the right environment. On the other hand, such a discovery would challenge the belief that the origin of life on Earth required a miraculous intervention on God's part."[121] After reflecting on the Christian doctrine of divine Incarnation as a human being on Earth, and this doctrine's multiple conjectures about and doctrinal formulations of the meaning to be attached to Incarnation that were developed over two millennia of Christianity, McMullin asks, in regard to Incarnation and the redemption of humans who had fallen into sin and how what had transpired on Earth might or might not be relevant to or for ETI: "[H]ow do or did they stand, morally, in the sight of their Creator? Do (did) they need the sort of redemption effected by the Incarnation on Earth? . . . [A]lien people might or might not be favored [by divine Incarnation in their world]. They might, for a variety of reasons, need it less (or perhaps even more) than we do."[122] He wonders, too, "How can we limit the ways in which the Creator of a galactic universe might relate to agents like ourselves on other distant planets?"[123] In his pondering, McMullin is close to Teilhard and Chalmers, though without taking the firmer stand evident in them.

In "The Evolution of Life on the Earth and Possibly Elsewhere: Reflections from a Religious Tradition," scientist and retired director of the Vatican Observatory George V. Coyne, SJ, ponders areas referenced by both Tarter and McMullin. He states that statistically there are "1017 Earthlike planets in the universe."[124] He asserts thereafter that "while religious belief may have played a key role in the inspiration of modern science, we now know that religious experience cannot be limited to that which science can discover. To use the concepts coined by Galileo, both the Book of Nature and the Book of Sacred Scripture can be sources of coming to know God's love incarnate in the universe. . . . [E]xperience of God exceeds the content of the Book of

---

121. McMullin, "Life and Intelligence," in Dick, *Many Worlds*, 157.

122. Ibid., 173.

123. Ibid., 172.

124. Coyne, "Evolution of Life," in Dick, *Many Worlds*, 180.

Nature.... Such experience also exceeds the Book of Scripture.... [I]t is mistaken to assume that rational processes exhaust the primordial experience of God, the source of both the Book of Nature and the Book of Scripture."[125] Concerning ETI, he states, "From the scientific evidence, presented in summary above, the existence of extraterrestrial intelligence must be taken as a serious possibility with all its consequences."[126] He affirms that a God who is good and who "saved" human beings would also "save" extraterrestrials and, as a next step to be taken when considering implications of extraterrestrial intelligent life and conduct, "theologians must accept a serious responsibility to rethink some fundamental realities within the context of religious belief." God's revelation to humans is present in the Scriptures, but "God has also spoken in the Book of Nature."[127]

Related to the latter ideas, Steven J. Dick observes in "Cosmotheology: Theological Implications of the New Universe," that most religious authorities do not think that extraterrestrial intelligent life encounters will catalyze a faith problem: "Internal to religion, flexibility seems to be the watchword, whereas those external to religion proclaim the imminent death of religions after such a wrenching discovery as extraterrestrial intelligence."[128] He suggests that humans should not expect ETI to have the same understanding of God expressed by humankind: "considering the divergence of human ideas of God, there is no basis for expecting convergence of theistic ideas by intelligences on other planets throughout the universe. Unless, that is, there is some scientific basis for it."[129] (Teilhard de Chardin and Thomas Berry have expressed similar views.) In words reminiscent of McMullin's speculations on the meaning of Incarnation and salvation, and Thomas Chalmers' ideas on intelligent beings elsewhere in the universe, he states that "we know nothing about good and evil in the universe in the context of extraterrestrial civilizations."[130] He concludes that "surely the history of God teaches us that the concept will persist, but that it ought to be adjusted to our knowledge of the universe. Surely history demonstrates that the true meaning of God is not grounded in any single human culture, but in the best elements of otherworldly thinking of all of them. To this body of thought we must now add the scientific world view, wherein the universe, or the multiverse, is large enough to encompass God."[131]

125. Ibid., 185.
126. Ibid., 187.
127. Ibid.
128. Dick, *Many Worlds*, 198.
129. Ibid., 203.
130. Ibid., 205.
131. Ibid., 207–8.

Unidentified Phenomena and Voyagers: Extraterrestrial or Extradimensional?

In the course of his extensive scientific investigation of UFOs and ET over more than fifty years, astrophysicist Dr. Jacques Vallee[132] concluded that theories about extraterrestrial intelligent beings' voyages were inadequate to explain how they traversed vast distances and how and why they disappeared instantly after accelerating to a velocity that seemed impossible to achieve under known laws of physics. He theorized that what was considered Extra Terrestrial Intelligence (ETI) might more likely be, in some events, Extra-Dimensional Intelligence (EDI). As he pursued his research, he became frustrated at efforts, particularly from the US government and fellow scientists, to stifle UFO narratives and ridicule UFO-curious scientists, credible witnesses, and members of the general public. He declared that "if there was ever a situation in science that called for the careful sifting and screening of data and for the questioning and testing of every hypothesis, it is the situation presented by the UFO phenomenon."[133]

Vallee presents a strong but not incontrovertible theory regarding exoEarth intelligent beings. Based on his analysis and interpretation of historical narratives about alien intelligent life, Vallee concludes that the beings described are not *extraterrestrial* but *extradimensional*.[134]

Theories about UAP and ETI or EDI have been suggested by scientists who have credible and justifiable scientific approaches to research into aliens from exoEarth—however "alien" being might be defined, and wherever "exoEarth" might be located.

Roswell and Malmstrom Updates

In regard to the extraterrestrial hypothesis, two credible witnesses have corresponded with the author about their respective experiences, as they reflected upon them decades afterward in dialogue with their current knowledge and experience.

Jesse Marcel Jr., when queried about the apparent lack of hard evidence about extraterrestrial intelligent beings, and why ETI has not made a

---

132. Dr. Vallee's academic background includes a Bachelor of Science degree in mathematics from the Sorbonne; a Master of Science in astrophysics from the University of Lille; and a PhD in computer science from Northwestern University. In 1961, he was an astronomer at the Paris Observatory; in 1962 he moved to the United States to work in astronomy at the University of Texas, Austin. When he went to Northwestern University, he became a close associate and lifelong friend of Dr. J. Allen Hynek, discussed earlier, who was then a scientific consultant to the USAF Project Blue Book.

133. Vallee, *Dimensions*, 4.

134. Vallee's data and theories are presented in depth in Hart, *Encountering ETI*.

significant, irrefutable public appearance in a metropolitan area in order to visibly and dramatically announce its presence, replied speculatively:

> I think that the extraterrestrials have a scientific curiosity about our civilization. They know we have passed the nuclear threshold and they may be curious about our ultimate survival. As you know we have had several close calls like the Cuban Missile Crisis and others. I was on board a troop carrier (APA 227) getting prepared to wade ashore for an invasion of Cuba in 1962 and I just don't think we could have come closer to all out nuclear war. As far as the extraterrestrials, they probably have more to fear from us than we from them. I have seen the horrors of combat firsthand; I don't think I would trust us either because we tend to shoot first and ask questions later.
>
> At this time, I do not think that there is anything that would prove to the skeptics that indeed UFOs are real. It would take a release of material from Roswell, etc. along with public scientific analysis that would prove to open minded skeptics the reality of our being visited by representatives of one or more intelligent civilizations. Those in our government who have control of these artifacts are not ready to release them.[135]

The Roswell experience sixty-six years ago of Colonel Dr. Jesse Marcel Jr., as a military officer and credible witness to UFO encounters, is complemented by that of Captain Robert Salas forty-six years ago in Montana.

Salas was the lieutenant on duty in the Oscar Flight ICBM control center when a UAP hovered over the center's silo and shut down the ten missiles, one by one. He, too, reflected on what interest extraterrestrials might have in regard to human civilization on Earth:

> When you get past the question of the reality of this phenomenon, you search for the answer to the question – Why are they here? Clearly they have announced their presence to all humanity. It is also clear to me that thousands of people have been abducted and subjected to probing and medical procedures related to genetics. During these abductions people, including small children, are given messages related to the future of the planet and humanity. I think the main message is that we humans seem to be bent on our own destruction by the way we are treating ourselves and our home planet. Nuclear energy and weapons is just one aspect of this and, because of my experience, that has been my main message whenever I speak or write.

---

135. Jesse Marcel Jr., e-mail message to author, October 12, 2012.

Further, we can only resolve these questions of the environment and our proclivity to be warlike through cooperative actions. There's the rub: How do we focus on taking concerted, coordinated actions to resolve the problems that threaten our very existence? We have apparently not been able to make much progress in that direction.

However, as I think you will agree, part of our inherent nature includes the superior qualities of love, caring, honesty, integrity, and compassion toward each other. If there is hope for us, these inherent qualities may save us. I am not a religious person, but I think that one of the reasons people practice religion is to see these qualities come out in ourselves and others. We all have to value them and visualize how they will help human community become one. I think this is what ET is hoping for us also. I think ET wants to see some results in this direction before making full contact.[136]

Jesse Marcel Jr. and Robert Salas had direct involvement with UFO/ETI incidents. They have seen physical evidence (Marcel: Roswell debris; Salas: a UFO above the silos control center and a simultaneous, inexplicable shutdown of missiles), and they have communicated with ETI either directly (Salas: through a telepathic message) or indirectly (Marcel: through his father's excited presentation of debris that provided UFO evidence). They are convinced that extraterrestrials do exist, that they travel in a technologically advanced but currently unknown way from their bases to Earth, and that perhaps they are interested in and hopeful of seeing human progress toward peace, social stability, and environmental responsibility.

ExoEarth Intelligent Beings: Extraterrestrial and Extradimensional

The perspectives of Marcel and Salas appear to contrast sharply with that of Jacques Vallee. However, they need not be incompatible. Vallee's research and expertise, and the cogency of his arguments, might catalyze acceptance of the data he presents as evidence that some extraterrestrials originate in another dimension of integral being. This would be especially possible, perhaps, if integral being is a *multiverse*, as he suggests, rather than a *universe*, as understood traditionally. Other researchers, such as J. Allen Hynek, and credible witnesses including corporate executives, politicians, military officers, and pilots, have remained convinced that they have viewed diverse types of evidence that indicate that, at least in their own experiences, some events might be explained better (or solely) as having originated extraterrestrially.

136. Robert Salas, e-mail message to author, August 24, 2012.

The variant views presented suggest three possibilities for serious reflection and scientific exploration. First: intelligent beings are *extraterrestrial* and travel at extreme velocities because they have developed extraordinary technologies, advanced far beyond current human comprehension, that enable them to traverse space in much shorter periods of time than thought possible previously, perhaps aided by wormholes or other unknown cosmic factors. Second: intelligent beings are *extradimensional*, and need not travel far because they exist as 'next door neighbors,' inhabitants in another dimension of materiality, and have developed a mode of interdimensional travel. Third, that intelligent beings are *transdimensional*: they have originated in the farthest regions of space, billions of years in their past and our future, and cross time and space using some unknown scientific knowledge and form of technology that enables them to travel rapidly from one place to another by entering an alternate dimension or cosmic phenomenon. Especially because of the limitations of current human scientific research and understanding, the relatively simple technologies that have resulted from it, and lack of sustained Contact and scientific information exchanges with ETI, the selection of a single possibility from among the three is presently not possible.

The integrated consideration of the range of phenomena should prove invaluable for comprehensive scientific research. There appears to be sufficient data for inquiry into both ETI and EDI, which might eventually establish that there is not a distinct divide between them—they might well be participants in diverse but complementary events. Regarding them as "both-and" rather than "either-or" would be a more scientifically fruitful research endeavor; it would engage scientists who have expertise in complementary fields of inquiry.

An Intelligent Life- and Human-Hospitable Cosmos?

In discussions about human origins and place in the universe, scientists agree that humankind evolved from primates, but disagree about whether or not that was inevitable. Some assert that the universe was "fine tuned" so that humans would evolve, whether the "fine tuning" had a divine origin or was a natural result of cosmic dynamics and biotic evolution; others see human emergence as purely a chance occurrence; some of the latter contend that humans are the only self-reflective conscious intelligence in the entire universe. If we assume that humans are alone in the cosmos, it puts a great deal of pressure on humans not to destroy either their Earth home (without which they could not survive), or each other in a catastrophic grand Armageddon. If we assume that there are other self-reflective conscious intelligent beings, we have "social" pressure to conserve Earth and coexist peacefully and justly with each other in order to be accepted among the family of intelligent beings. In either case, humans have a significant socioecological responsibility to care for planet Earth and all its biota, and to be compassionate toward and solicitous of the human species as a whole and in its diverse parts.

Intelligent beings, because they can learn the intricacies and appreciate the grandeur of the cosmos, are the cosmos reflecting upon itself, integral being's focal points. Consequently, if we extinct ourselves directly through wars and social injustice, or indirectly through ecological irresponsibility, the cosmos either ceases to be self-reflective because we, the single intelligent species, are gone, or loses a particular human perspective that can be integrated with the insights of other intelligent beings to formulate a holistic understanding of cosmic complexity and diversity.

In all cases, humankind is provided with a choice to embrace cosmic being or anthropocentrically attempt to pursue individual species advantage. If humankind decides to responsibly explore and become part of cosmic being, then humans, Earth, and all biota on Earth and in the heavens will continue their evolution, interdependent integration, and community interrelationship through eons to come.

In *At Home in the Cosmos*, David Toolan expresses this well: "Quite literally, we are the fallout of the stars and the lucky outcome of a succession of contingent transformations that have merged out of chaos. . . . And our role in this mind-boggling creation? Unless there are extraterrestrials, we are the only ones in the cosmos who will be able to tell its story and say what it shall mean."[137] Our role too, is to find continually our dynamic niche in a cosmos *in statu viae* (in a state of ever-becoming), a sense of place across places and in every place, and ever-developing and closer apprehensions of ultimate meaning as we explore and are enlightened in our being and our becoming amid cosmic dynamics.

As has been obvious in this chapter, there are numerous credible witnesses who assert strongly, even in the face of ridicule, public embarrassment, or professional retribution, that they have been part of some event in which, from near or far, they encountered ETI or at least a UAP that might have indicated ETI's presence. There are too, particularly over the last quarter-century, noted thinkers who have discussed and debated some of the implications of Contact between terrestrial intelligent life and extraterrestrial intelligent beings. The issues of Contact and of the possibility that UAP are operated by intelligent beings have gained respect in academia and among the general public. The information provided here as part of the thought experiment—as if events narrated and investigated had actually taken place—should stimulate serious discussion of not only the possible or

137. Toolan, *At Home*, 177.

probable existence of intelligent life elsewhere in the cosmos but, if so, how that existence and subsequent TI-ETI interaction might affect us now and in the future: at home on Earth, or in new homes in space.

PART IV
TERRA COSMICA

10

Cosmic Coexistence

Cosmic Consciousness and Cohesion

The cosmos as a whole and in its intertwined existents–elements, energies, entities, events, and entropy–manifests dynamic integral being. It came into being at an initial moment that, according to physicists' calculations, occurred some 14.82 billion Earth years ago. The cosmos continues to come into being: it is active, not static. After its primordial moment it was not complete "as is," as a "being," but has continued to complexify, to "become." The cosmos is a being-becoming existent, a complex that is understood only partially, tentatively, and theoretically, on the whole. The cosmos as it emerges and unfolds is at once a dialogic and dialectic relationship of necessity and contingency, with seeming chaos thrown into its dynamics.

In order for cosmic interspecies peace to result from Contact with dynamic integral being, humankind must experience conversion from the anthropocentrism and androcentrism that have historically characterized most human cultures' self-designated terrestrial and even cosmic place and role. Human-centeredness and "man"-centeredness must be replaced by a humble sense of self and species that is localized within, and interdependent and interactive with, other species, spaces, and space. In the cosmic context of the vastness and complexity of integral being, such self- and species-humility and relation must be accompanied by and evolve with a reasoned, tentatively formulated, but at times dynamic understanding of how terrestrial intelligent life and extraterrestrial intelligent life can relate to each other, to other species (complex and simple, encountered or unknown), and to abiotic being (planetary places or, perhaps, intelligent beings whose *be*-ing—existence—does not correspond with humans' understandings of "life") with a relational consciousness and conscientiousness. The latter relationality should be expressed and evident, for greater material implementation and cosmic consociation, in practices of cosmosocioecological ethics. For humans, that initially will flow from the theory and practice of Earth socioecological ethics, and subsequently be adapted to, from, and in diverse space milieus.

Humankind's acceptance of having a less-than-centric place in the cosmic commons enables human consciousness to be open to relation with other intelligent beings from other places in space, and to act accordingly in the dynamic cosmos context. This provides a base for accepting extraterrestrial others, no matter how unique and diverse in form their materiality might be. It is also a necessary condition for and prelude to peaceful cosmic coexistence.

But, can or will humans know for certain that "we are not alone"? Unless a person or group has had a "close encounter" of the first, second, or third kind (as described below), such certainty is hard to come by for many people. Even narratives by seemingly credible witnesses fall short of providing grounds for many people to accept that ETI has appeared on or near Earth, or that ETI even exists. Currently, however, despite US government efforts in particular to dissuade them otherwise, a majority of people in the United States and abroad accept the testimony of witnesses, and their own observations (when these have occurred), that UFOs and ETI exist and, in fact, have interacted with Earth places and people through direct Contact on Earth or by communication from Earth's atmosphere.

Close Encounters

In the previous chapter, narratives of sightings of UFOs/UAP and of actual Contact with ETI were presented. The next part of our "thought experiment" to help us to consider seriously the impacts and implications of terrestrial-extraterrestrial intelligent life Contact is to discuss a scientific approach to both UFO events and to the possibility that UAP whose maneuvers appear to violate known laws of physics or capabilities of meteorological phenomena are controlled by intelligent operators (internally or externally), whether or not the source of the intelligence might be understood to be extraterrestrial. The ideas and analysis of Dr. J. Allen Hynek, highly regarded as the top scientist to have studied and analyzed UFO data in depth, will be discussed. Hynek, who proposed the three categories of "close encounters" that are still used today, decades after their first formulation, is a particularly fascinating, thought-provoking, and credible person. He was an accomplished astrophysicist and astronomer who began work in the field of UFO studies as a civilian scientific consultant to the US Air Force. He investigated perceived UFO phenomena and claims presented regarding Contact with ETI, and through more than two decades served as the official scientific expert who denied the veracity of any reports about UFOs and ET. Over time, however, after probing further and reflecting extensively on the cases and evidence that he had analyzed and, at times, ridiculed, he altered his longstanding position on UFOs.

J. Allen Hynek: Enter the Scientific Method

Josef Allen Hynek (1910–86), known as J. Allen Hynek, was an astrophysicist and astronomer. He received his BS (1931), and his PhD (1935) in astrophysics from the University of Chicago, where his principal doctoral studies were undertaken, with direct application of theory to practice, at the university's Yerkes Observatory in nearby rural Wisconsin. His professional credentials included professor of physics and astronomy, Ohio State University; research scientist on satellite tracking (he developed the system used by the US to track satellites, which was subsequently copied by the USSR), Smithsonian Astrophysical Observatory; and professor and chair of the Astronomy Department, Northwestern University. He authored numerous scientific books and articles on astrophysics (and, later, on UFOs), and wrote the astronomy column for *Science Digest*. He was highly regarded as an accomplished professional before, while, and after he worked with the US Air Force.

Hynek was for twenty-two years (1947–69) a consultant and principal astronomy expert for successive US Air Force projects focused on UFOs (Project Sign, 1947–49; Project Grudge, 1949–52; Project Blue Book, 1952–69). These projects had the supposed function of determining whether or not people's perceptions of UFOs were misidentifications of natural phenomena (e.g., lunar reflections on, or a Venus sighting from, a pilot's cockpit), human activities on Earth or in the heavens (e.g., weather or military satellites), or actual UFOs under intelligent control. In reality, the principal functions of the projects (as acknowledged by Hynek after he resigned from his position) were to collect and store data on UFOs; deny that UFOs exist, or that they had appeared in US skies; dissuade people from thinking that they had seen a UFO—even when (and especially when) the Air Force and the US government knew that a UFO event had occurred at the time and place witnesses stated that it had; fabricate false explanations for UFOs when they were known by the Air Force to have been actual aerial anomalies or perhaps space-originated vehicles; and use ridicule in press conferences and news releases, among other places, so that people became concerned about possible professional and personal adverse consequences if they acknowledged they had viewed a UFO even from afar. Consequently, highly respected, tenured senior professors and those of lower academic rank feared to mention— even to extended family members, friends, and coworkers—that they had observed unidentified aerial phenomena, or that they were curious about UFOs.

In academia, where the quest for knowledge and for truth are supposed to characterize any personal intellectual quest and institutional enterprise, a consequence of government policies regarding UFOs was that university professors denied even to themselves what they had perceived, accepting either that (a) what they had seen could not have occurred, despite visual contact and the analysis thereof that their mind had consciously processed precisely at the moment of sight, because they chose to believe or came to believe government assertions rather than their own eyes, or (b) they had to self-censor because they feared that, if they relayed their narrative to others, negative consequences would result: they would not receive promotion or tenure; they would experience ridicule from colleagues who unreflectively followed or consciously followed, for similar reasons, the official US government line that it was absurd for intelligent people to "believe" in extraterrestrial intelligent beings; or they would be fired for appearing to take seriously the theoretical existence, let alone an actual appearance, of spacecraft possibly controlled by ETI. In the United States, correspondingly, there was neither funding of nor even application for grants to do serious research into UFO/UAP events. The operative, prevailing "wisdom" has been that government officials' statements—made by military officers and press officers, few of whom (if any) had a scientific degree or even background in the field upon which they expounded with great confidence—are supposed to suffice to explain UFO occurrences, which scientific, social scientific, and humanities faculty would otherwise find exciting to consider. Hynek stated that his task as the official Air Force scientist was to be a "debunker," and he relished this role for most of the time that he had it.

Eventually, however, Hynek came to realize, after poring over thousands of reports, that while about 95 percent of events could be explained as having originated in observations of meteorological phenomena, other natural causes, or aviation or other human technological sources, the remaining 5 percent could not be so explained. He realized, too, that in other nations' agencies and among their scientists, serious documentation of, investigation into, and scientific study of UAP did occur. Hynek took a personal poll among US astronomers known to him in which he asked, in confidence, if they had had any sightings of unidentifiable aerial phenomena while doing their professional work. He was surprised to learn that more than 11 percent of the astronomers had observed objects not explainable from their area of expertise—a percent higher than that of the general public. This happened despite their expertise, which they used to discard meteorological events and human-caused incidents. Individual scientists had been afraid to state this in their research reports, and even less so publicly or even to other faculty or government and corporate research scientists, fearing especially retribution in the form of ridicule from colleagues and professional reproof from their supervisors, which would have imperiled their career. Astronomer Dr. Clyde Tombaugh, who had discovered Pluto, was a major exception to scientists' self-censorship; he openly discussed his sightings of anomalous aerial objects and events, and speculated about UFOs and who or what might control them.

Hynek became increasingly dissatisfied with the Air Force's refusal to follow a scientific method to investigate UFO reports: the USAF always chose to reject them out of hand—emphatically and even coercively, if necessary. He discusses his experiences extensively in two books, *The UFO Experience: A Scientific Inquiry* (1972) and *The Hynek UFO Report* (1977).

During his years as a US Air Force consultant and afterward, Hynek continually pressed for a scientific rather than political analysis and discussion of UFO phenomena. In *The UFO Experience*, he presents and discusses his own efforts to do such scientific study. As noted earlier, he reports in this book that approximately 95 percent of UFOs are attributable to natural, usually meteorological, phenomena, and sometimes to human sources. The "unidentified" becomes "identified" when any of these theorized causes of a perceived encounter with a UFO are established to be factually accurate. The other 5 percent are true "UFOs," that is, "unidentified" flying objects whose nature and maneuvers cannot be understood in terms of currently available (to humans) science and technology. In order to ensure that a pre-screening of all UFO reports results in the weeding out of the 95 percent for which known objects will probably be identified as the cause of a phenomenon, Hynek provides the following definition of "UFO":

> *We can define the UFO simply as the reported perception of an object or light seen in the sky or upon the land the appearance, trajectory, and general dynamic and luminescent behavior of which do not suggest a logical, conventional explanation and which is not only mystifying to the original percipients but remains unidentified after close scrutiny of all available evidence by persons who are technically capable of making a common sense identification, if one is possible.*[1]

The definition does not permit an individual, group, the press, or the general public to decide whether what is unidentified should remain in that somewhat mysterious category if another explanation can be found, nor to declare automatically that unidentified objects are under intelligent control,

---

1. Hynek, *UFO Experience*, 10 (Hynek's italics).

in particular (because of a general public perception or UFO buffs' enthusiasm) control by extraterrestrials. Finding "logical, conventional explanations" should be the first order of business, Hynek asserts, when UFOs are perceived. Technically competent people (from diverse fields, as needed) would provide data assessments in each case. Most UFO reports are about natural or human-caused events which competent professionals can identify upon careful consideration of data: natural phenomena or human-related activities with which people either are not familiar or have not perceived in settings identical or similar to those in which the events occurred. If "close scrutiny" cannot identify an object or event with some certainty, it remains an "unidentified flying object" until new data comes along. Data gathered from prescreened narratives that have been verified as perceptions of UFOs might then be compared with other incidents in the body of knowledge that were compiled previously, on different occasions, in order to find similar patterns or events. This, in turn, might lead to the development of innovative categories that previously would have been dismissed as "strange" or "highly unusual." If cases are considered separately, in isolation and with no cross referencing, this might be an accurate assessment; however, if the "strange" characteristics of particular unusual actions occur multiple times, they become, in a sense, "ordinary" descriptions of like events.

UFO Event Categories: Distant Sightings to Close Encounters

In *The UFO Experience* (1972), Hynek devised six categories[2] to describe types of unidentified phenomena events and the observing reporter's proximity to an event. (Hynek observes several times that there are probably significantly more UFO sightings than have been reported, but that many, if not most, of them are not reported in the United States because official government statements declare absolutely and unequivocally that UFOs do not exist.) Distant objects were classified as nocturnal lights, daylight discs, or radar-visual events. Objects near the observing subject were placed in the category of "close encounters," a nomenclature that became especially known publicly because it provided the title for the 1977 film *Close Encounters of the Third Kind*, starring Richard Dreyfus, for which Hynek served as a technical consultant (and in which he has a brief cameo near the end). The "close encounters" could be classified in one of three types: close encounters of the first kind (CE-I), in which the observer sees an object less than five hundred feet away; close encounters of the second kind (CE-II), in which physical evidence of a UFO occurrence is found at the site where the UFO was thought

---

2. Ibid, 28–31.

to have landed; and close encounters of the third kind (CE-III), in which humans come into direct Contact with extraterrestrials.

Hynek elaborated on the close encounter categories and provided examples of them in his second book on the topic, *The Hynek UFO Report*.

On CE-I, Hynek wrote:

> Here we have a close encounter with a UFO but there is no interaction of the UFO with either the witness or the environment, or at least none that is discernible. The encounter must be close enough, however, so that the UFO is in the observer's own frame of reference and he is able to see details. The chance, therefore, of this sighting being a misidentification of Venus or a conventional aircraft, etc., is quite small, particularly if the sighting is made by several persons.[3]

Turning to CE-II, Hynek stated that

> Here the UFO is observed interacting with the environment and frequently with the witness as well. The interaction can be with inanimate matter, as when holes or rings are made on the ground, or with animate matter, as when animals are affected (sometimes becoming aware of the presence of the UFO even before human witnesses). People, too, can be affected, as in the many reported cases of burns, temporary paralysis, nausea, conjunctivitis, etc. But in order for a CE-II to have taken place, the presence of the UFO must be established at the same spot in which the physical effects are noted. That is, if a burnt ring on the ground is noted, it must be at the exact place where the UFO was sighted hovering, or if an automobile ignition system is interfered with, such interference must have occurred at the time and place of the UFO sighting.
>
> The observed physical effects in these cases (often called "physical trace cases") must not be explainable in some other obvious way. That is, if holes in the ground ("landing marks") are found, these marks must be unique, and not like marks found elsewhere in the vicinity.
>
> Close Encounters of the Second Kind are of particular interest to scientists who can, in a sense, bring the UFO "into the laboratory." Burnt grasses, samples of disturbed soil, etc., can be tested with a view toward determining what caused the burn, what pressures were necessary to produce the imprints on the ground, and to finding what chemical changes occurred in the soil samples by comparing the affected soil with control samples

3. Hynek, *Hynek UFO Report*, 19–20.

> from the vicinity.... A catalogue of over eight hundred cases in which the UFO was both seen and left physical traces has been compiled.[4]

Finally, in terms of CE-III:

> Here there is not only a close encounter with the UFO, but with its apparent "occupants" or "UFOnauts." Close Encounters of the Third Kind bring us to grips with the most puzzling aspect of the UFO phenomenon: the apparent presence of intelligence other than our own, intelligence we can recognize but not understand. Hundreds of close Encounters of the Third Kind have been reported all over the world in the past decades. A catalogue of over one thousand cases has been compiled....
>
> In Close Encounters of the Third Kind, where the occupants make their presence known, we find reported creatures who resemble humans but are predominantly shorter and slimmer, capable of communication in their own way and on their own terms. Their interaction with humans has been reported to be largely impersonal, neither overtly friendly nor hostile.
>
> Clearly, Close Encounters of the Third Kind hold the most fascination for us because they bring into focus most sharply our fear of the unknown, the concept of other intelligences in space, and the possibility of intelligent contact with such beings, with all that such contact might imply for the human race.[5]

Hynek condensed the three close encounter categories and provided a special focus on a particular place, New York State's Hudson River Valley and areas in surrounding states, in the book he coauthored with Philip J. Imbrogno and Bob Pratt, *Night Siege* (discussed in chapter 9).

The close encounter categories continue to be used today among scientists and others who study reports of purported contact with UAP. The three types of categories provide a quick identifier of the type of encounter described, and facilitate cross-event comparisons and research.

Hynek's writings provided a previously lacking depth and probing analysis based on the scientific method that he used as he reflected on and elaborated several topics. In *The UFO Experience*, he describes a "credibility index" by which he judges event reporters' degree of believability in what they say has occurred.[6] He notes that there are variations in how multiple reporters describe an event, just as there are discrepancies in descriptions of

---

4. Ibid., 20–21.
5. Ibid., 21–22.
6. Hynek, *UFO Experience*, 18–20.

witnesses to an accident, fire, or robbery; but similarly, he observes, that just as there is no doubt in the latter events that it was indeed a fire and not a bank robbery being described, so, too, are credible witnesses all describing a UFO event.[7] He comments, too, on the "strangeness" of events or parts thereof:

> Still, there exist UFO reports that are coherent, sequential narrative accounts of these strange human experiences. Largely because there has been no mechanism for bringing these reports to general attention, they seem to be far too strange to be believed. They don't fit the established *conceptual framework* of modern physical science. It is about as difficult to put oneself into a "belief framework" and accept a host of UFO reports as having described actual events as, for example, it would have been for Newton to have accepted the basic concepts of quantum mechanics.... [T]he strangeness spectrum of UFO reports is so narrow that ... a *definite pattern* of strange "craft" has [been reported].... If UFOs indeed are figments of the imagination, it is strange that the imaginations of those who report UFOs from all over the world should be so restricted.[8]

In chapter 1 of *UFO Experience*, "The Laughter of Science," Hynek notes how effective US government pressures have been on the academic community, particularly by government use of ridicule of any faculty member who suggests interest in UFOs, a ridicule mirrored in the media and in attitudes of members of the general public (although with diminishing use as a majority of people have come to accept that there are both UFOs and ETI). He comments:

> The scientific world has surely not been "eager to find out" about the UFO phenomenon.... The almost universal attitude of scientists has been militantly negative. Indeed, it would seem that the reaction has been grossly out of proportion to the stimulus. The emotionally loaded, highly exaggerated reaction that has generally been exhibited by scientists to any mention of UFOs might be of considerable interest to psychologists.... [G]iggles and squirming suggest a defense against something the scientists cannot yet understand.... [S]uch exhibitions by mature scientists are [perhaps] expressions of deep-seated uncertainty or fear.[9]

Hynek notes, too, regarding scientists' reactions:

7. Ibid., 20.
8. Ibid., 23.
9. Ibid., 6–7.

The facts are not strictly scientific. Yet the data nonetheless form a fascinating and provocative field of study for those whose temperaments are not outraged by the character of the information. And it should be remembered that there are those whose fields of study abound with equally "unsatisfactory" data. Anthropologists, psychologists, and even meteorologists deal daily with evidential and circumstantial data that must be fitted together like pieces of a jigsaw puzzle.[10]

In a related discussion of scientists' responses to and ridicule of UFO event narratives, Hynek declares that the "history of science has shown that it is the things that *don't* fit, the apparent exceptions to the rule, that signal potential breakthroughs in our concept of the world about us. And it was these cases that should have been studied from many angles."[11] Returning to the practice of ridicule, he cites a 1953 article in which he wrote that "ridicule is not a part of the scientific method, and the public should not be taught that it is.... The steady flow of reports, often made in concert by *reliable* observers, raises questions of scientific obligation and responsibility."[12] He comments further: "It should be emphasized that in science one never knows where inquiry will lead—('if we know the answers in advance it isn't research')—that a primary aim of science is to satisfy human curiosity, to probe the unknown, and to open new paths for intellectual adventure."[13]

In his chapter on CE-III, Hynek notes that claimed encounters with ETI are "the most bizarre and seemingly incredible aspect of the entire UFO phenomenon." He states that he would "gladly omit" such consideration, but in the interests of scientific integrity cannot do so: "one may not omit data simply because they may not be to one's liking or in line with one's preconceived notions." He says, too, that he "shares a prejudice that is hard to explain," wondering "Is it the confrontation on the animate level that disturbs and repulses us? ... Encounters with animate beings, possibly with an intelligence of different order from ours, gives a new dimension to our atavistic fear of the unknown. It brings with it the specter of competition for territory, loss of planetary hegemony—fears that have deep roots."[14]

Complementarily, when discussing Air Force projects' internal divisions regarding the nature of UFO events he comments:

10. Ibid., 33–34.
11. Ibid., 194.
12. Ibid., 207.
13. Ibid.
14. Ibid., 138–39.

> When the mind is suddenly confronted with "facts" that are decidedly uncomfortable, that refuse to fit into the standard recognized world picture, a frantic effort is made to bridge that gap emotionally rather than intellectually (which would require an honest admission of the inadequacy of our knowledge). . . . When we are faced with a situation that is well above our "threshold of acceptability," there seems to be a built in mental censor that tends to block or to sidestep a phenomenon that is "too strange" and to take refuge in the familiar. . . . The history of science is replete with "explainings away" in order to preserve the *status quo*.[15]

In *UFO Report*, Hynek returns to several themes presented in *UFO Experience* but also includes extensive illustration and elaboration of "encounters" events cited in the Air Force Blue Book project. In *UFO Report*, Hynek refers back to the poll of astronomers he took in which 11 percent of his astronomer respondents answered affirmatively when queried privately about whether they had viewed unidentified objects. He supplements his own data by citing a Stanford University confidential survey of professional astronomers, the Sturrock survey of 1977, which found as he did that some scientists did not want to be identified even if it were only as being among those who had completed the questionnaire: only 34 of the 1,356 members of the American Astronomical Society (52 percent of the total membership) who responded to the poll (48 percent did not) were willing to sign the questionnaire, even though all had been assured that responses would be held in confidence.[16] Later in *UFO Report*, he states that 4 percent of the astronomers had stated that they had seen something in the sky that they could not explain.[17]

Reflecting on his change of mind about UFOs, Hynek observed:

> The transformation from skeptic to—no, not believer because that has certain "theological" connotations—a scientist who felt he was on the track of an interesting phenomenon was gradual, but by the late '60s it was complete. Today I would not spend one additional moment on the subject of UFOs if I didn't seriously feel that the UFO phenomenon is real and that efforts to investigate and understand it, and eventually to solve it, could have a

---

15. Ibid., 170. As examples of "explainings away," Hynek cites reactions to heliocentrism, the discovery of fossils, and the suggestion that meteors were "rocks from space."
16. Hynek, *UFO Report*, 14.
17. Ibid., 68n.

profound effect—perhaps even be a springboard to a revolution in man's view of himself and his place in the universe.[18]

During the course of his work on the succession of Air Force UFO projects, Hynek came to realize that the dominant, controlling group in the Air Force had decided *a priori* that UFOs did not exist. Consequently, their unalterable policy would be that UFO reports *"had to be nonsense."* Their perspective and policy effectively were expressed, he states, in the guiding principle "It can't be, therefore it isn't"; everyone "learned to follow suit or else."[19] Among critics, therefore, Blue Book was derided as the "Society for the Explanation of the Uninvestigated."[20] Hynek reports that after he had reviewed Blue Book files later, of the 13,134 investigations (which included meteorological explanations of events initially reported as UFOs) 354 were described in the official files as "unidentified"—even though, speaking to the general public and in Congressional presentations, the Air Force had claimed that there were no "unidentified" cases in the files.[21] He also discovered that reports made by credible witnesses, who had queried him about the event which they experienced and which they had described for Blue Book files, were missing from the files. He reports that in September-October 1973, there were seventy CE-III events in the United States that had been reported.[22] Separately from the files, astronaut and USAF pilot Donald K. ("Deke") Slayton described in a letter to Hynek his sighting of a UFO disk that, when he closed on it from behind at ten thousand feet, accelerated out of sight.[23]

In this book, too, which discusses numerous Blue Book files' CE-I, CE-II, and CE-III cases in detail, Hynek ponders people's fears about acknowledging the existence of ETI, let alone encountering ETI, in CE-III events:

> Why should it be more difficult for us to accept encounters with "creatures" than with "craft"? Probably because once we dare to admit that beings alien to ourselves exist, we are forced to face our deepest fear of the unknown, along with our more basic and specific fears of competition and hostility. But, as in the other types of UFO experiences, we cannot ignore the reports which *do* exist, for they are made by seemingly credible persons and are widespread.[24]

18. Ibid, 17.
19. Ibid., 23.
20. Ibid., 35.
21. Ibid., 244.
22. Ibid., 206.
23. Ibid., 281–82.
24. Ibid., 189.

In 1978, Hynek presented a statement to the United Nations General Assembly[25] urging the establishment of a UN agency that would assemble data from around the world regarding UFO phenomena, and facilitate an international exchange of scientifically generated information. In his statement, Hynek noted that the UFO phenomenon was worldwide, with reports having come from 133 countries, and that a majority of the general public considered such reports to be accurate (a Gallup poll found this to be the case for 57 percent of Americans, more than one hundred million people); many of these reports have been made by "highly responsible persons," Hynek said, including "astronauts, radar experts, military and commercial pilots . . . officials of governments, and scientists, including astronomers!" He asserted as a scientist, speaking for himself and many colleagues, that "the UFO phenomenon whatever its origin may turn out to be, is eminently worthy of study." He describes a French government report on UFOs which concluded that most narratives studied "involved a material phenomenon that could not be explained as a natural phenomenon or a human device." He notes that he has received UFO reports directly and confidentially from US scientists who were "associated with large and prestigious scientific organizations . . . [who] . . . are silent or even officially derisive about the UFO phenomenon." As a consequence, scientists who have "intimate knowledge" about UFO phenomena "are restrained by organizational policy to remain officially silent." Hynek noted that UFO reports are often "met by ridicule and derision by persons and organizations unacquainted with the facts" (and likely, too, given what Hynek states elsewhere about his own role as a debunker, by organizations and persons well acquainted with the facts). In concluding his presentation, Hynek observes: "I began my work as Scientific Consultant to the US Air Force as an open skeptic, in the firm belief that we were dealing with a mental aberration and a public nuisance. Only in the face of stubborn facts and data similar to those studied by the French commission . . . have I been forced to change my opinion."

Hynek is an important figure in and advocate for the scientific study of UFO phenomena, not only because of his impressive professional credentials but because of the sharp shift he made in his perspective about and approach to UFO reports and data. He remained until his death a staunch advocate of serious scientific study of UFO phenomena based on the scientific method of research, was involved in numerous investigative journeys to sites of reported UFO encounters, and founded the Center for UFO Studies (CUFOS).

---

25. The J. Allen Hynek statement to the UN General Assembly is available online: http://www.ufoevidence.org/documents/doc757.htm. The quotations that follow are from this statement.

J. Allen Hynek is a highly important voice in the UFO/ETI discussion because of his science-based approach to ordinary and extraordinary phenomena. He is particularly important as a resource to encourage scientists and other academics to seriously investigate UAP. His definition of "UFO," his method and template for analyzing particular narratives, and his organization of stories into specific categories should satisfy the honest skeptic that intensive and transparent efforts are being made to analyze with an open mind—part of a true "scientific method"—and with the best scientific techniques the UFO phenomena that have been pre-screened by his process.

Encounters and the Thought Experiment

The intellectual exploration of UFOs and ETI as a "thought experiment" receives an assist and a focus from the depth and breadth of the scientific body of work produced by J. Allen Hynek. His effort provides an excellent base for a focused consideration of plausible accounts of extraterrestrial visits. As Hynek notes, to come to ascertain the meaning of Unidentified Aerial Phenomena and the intelligent beings who operate them requires a twofold process—understanding UFOs and understanding ETI. In response to his assertion, it can be suggested that this process could be explored through three questions related to the work of Hynek and his successors: What explanations might be suggested for UFO phenomena capabilities and origins when UFOs are perceived as nocturnal lights or daytime discs (and other shapes), in radar-visual sightings, and in the categories of "close encounters"? What intelligence controls them—is it ETI? Have there been confirmed visits by ETI? The search for responses to these questions should excite scientists in multiple disciplines of the physical sciences, social sciences, and humanities. When that happens, even in the face of ridicule and other adverse reactions, humankind will make a giant leap forward in understanding the cosmos and take great strides toward finding their niche and meaning in it.

Absence of Evidence and Evidence of Absence

In reacting to reports of UAP, scholars, scientists, and other members of the public ask to see evidence of UFO encounters, a request made more demandingly when ETI is claimed to have been viewed or personally engaged. "Evidence" as presented by the testimony of even credible witnesses, including scientists studying unusual phenomena, is dismissed by skeptics (some of whom Hynek described as individuals who did not bother to consider seriously any information regarding extraterrestrial phenomena) who want "hard" evidence, such as some object left behind that is clearly not of Earthly origin. Suggestions about the possible existence of ETI, provided by scientists from diverse disciplines who have analyzed pertinent data derived from scientific analysis of reports, is insufficient for these skeptics—even when it is presented by astronomers who have viewed anomalous objects and events. The skeptics believe that if the type of "hard" evidence they require—and they reserve the right to judge for themselves whether what is presented is or is not "hard" or even "evidence"—is absent, then what is thought to exist does not exist: the absence of hard evidence is evidence of the absence of UAP and ETI. This attitude can be linked with assertions about the "impossibility" of traversing vast distances of the observable universe—for example, to solar systems of the nearest stars beyond the sun: such skeptics do not realize (or are too anthropocentric to consider) that "impossible" might describe only what might not be accomplishable now by humankind, given the present state of humans' knowledge of physics and engineering and of humans' technology. It is both anthropocentric and anthropomorphic for skeptics (academics and others) to state that distinct, probably older civilizations (given that Earth has existed for less than one-third the life of the universe) would not have advanced technologically far beyond human civilizations' creativity and humans' current scientific theory and technological acumen. Such a "certainty of impossibility" might be, too, part of the government effort to derail serious research into extraterrestrials' existence. It is ironic (and somewhat unnerving) to see scientists accept the word of government officials, who often lack scientific degrees or even minimally adequate scientific training, when they state unequivocally that UFOs and ETI do not exist, but would not accept such statements on other science-related topics when the person expressing them lacks credentials in the field about which they speak.

Faith in Beings Unseen

The phrase "absence of evidence *is not* evidence of absence" is used by those who affirm the existence of an "unseen" divine Being or of "unseen" extraterrestrial beings, as a way to respond to those who deny divine or ETI existence because "there is no evidence." Their adversaries use a variation of the same phrase: "absence of evidence *is* evidence of absence." In reality, "absence of evidence" ultimately neither confirms nor denies the existence of the "what" or "who" to which the speaker refers. If something seems to be absolutely "absent" or nonexistent, it is possible that the wrong types of evidence might have been sought, or the tools to find adequate and useful evidence were lacking. The work of the SETI Institute provides a parallel: the lack of a response from space does not provide "evidence of absence"—proof that intelligent beings do not exist—or "evidence by absence"—the assumption that since intelligent beings do not respond to human technology it means that they do not exist, do not want to be Contacted, or have not reached the stage of technology by which they would detect and respond to signals from space. However, intelligent beings might be so advanced, in fact, that they have progressed far beyond such technology long ago and no longer have it available; or, that they could not be bothered to respond to what is viewed as a "primitive" technology—they might (or might not) respond to more advanced technology; or, that they do not care to respond to radio signals, preferring, for whatever reason, not to be Contacted (a course of action that, as seen in the Introduction, Stephen Hawking has suggested that humankind on Earth should follow, so as not to be detected by hostile extraterrestrials).

The statement "there is no evidence"—expressed also interrogatively as "Where's the evidence?" by skeptical scientists, religious leaders, and politicians, among others—restricts or constricts "evidence" to that with which they are familiar, or otherwise have the capacity to detect and/or understand. The phrase "absence of evidence is not evidence of absence" can express, by contrast, a deeper, more complete openness and understanding, embodied in a willingness to keep searching, using ever-improving search engines and techniques (including, for some researchers, going beyond mere technology to telepathic, intuitive, or other types of evidentiary probes). This kind of willingness would indicate that they are open to new data and new truths (to the extent that "truth" can be subjectively or objectively known in this space and time), and are willing to learn that nothing that exists confirms or negates their belief or their prior belief and assertions. For both sides of the "evidence" phrase, it could mean that there is nothing or no being that exists *in the way* or *with the characteristics/be-ing* that they seek, because they have pre-determined what such a being should be.

In the case of divine Being, people (pro and con) often limit who/what that Being is to their finite understanding, conceptions, beliefs, doctrines, speculations, or projections at any particular historical moment—despite the (limited) extent of their intellectual or material (including technological) (in)capability of discerning something that is Other than who or what they are, or other than what their species is.

In the case of extraterrestrial intelligent being, similarly: humans project—based on who they are, what their culture is, what their scientific or religious perspectives and perceptions are, and what technology they use or have available to use—certain understandings or beliefs regarding that which they seek, in order to affirm and find "proof" for their preconceived notions, and thereby limit evidence of its existence to what they are able to discern or want to discern despite their intellectual and technological limitations at the current historical moment. Civilizations that have had billions of years more than humans to evolve and complexify should by now far surpass humans in science understandings, technological capabilities, and overall intellectual abilities. This could include extraordinary (in human perspectives) means of interstellar, interdimensional, or extra-native dimensional travel, communication, etc. It could include, too, the extent of their use of myriad particular capabilities they have based on their species' unique evolution and corporeality, such as, for example, a capacity for telepathic communication via the brain rather than speech communication via mouth and ears.

UFO Skeptics and Evolution Skeptics

UFO skeptics' reasoning and cultural conduct somewhat mirror some people's—especially some religious peoples'—response and reaction to biological data and theories of evolution. Creationists, for example, continue to declare that all creatures were individually created and directly placed on Earth by God several thousand years ago. Some have amended their position in light of what they accept as being overwhelming evidence regarding select species' evolution, and acknowledge that limited biological evolution has occurred; they do not include in their general acceptance of evolution a similar acceptance of human evolution from primates. Other people, by contrast, accept evolutionary theory as firmly established scientifically even if it needs to be modified over time because of new data and understandings, resulting in revised or even alternative evolutionary theories. They think that all life has evolved, and that eventually increasing scientific evidence will determine that this is the case. However, here, too, some deny the existence of similar evolution on habitable planets elsewhere in the universe.

In physics, creationist-like beliefs and assertions that reject all evolutionary biology are mirrored or at least complemented by UFO skeptics' 100 percent outright rejection, without any open-minded consideration, of any possibility of ETI existence. Scientists might say, for example, that "the physics aren't there." Over centuries, Isaac Newton's physics have been modified or disproven, in part, with advances in the field, the most obvious current examples of which are Albert Einstein's theories of relativity—parts of which today are in question—and contemporary theories of quantum physics. However, not all of Newton's ideas have been rejected; some continue to be used. As with absolutist creationists in regard to biology and evolution, UFO skeptics—using as their basis current physics knowledge and the limitations of current human technological developments—reject any assertions, however plausible, of sightings of extraterrestrial spacecraft, let alone of descriptions of close encounters of the third kind.

While some creationists have modified their absolutist position in light of scientific data, and now accept limited biological evolution, ETI skeptics—even some scientists, at least publicly—will not consider a similar modification: they reject all UFO reports as inaccurate, without exception. Some, ironically, accept religious faith claims of divine action without any similar proof. They believe that credible biblical witnesses and perhaps even some contemporary teachers and pastors are relating actual occurrences of events; they accept these "on faith." Not so, however, with credible witness reports concerning UFOs/ETI.

As time marches on, and despite ever emerging-weaknesses in their original theories as new scientific data become available, Darwin's and Newton's essential ideas continue to serve a research function in their respective fields of biology and physics. Regarding biological evolution, while creationists grasp at statements from those who deny evolution, and claim that new scientific theories destroy the concept of evolution or at least debunk it, particularly when scientists find a way to amend the theory (but not to reject its basic tenets), the preponderance of evidence remains on the side of the evolutionists.

Similarly, with physics and ETI: skeptics start with a preconceived notion that there is no such thing as intelligently controlled spacecraft, occupied or not; consequently, they declare that 100 percent of UFO sightings and ETI close encounter reports are, variously: spurious; misunderstood natural phenomena; fabricated for notoriety, personal financial gain, or because of some mental issue; etc. Such skeptics have no evidence of their absolute and universal claims. Advocates of the position that ETI does exist, by contrast, are not absolutists who claim that every or even most narratives are accurate accounts of an ET presence: in fact, they acknowledge facts noted earlier, such as the scientific assessment that about 95 percent of ET claims are errors, and only about 5 percent are not explainable by current knowledge. The latter 5 percent need and merit further investigation—especially by qualified, inquisitive scientists seeking credible UFO data, based on current scientific facts or at least on plausible theories that relate to what is said to have transpired. And, as Hynek observes, probably there have been numerous other UFO-related events that have not been reported because of peoples' fears of ridicule or professional punishment for even suggesting the possibility of UFOs that are intelligently controlled. There have been, too, credible witnesses' detailed reports which have been "lost" from government files—in both the United States and the United Kingdom.

ETI Consciousness: Recurring Cosmos Occurrence?

Australian astronomer Paul C. W. Davies conjectures that consciousness is a natural development in evolution, a product of operative natural laws.[26] To date, there continues to be debate over this idea, and no clear consensus regarding how and to what complexity biota evolve, nor whether or not "natural laws" are at work. The idea does suggest several questions worth pondering, even if responses to them cannot be formulated in a manner that would effect general agreement: What distinct-from-humans physical form; modes of social organization; sciences; technological development (if any); spiritual or religious organizations (if any); philosophically expressed ideologies; means of communication: spoken or written language, telepathic transfer, electronic media; economic system; natural goods use and exchange arrangements might other intelligent life have? What was needed or useful for the evolutionary development or adaptation of diverse species of ETI in distinct settings? What types of mutations might have occurred previously, are occurring at present, or can be enabled to occur in new settings to find a new niche, given that diverse ETIs might have distinct genetic constitutions and complexity, as ETI engages with and relates to diverse other biota and planetary, celestial body, or spacecraft contexts?

The Lord's Prayer Redux

In a Christian perspective, the Lord's Prayer (popularly called the "Our Father") could assist Christian laity, clergy, and institutional leaders to consider that ETI might provide insights helpful for improving human consciousness and conduct.

Earlier discussion of the prayer (chapter 2) noted that English translations of the Latin *caelum* (which is itself a translation of the Hebrew of the Hebrew Scriptures and the Greek of the Christian Scriptures) use "heaven" in the prayer and throughout the Bible when referencing a divine abode, and "heavens" when referring to the skies above. Over the years that translation has proved helpful to exhort Christians, as responsible followers of Jesus, to implement social justice among and within human communities, and to take better care of Earth and of all their relatives in the biotic community. All that is, is God's creation; all that exists began at the singular divine creative moment when it emerged in seminal form from divine being; all that emerged in that singularity was pronounced "very good" by the Creating Spirit, and has unfolded and evolved since that moment in the creative freedom granted to it, and in the solicitous love that permeates it. Humans who see themselves

---

26. Cited in Michaud, *Contact*, 79.

as "images of God" can do no less than to care for and about *all* creation as does the immanent-transcendent Creator.

The "heaven" translation helped to teach that since in God's heaven there is neither social injustice, nor warfare, nor degradation of God's place, when Christians pray, "Your will be done on Earth as it is in heaven," they express a hope and a commitment: they *hope* that, in the future, present injustices perpetrated among humans within their own species, and between humans and other biota and humans and Earth, will no longer exist; and they *commit* to act as divine "images" and work under God's guidance to do their part to eliminate the very ills and evils that they hope will not exist "on Earth" any longer.

The literal translation of *caelum* as "heavens" could have a similar impact on Christian thought and action. In a reversal of the apparent reversal of the prayer begun in earlier centuries of the Common Era, in the fifteenth century and thereafter Christian conduct, in accord with Discovery Doctrine, effectively reflected Christian belief that what Europeans were doing against peoples beyond Europe was God's will. This interpretation of the prayer as used in Discovery Doctrine practices on Earth might become operative on distant celestial bodies.

On TI-ETI Contact, Christians might realize, alternatively, that other intelligent beings have already learned to coexist in interrelated and integrated communities on their home planet and wherever they travel. They might not be "perfect," but perhaps they are closer to Christian (and other religions') "perfection" ideal than is humankind. Perhaps God's will *is* being done elsewhere in the cosmos, in "the heavens." Therefore, to pray, "Your will be done on Earth as it is in the heavens" would mean that humanity could see that ETI from the "heavens" is benevolently bringing a living example of acting according to the divine will. In doing so, they are providing humankind with a way to live in accord with the teachings of the ancient Hebrew prophets and of Jesus, to enact and embody the consciousness and conduct taught by Jesus in the Last Judgment story (Matthew 25), and to follow the practical example of the Christian community (described in Acts 2 and 4) who "shared all things in common."

In previous centuries, Christian thinkers and religious leaders denied the possibility of a plurality of worlds and of the existence of other intelligent beings. Augustine denied categorically and condemned the idea that there were humans, intelligent Earth beings, beyond Eurasia.[27] Augustine's dogmatic denial that humans existed beyond the regions of Earth with which he, a very learned man, was familiar was negated definitively a millennia later when European explorers came upon indigenous peoples in the Ameri-

---

27. Ibid., 15.

cas. However, seeds of acceptance of the idea and reality of other intelligent beings were planted by, among others, Nicholas of Cusa who stated in 1440 that there exists a plurality of worlds; without doubt, humans exist on other celestial bodies.[28] As noted earlier, Giordano Bruno was burned at the stake by the Inquisition in 1600; 40 years after publication of Copernicus's work, he taught that the stars were suns and were centers of their own solar systems, and that divine Incarnation was not unique to Earth.[29]

Despite condemnations, however, speculation continued—but the speculators dared not do so publicly. Even a scientist as eminent as Teilhard de Chardin suggested, in an essay unpublished during his lifetime, and despite his conscious efforts to be ever faithful to Catholic doctrine, that there were likely other intelligent beings in the cosmos. As such considerations became more public, and were stimulated by credible reports about UFO sightings, other citizens were energized to go public about their experiences.

Once scientists, social scientists, philosophers, and religious thinkers, among others, began to discuss the ETI issue at length and in depth, including in conferences and workshops sponsored by the Vatican, the acceptability of such discourse was assured—despite the efforts of US government and other officials to prevent speculation by denying funding for serious scientific work, deriding those who suggested that it be undertaken, and lauding and funding others, especially academics, who made public pronouncements, gave lectures, and provided other communications with the public at large in concert with government policies and practices.

Even with the drawbacks of doing so in this intellectual and social context, some people have given serious consideration to theoretical explorations of the implications of the existence of extraterrestrial intelligent beings, and their possible motivations for visiting Earth.

ETI Intent in Earth Visits: Peaceful Coexistence or ... ?

Humankind has noted an apparent and periodic extraterrestrial presence for millennia with a mix of curiosity, awe and, very likely, concern. Sightings are indicated in, and illustrated or described in varying levels of detail, by primitive art, Renaissance art, and indigenous peoples' narratives.[30] As Michael A.

28. Ibid., 12.
29. Ibid., 14.
30. Phillip Deere, a Muskogee ("Creek") spiritual leader, healer, oral historian, and human rights activist, recorded for me a quarter-century ago the millennia-old migration story of his people. He had an extraordinary memory, and narrated the story for almost four hours—without notes of any kind. In his narrative, he related how his ancestors migrated from west coast to east coast because the Creator had instructed them to go to where the sun rises; when they arrived at the Atlantic in what is now the southeastern

G. Michaud observes in *Contact with Alien Civilizations*, "The idea that intelligent beings exist beyond the Earth has been part of the Western intellectual tradition for more than 2000 years."[31]

People have wondered even more extensively in recent decades, when reports by credible witnesses about UFO sightings or ETI experiences began to multiply, not only about what the human place or niche is in the cosmos, but what the ETI intent is in regard to Earth and humans.

Narratives to date generally have described benevolent ETI. When fired upon with rockets or cannons from Earth military aircraft, such as happened in Peru and possibly elsewhere, spacecraft absorbed the projectiles into themselves (without apparent damage), or exited the scene at very high speed; they did not return fire. An ETI team shut down US nuclear missiles on at least two occasions in Montana, and on at least one occasion in the former USSR. An unconfirmed report states that near Big Sur in California a UAP used a focused laser beam to destroy a missile with a dummy nuclear payload that had just been launched—in a training exercise for a possible real launch of an armed missile.

As noted earlier, there might well be various species of ETI roaming the heavens, as indicated by the diverse types of spacecraft described and photographed. Perhaps those with less patience and less benevolent or even malevolent intentions have not made their presence known so dramatically. ("Abduction stories," which are not considered in *Cosmic Commons* because of varied degrees of credibility among the people who claim to have been abducted, might reveal different types of ETI: some people seem awed by the experience, rather than frightened, while others are frightened, and concerned about what might have been done to them, including implanting something in their body: a credibility concern here, however, is that it is possible that the US government, which has sent to UFO conferences and other platforms, on some occasions, absurdly attired people to attract attention and to ridicule the UFO idea, and on other occasions has used serious, including scientific, academics to dismiss consideration of UAP and ETI, has also arranged for some abduction narratives to be used to caricaturize events and distract attention from and cast doubt upon all UAP and ETI accounts.[32])

United States, they knew that they had come to the right place. They were settled there, in towns with agricultural lands (corn was a principal crop), when they discovered the first European explorers entering into their territory. During the journey across the continent, possibly when they were three-fourths or more on their way, they came upon some people who said they had arrived in "boxes from the sky." Phillip paused in his narration and said to me, without any surprise or concern, "Maybe they came from another planet"—and resumed recording his migration narrative.

31. Michaud, *Contact*, 1.

32. This strategy has been effective. As noted earlier, academicians and other citizens

Philip Corso, retired Army colonel who served in the Pentagon and was given responsibility to analyze data regarding Roswell and other ET events, materials, and activities, has a negative perspective on extraterrestrials, based on his assessment of their motives, which is based in turn on his evaluation of terrestrial-extraterrestrial engagement via long-distance encounters and direct Contact moments that he finds described in top secret US government files. Using the National Security Council's designation of ETI as an "Extraterrestrial Biological Entity" (EBE), Corso concludes from his analysis, and declares unequivocally, that "EBEs weren't just friendly visitors looking for a polite way so say 'Hello, we mean you no harm.' They meant us harm, and we knew it."[33] This despite his statement as fact that astronaut Gordon Cooper, while a military pilot, "scrambled with other Sabre Jet fighter pilots to intercept a formation of UFOs flying over his base, but when his fighter group got too close, the formation of UFOs flew away."[34] He also stated—referring to the US military's proposal to construct a lunar base—that the United States "was determined to stake out its territory and defend the moon"—where, he said, EBEs already had a base.[35] Unwittingly, Corso was expressing well a modern formulation of the Discovery Doctrine. In light of his comments, Neil Armstrong's act of planting a US flag on the moon has far more implications that merely celebrating "one small step for a man, one giant leap for mankind." In a similar Discovery vein, Corso declares that the US lunar successes "demonstrated that if there were any deals to be made, any proxy relationships to establish, the Soviets were not the ones to deal with."[36] In fact, in an appendix to his book, a declassified 1999 "US Army Study for the Establishment of a Lunar Outpost" states in its Summary regarding lunar successes other than a military base,

> these accomplishments will not have the same political impact that a manned lunar outpost could have on the world. In the still vague body of fact and thought on the subject, world opinion may view the other applications similar to action on the high seas, but will view the establishment of a first lunar outpost as similar to *proprietary rights* derived from first occupancy.[37]

In this statement, Discovery Doctrine is well evident. Just as European seafaring nations established their firm claim to have Discovered new

---

fear public ridicule, and fear for their jobs, even when they have observed something extraordinary.

33. Corso, *Day After Roswell*, 132.
34. Ibid., 135.
35. Ibid., 170.
36. Ibid., 171.
37. Ibid., 316. Italics in original.

territory by establishing a fort in that territory, so, too, would the United States establish its firm claim to the moon by constructing a lunar military base the military named "Horizon."

Corso's military analysis complements the scientific comments of Stephen Hawking extracted from his own evaluation of ETI existence and mission probabilities. These were projected in part from computations by his "mathematical mind," and in part by his androcentric, anthropocentric, and anthropomorphic attribution of human conduct and consciousness to ETI. Hawking, as noted previously, discoursed eloquently and directly regarding reasons why SETI should cease to seek Contact, and humans should strive to evade notice by or Contact with ETI in general because extraterrestrial intelligent beings likely have destroyed their planet of origin and now roam the universe trying to locate territories they can inhabit or natural goods they can extract. Either objective would be easily accomplished, given that their vastly superior technology would have begun to be developed far in advance of humans' evolutionary emergence.

Robert Salas, a retired USAF captain who was in the Oscar Flight control center module when the flight's ten missiles were shut down one by one near Great Falls, Montana, has a positive perspective on ETI, in contrast to Corso. In responding to the query, "Why are they here?" Salas states:

> Clearly they have announced their presence to all humanity. . . . I think the main message is that we humans seem to be bent on our own destruction by the way we are treating ourselves and our home planet. Nuclear energy and weapons is just one aspect of this and, because of my experience, that has been my main message whenever I speak or write. Further, we can only resolve these questions of the environment and our proclivity to be warlike through cooperative actions. . . . I think ET wants to see some results in this direction before making full contact.[38]

The first part of Salas's assessment corresponds with the reason Stephen Hawking provided (see Introduction) to justify humans' need to have, in the very near future, a lunar base and a Mars colony: we are destroying ourselves and Earth. The last part expresses a hope, based on their prior known actions, that ETI has peaceful intentions and perhaps even is nudging violent elements of humankind away from their propensity to self-destruction.

It is possible, of course, that both Corso and Salas have correctly assessed ETI: they might be describing respectively distinct species of ETI who have diverse places of origin in the vast cosmos. These distinct intelligent species' conflicting consciousness and their conduct corresponding to their

---

38. Robert Salas, e-mail message to author, August 24, 2012.

respective perspectives might have brought them into conflict in the past. However, since different types of UFOs have been reported, they might have come to a truce, based on the respect each has for the other's technological and even military-technological capabilities.

If the latter is the case, the fact that these contradictory and conflictive ideologies have prompted both types to become interested in Earth presents a chilling prospect: that competing ETI species might act just as the US and the USSR did during the Cold War: they hoped, initially, to limit a nuclear war just to their respectively controlled nations in Europe, whose peoples and places would be devastated by the nuclear arsenal unleashed by the two superpowers; Western Europe would be the superpowers' proxy "nuclear sacrifice zone," not a pleasant prospect for Europeans. Similarly, in ETI species' struggle for hegemony and goods acquisition, Earth would become a "sacrifice zone."

A more positive scenario is possible: that all ETIs have evolved to become socially cohesive, stable societies, and accept and collaborate with other ETIs with whom they come into Contact. In this case, Corso's aggressively military perspective (he states repeatedly that his Pentagon mission was to battle communists and extraterrestrials concurrently) would have been, then, an overreaching analysis and projection of ETI conduct. Several credible witnesses, in fact, have suggested the possibility that the US government has fabricated or exaggerated ETI conduct to cast it into a negative light—even though ETI has not retaliated when air force personnel from diverse nations around the world have fired on UAP that appear to be operated by intelligent beings.

Humans should take precautions if they expect the worst case scenario to be true, but simultaneously develop prescriptions for dealing with the best case scenario for which they hope. Since at least one species shut down nuclear missiles in underground silos in both the Soviet Union and the United States, ETI obviously have the technology to attack and subdue human populations and colonize and control Earth—which, science fiction movies notwithstanding, they have not done. Humankind should continue to develop collaborative conduct such that Contact will prove to be not a conflictive event, but a moment in which to construct a wider cosmic community.

In the latter case, cosmic peaceful coexistence becomes possible. A shared cosmic consciousness will effect a cohesive cosmic community, and conduct that corresponds to it.

Commons Good and Commonweal: On Earth and in the Heavens

The *commons good* has been understood traditionally as the good of the Earth environment that is the home and habitat of all life. Earth has the capacity and the fruitfulness to meet the needs of the entire biotic community—when its natural goods are used responsibly in place, or removed—dis-placed—to meet human needs, and shared equitably. Humankind has now initiated and, assumedly ETI has developed extensively, the capability to explore beyond their home planet, onto nearby lunar bodies, home solar system planets (note the successes of the various Mars rovers that are or have been operative, Curiosity being the most recent at this writing), and eventually (or, actually, perhaps, for ETI) distant places in the cosmos. Consequently, a more inclusive *cosmic commons good* should be incorporated in theory and practice, in consciousness and conduct, by humankind—and by ETI, whenever or wherever ETI is Contacted.

Similarly, the terms *commonwealth* (several US states call themselves "the commonwealth of" rather than "the state of") and *commonweal* have been used extensively to describe an area or place on Earth in which the common well-being of all residents is sought, and laws and legislation are used to provide government or government agency efforts to initiate or ensure this type of mutually beneficial association. Today, again in light of actual or pending space voyages, a *cosmic commonweal* should be conceived and efforts dedicated to realize ("real-ize"; "make real") its concrete establishment and continuance.

The words of the second paragraph of the US Declaration of Independence, signed on July 4, 1776, by members of the Continental Congress, might prove helpful in this regard, as amended in order to promote a *cosmic commonweal*:

> We hold these truths to be self-evident, that all intelligent life is created equal, that they are endowed by their Creator with certain inalienable rights, that among these are life, liberty, social equality, community well-being, commons well-being, commons goods equitable sharing, and the mutually and socially responsible pursuit of happiness.—That to secure these rights, intraspecies and interspecies governments are instituted throughout the cosmos, deriving their just powers from the informed consent of the governed.

Obviously, as with the original Declaration of Independence, this is a hopeful, even idealistic proposal and paragraph. It is, however, *utopian* in a positive sense: a vision of what is not yet, but what might be—if terrestrial and extraterrestrial intelligent life were to accept it or something equivalent to it as an ideal toward which to strive. Consequently, it would provide a vision to inspire collaboration. If achieved, or even in the process of striving to achieve it, gradually it would make this the future cosmic social reality, into which newly encountered intelligent beings would be invited and integrated in order that they, too, might share in its transcendent vision and immanent, material implementation.

The *cosmic commonweal*, then, would be an interspecies, inter-intelligent life concept and evolving construct. In conceptual theory and as an ideal practice toward which intelligent beings should strive it might be codified in a Cosmic Charter, as proposed in the next chapter. It might come to apply to Contact and colonization on Earth and in the heavens. In current practice and as a possible paradigm on Earth for what might be possible universally, the cosmic commonweal in a terrestrial form and application would mean that current ideologies, political and economic arrangements, and present and future distributions of Earth's natural goods would be significantly altered, or at least partially amended, to provide for the common good of all humankind.

The commonweal would be a way and a place that ensures that all people have at least the material necessities of life; that the common good of all biota is promoted such that their habitats are conserved and their evolutionary possibilities continued; and that the well-being of Earth, our home planet and our shared habitat upon which we and all biota depend for our very existence, is promoted, effected, and sustained intergenerationally.

The "dialogic relation" suggested earlier between present and future, and between Earth and other contexts, would be initiated terrestrially now in anticipation of or at least with hope for it being complemented in exoEarth settings. The effort to promote the cosmic common good in a cosmic commonweal would be complemented by and implemented by a theory of *cosmosocioecological ethics* and practice of *cosmosocioecological praxis ethics*, both as cosmic extensions of Earth-born and borne socioecological ethics, and when integrated with extraterrestrial perspectives and practices in Contact contexts would provide a local and potentially universal *cosmoethics*.

Cosmic Existence and Coexistence

The ideals, principles, and proposals for human conduct on Earth that have been gradually developed internationally, and presented in United Nations documents and the Earth Charter (as discussed in earlier chapters), if realized or at least partly embodied not only in human consciousness but also in humans' conduct, individually and nationally, would provide a foundation for a process toward utopias in other worlds, and stimulate steady progress toward realizing them.

A dynamic Cosmic Charter developed terrestrially, if evolved and implemented through collaboration with other intelligent beings, would provide an initial foundation upon which commonweals and extraterrestrial utopias might be built, develop, stabilize, and endure in order to provide for the relational well-being of all life. The next chapter explores this possibility.

11

Cosmic Charter

Constructive Consultation and Consociation

The extraordinary event of interpersonal Contact between terrestrial and extraterrestrial intelligent life in a dramatic, public way, if it occurs and is acknowledged, would have substantial impacts on the distinct species involved. As discussed previously, on Earth it would stimulate discussion of differences in perspectives and experience especially in four areas: ecology, ethics, economics, and ecclesiology/religion/spirituality. It might well alter the ways in which we humans understand or envision ourselves not only in our particular Earth home in our specific solar system, but in the immense cosmos.

Prior to such Contact, it would be well for humankind to consider their current and projected interbiotic, interplanetary, intergenerational and, perhaps, interdimensional responsibility. How should humans relate to other members of the cosmic biotic community, whether intelligent or less complex, whom they have not encountered previously? What might biotic rights be when considered in a cosmic context? How would "natural rights" be explained and elaborated interculturally and intercontextually? In such considerations and speculation, it would be helpful to develop proposals for human conduct, expressed in initially formulated principles, which would evolve in context if there were a constructive consultation with ETI on Contact, or as soon as possible after Contact. Efforts to consociate, probably tentative at first as each species seeks to understand the other and weigh concepts and principles that might not have been previously considered, should prove to be beneficial to all parties. As intelligent beings, all would be conscious of their interrelationship within integral being, however understood and expressed.

The communicants might explore together their respective understandings of *ecological ethics*, as expressed in relationships among biota, including among intelligent communities of biota; note how they correspond to and complement each other; and from that foundation formulate an innovative, interrelated ecological ethics. Through a similar process involving their respective understandings and experiences of *social ethics*, each party could inform the other, in a mutually respectful and insightful manner, what they have come to accept and strive to live by in their respective approaches to justice; consequently, the parties involved could collaboratively interrelate and integrate their congruent and complementary ideas, values, and principles. The collaborative consideration of species' thinking on these two key issues should result in a preliminary proposal for a *cosmosocioecological ethics* which would become operative in context as *cosmosocioecological praxis ethics*.

In the consociation and collaboration, issues of species regard and species collaboration would take on new meaning. Intelligent beings would directly acknowledge the intrinsic value of each other and of all other biota, including those not previously encountered by either TI or ETI, and be able to formulate a responsible, judicious assessment of the potential instrumental value of some beings (abiotic and biotic) that provide natural goods needed for intelligent species' sustenance, survival, and social well-being.

A common exploration of a possible Cosmic Charter could be developed on present or future common ground (literal and figurative). It would present a dynamic and evolutionary utilization of terrestrial and extraterrestrial socioecological ethics in ongoing dialogue, and include engagement with the respective intelligent beings' communities' social ethics (originating in their respective places of origin, and developed further during their socioecological experiences in diverse contexts). Ecological ethics analyzes abiotic and biotic contexts and needs, and expresses and guides commitments to cosmos and community in material and intellectual relational environments. Social ethics advocates for the basic well-being of all members of the human community, with particular concern for the most vulnerable. Socioecological ethics relates principles to context, prioritizes and applies principles in context, and reformulates principles, as necessary, for changing or changed physical or philosophical contexts.

Spirit and Science: Collaborative Consultations in Cosmic Contexts

Spirit and science might serve as bridges between terrestrial and extraterrestrial intelligent life dialogue and developments that result from exchanges of these components of culture, however understood by significant communities of a particular culture as a whole, or in its individual members. Challenges presented by Contact could stimulate spirit and science—in ways reflective of their current collaborative efforts on Earth to address Earth's ecological problems—to negotiate contrasting perspectives and develop complementary strategies and cooperative projects, based on shared values. This collaborative dialogue could provide a theoretical foundation upon which a Cosmic Charter would be conceived and concretized, or continued to be developed and disseminated.

A Cosmic Charter

In order to promote planetary and interplanetary well-being, people need to develop a new sense of their place in and relations with their Earth home and in the expansive and still expanding cosmos. In so doing, humans would learn to respect both who they are and where they are going, and become responsible explorers and inhabitants throughout space and through time. People would be educated and stimulated to be socially and ecologically considerate, to be respectful during terrestrial-extraterrestrial engagements, and to develop—before and during explorations—religious or spiritual reconciliation with the implications and impacts of Contact.

Humans as a whole do not know with certainty whether or not aliens have come among them already–the Roswell Incident in New Mexico and Hudson River Valley sightings in New York, among others, are still being debated. People do know that they live in an extraordinarily dynamic and beautiful creation–the wonder of the Hubble telescope's images is not debated, nor is the intricacy and diversity of life on Earth, some of which is viewed only through microscopes.

In preparation for space travel and for ETI encounters in space or on Earth, humankind could develop a Cosmic Charter preemptively, in order to promote changes in human consciousness and conduct—for the mutual benefit of TI and ETI.

A Cosmic Charter, as originally developed and elaborated on Earth, would advocate for interbiotic, interplanetary, and intergenerational responsibility. It would incorporate the best of current international thinking on three key ethical issues: social justice and well-being; Earth and exoEarth ecological well-being; and terrestrial-extraterrestrial mutual respect and accommodation. As discussed earlier, this integration would provide a foundational and dynamic *cosmosocioecological ethics*, its activation in specific space contexts as *cosmosocioecological praxis ethics*, and the integration of respective TI-ETI theories and practices as *cosmoethics*. A Cosmic Charter, guiding Contact and catalyzing cooperation in cosmic contexts, would facilitate the acknowledgment and acceptance of biota's inherent intrinsic value and natural rights. A corollary understanding would be that out of necessity, in some places and times, some biota might come to have, because of need, instrumental value for other biota; predator-prey relationships would be accepted as continuing naturally during the evolution of biota. (Intelligent beings would not apply such an understanding to intelligent biota.) Similarly, the inherent intrinsic value and natural rights of abiotic beings would be acknowledged and accepted as they are in place; however, because of a pressing need they too, in times of need, would be judged to have instrumental value and extracted (e.g., minerals) or diverted (e.g., a river) from their origin place.

The Cosmic Charter, as an intentionally dynamic document, would contain initially formulated principles that would adapt to, adapt from, and integrate the perspectives and perceived needs of terrestrial and extraterrestrial intelligent species involved in the dialogue as it evolves in diverse abiotic and biotic cosmic contexts. Intelligent life from Earth and elsewhere would collaborate to refine the document as needed, over time, to address unforeseen circumstances and needs. Such collaboration would include integration of whatever *cosmosocioecological praxis ethics* had been developed and used previously and independently, by terrestrials and extraterrestrials, and incorporate new thought and theory resulting from the dynamics of their interaction. The collaboration would assume a continuing dynamic and evolutionary aspect of Charter secondary principles, while seeking to ascertain and retain apparently universally applicable and accepted core principles that transcend the particularity of cultures and species, but are contextually relevant for and potentially can be immanent in every social setting.

The development of a Cosmic Charter requires contextual flexibility of ethical principles in diverse extraterrestrial contexts. Engagement with ETI might require, for example, that some human and ETI ideas and principles be adjusted to context. An ongoing *praxis ethics*[1] approach would provide a means whereby humans (and ETI) might resolve variant perspectives and principles as they negotiate acquisition and development of natural goods on celestial bodies.

The concepts of a cosmic commons and a Cosmic Charter suggest a realistic assessment of human cosmic aspirations, and ethical principles and an ethical process to guide space exploration, acquisition of natural goods from

---

1. *Praxis ethics* is the dialogic relationship of theory-practice-theory . . . and principles-practice-principles . . . in context. In socioecological—and cosmosocioecological—contexts, possible principles and their application as choices in concrete situations are proposed. While courses of action might be considered theoretically beforehand, decisions on how to activate, prioritize, and use them or revise and use them occur in concrete situations, in context.

other worlds, and encounters with alien species. They suggest, too, ways for people to react to and relate to potential and actual Contact with intelligent life, in order that a sense of cosmic displacement, such as was engendered by both the Copernican and Darwinian revolutions, will be avoided—or at least mitigated.

Humankind's original, terrestrial integration of "cosmic commons," Cosmic Charter, and cosmosocioecological praxis ethics would incorporate, innovatively, considerations of *time* (present-future relationships), *space* (human exploration of the cosmos), *relation* (human-alien intelligent biota interaction), and *place* (terrestrial-extraterrestrial ecological and economic impacts of human colonization of other worlds: extraction of natural goods; alteration of abiotic places; and intrusion in biotic evolution).

The integration would include reflection on projected socioecological impacts of terrestrial-extraterrestrial engagement; suggestions for a present-future-present dynamic that would promote ecological and social responsibility on extraterrestrial and terrestrial contexts; a dynamic praxis ethics, originating terrestrially but adaptive to extraterrestrial contexts; and collaboration of humanist scientists, members of faith traditions, sociologists, environmentalists, and ethicists to process impacts of space exploration and Contact, and to propose dynamic ethical principles and practices to minimize or mitigate harm.

A Cosmic Charter, in order to be both internationally and interculturally accepted (and religiously neutral, comparable to the Earth Charter and official UN instruments in this regard), must be developed collaboratively. Input must be invited from all faith traditions—Christianity, Judaism, Islam, Buddhism, atheism, et al.—despite the degrees of fundamentalist orientation and absolutism characteristic of segments of each of the preceding. Currently, for example, Christian fundamentalists claim that the Bible is literally, scientifically, and historically factual "truth," and deny biotic evolution; and atheist fundamentalists claim that materiality is the only "reality," and that anything that cannot be verified, quantified, or falsified does not exist. The narratives of religious faiths ordinarily express belief in and engagement with a spiritual dimension of reality, which is permeated by the presence of a transcendent or transcendent-immanent Spirit (who has distinct names in diverse cultures, and whose characteristics are expressed in understandings and doctrines particular to each of those cultures and belief systems). The essence of these beliefs might be infused in some way, *sans* specific doctrines, into a Cosmic Charter such that both believers in religious faiths and believers in atheistic faiths might recognize traces of their transcendent, "metaphysical" affirmations—not as a watered-down "least common denominator," but as a subtly infused most common integrator. In terms of ecological issues and concerns, theists and atheists have much in common, as has been affirmed or at least indicated by, among others, Harvard emeritus biologist Edward O. Wilson and other secular humanists; Orthodox Patriarch Bartholomew I of Constantinople, the "Green Patriarch"; Rabbi Art Waskow of the Shalom Center; and others.

The focus and content of the Cosmic Charter would not be, of course, specifically "religious" even in transcendent moments. Humankind does not know, nor can anticipate, whether or not extra-terrestrial intelligent beings have some form(s) of religion or spirituality. Since religions are born in particular cultures in a specific era (although they may come to be embraced by people of other cultures and times), ETI's understandings and beliefs can be expected to be significantly different from those of Earth peoples—with, perhaps, a shared belief in or experience of, in some way, the presence of a Spirit immanent in and throughout the cosmos.

As suggested previously, international agreements and documents initiated, developed, and promulgated on Earth, especially those emerging (directly or, as with the Earth Charter, indirectly) from United Nations member states, would be especially helpful as foundations for, and incorporated in some way into, a Cosmic Charter. At particular historical times the states have carefully crafted statements to which peoples of diverse cultures, ethnicities, social classes, and nations could all agree, and commit themselves to effecting immediately or gradually, in whole or in part, within their own national borders and, with the assistance of UN agencies, globally. They have provided a foundation and a starting point for human international development and acceptance of a complementary cosmos instrument.

The Cosmic Charter, once it becomes a foundational but not fixed document, will provide a firm and visionary elaboration of a consciousness and conduct that will enable TI and ETI to consult together to explore ways of constructing, in theory and practice, mutually agreeable steps, however tentative, experimental, and transitional, for their consociation in a transcosmic community and eventually, perhaps, a pancosmic community within integral being.

It is possible that ETI includes both malevolent and benevolent interstellar explorers, colonizers, and industrial representatives with military accompaniment. Until sustained direct Contact occurs with either or both types of ETI (should contrary characteristics and intentions exist) humankind will not know with whom they are interacting. An Earth international community-created Cosmic Charter would still be an important document to develop, in the face of the current uncertainty regarding ETI intentions. If both types exist and are encountered, humankind might strive to form relationships with representatives of benevolent ETI, for mutual benefit—including joint defense efforts against malevolent ETI; these relationships will have a firm foundation, in human eyes, if the Charter has been formulated and begun to be implemented in terrestrial settings and on such celestial bodies, if any, where humankind has become present.

If only malevolent ETI is encountered, and humans already have begun to use Charter principles and proposals to better Earth's socioecological conditions and to think and act on celestial bodies in a manner geared toward avoiding the harmful consciousness and conduct that were detrimental to human community and Earth, the effort still will have been worth making. If only benevolent ETI is encountered, then the kind of collaborative process theorized and proposed for interaction will be facilitated on Contact.

A Cosmic Charter, then, could help humanity now and into the future, in Earth and exoEarth milieus, by formulating and implementing principles, policies, and practices that effect at least the beginning of change in human consciousness and conduct, as has been suggested numerous times in *Cosmic Commons*. Change is necessary for the responsible exploration and colonization of celestial bodies—whether or not ETI is encountered.

International Earth Bases for a Cosmic Charter

Over the past sixty-plus years the United Nations—the international body formed in the aftermath of World War II to promote world peace and human rights—has developed and promulgated through its member nations several documents that have a significant and substantive bearing on Earth and exoEarth milieus. The principal ones to be explored and edited to derive principles for a Cosmic Charter that embodies pacific, humanistic, and socioecological concerns are in the areas of human rights and responsibilities vis-à-vis humankind; human responsibilities vis-à-vis Earth; and human responsibilities in extraterrestrial contexts (for which case incorporation of supplementary concepts and principles will be suggested, to become operative in the event of Contact). The Earth Charter complements UN documents: it, too, was internationally developed and proposed to be a UN initiative and statement, in the vision of its originator, Maurice Strong, and of the core of people who developed and disseminated it with him. Many of these documents' Earth- and human-based principles and proposals can be adapted to extraterrestrial places and shared with ETI in the form of a dynamic, continually contextually developed and adapted Cosmic Charter.

International Human Social Rights and Responsibilities: Cosmic Inter-Terrestrial Adaptation

Adaptation to and from context should characterize a dynamic Cosmic Charter. The word *inter-terrestrial* is used here. Although *terra* (Earth) literally and linguistically means having to do with humans' planet of origin, it will be used in this particular conceptual setting to indicate the home planet of any species, terrestrial or extraterrestrial, regardless of its cosmic setting. It should be remembered that in outer space, whenever humans land someplace (and even in their journey to get to that place) they have been themselves—from the time they left Earth (and while they were based on the moon, Mars, Europa, or elsewhere after having left Earth)—*extraterrestrial*, that is, away from Earth. Should Contact be made with ETI inhabitants of and on a planet other than Earth, for example, humans would be the "extraterrestrials" both in terms of what they had become upon leaving Earth, and of what they become when arriving elsewhere. Humans would be "aliens" to native inhabitants—much as the European explorers who were discovered by native peoples who saw their arrival on Turtle Island were aliens (foreign to this place) and extraterrestrials (originating in a place other than the natives' Earth place, the natives' own homeland). Social rights and responsibilities initially developed to guide, and to offer ideals for, human terrestrial conduct could be extrapolated and adapted for and to exoEarth contexts, as human extraterrestrial rights and responsibilities.

The UN Universal Declaration of Human Rights→Universal Declaration of Intelligent Being Rights

In the UN Universal Declaration of Human Rights (DHR) numerous principles can be adapted with little alteration in order to be applicable to and for intelligent beings' interspecies Contact[2] as a Universal Declaration of Intelligent Being Rights (DIBR). (Parenthetical references in the following indicate paragraphs and articles in the original documents issued by the UN.)

The preamble of a Universal Decaration of Intelligent Being Rights, adapted to exoEarth contexts, would declare that cosmic freedom, justice, and peace are founded on recognition of the dignity and "inalienable rights" of all "intelligent beings" (par. 1). On distinct worlds in space, disregard for intelligent species' rights could lead to "barbarous acts" (par. 2). If rebellion and war is not to be TI's and ETI's only perceived recourse to respond to these acts, their intrinsic rights must be codified in and safeguarded by laws

---

2. The original expressions of these principles and of those in other UN documents and the Earth Charter were elaborated in chapter 3, "Terrestrial Transformation."

(par. 3), and by fostering "friendly relations" among TI and ETI (par. 4). TI and ETI beings must affirm faith in intelligent species rights, including "equal rights" among people of diverse genders, sexual characteristics (anatomically, psychologically, and socially), and sexual orientations—"men" and "women," or "male" and "female" might not apply in exoEarth settings (par. 5). There must be "universal (literally!) respect" for all intelligent beings' rights (par. 6), based on a "common understanding" of who they are in themselves and in relation to each other(par. 7). The Preamble's concluding paragraph should be amended to a "common standard of achievement for all intelligent beings" and their political entities, however constituted (they might have evolved, as intelligent beings *per se* or culturally, beyond nation-state forms of political and social organization).

The DIBR's thirty following Articles present additional human insights and principles that might be bases for interspecies collaborative development of a mutually acceptable Cosmic Charter. This, in turn, would adapt to, adapt from, and be adapted by such other intelligent species as might be Contacted over eons of time in extensive cosmos space. Some species to be Contacted in the future might emerge and evolve to be intelligent beings on worlds as yet unborn or barely born today, worlds that will eventually orbit currently unborn stars. Consequently, the DIBR principles would include: "all intelligent beings are free, and equal in dignity and rights" (Art. 1); all intelligent beings are entitled to all the DIBR's rights and freedoms, regardless of distinctions among species who have diverse characteristics complementary with those currently expressed in human terms as "race, color, sex, language, religion, political or other opinion, national or social origin, property, birth, or other status," or based on any political or economic structural differences (Art. 2); rejection of slavery or any other unjust practice of intrapecies or interspecies oppression of any intelligent species (Art. 4), including torture (Art. 5); and equal status in formulation and enforcement of legal documents that promote social justice in any terrestrial or extraterrestrial context (Art. 6, 7). So, too, should all intelligent species, once they have worked out planetary cohabitation arrangements, have freedom within their own borders (Art. 13) and, it should be added, across "national" and other political boundaries in planetary contexts; and, to seek asylum in locales external to their own places of origin should the need arise (Art. 14).

The "right" of property ownership, individually or "in association with others" (Art. 17) would be problematic in extraterrestrial cultures that have communal property, just as it was for indigenous cultures on Earth—on every Earth continent—prior to European monarchies' elimination of the practice through internal imposition of Inclosure Acts, and external imposition of a self-serving Discovery Doctrine when they came upon indigenous territories, agricultural lands, and hunting, fishing, and sacred sites. The "association" qualifier might be acceptable, depending on how "association" is defined: many people in the United States and abroad are appalled and angered that US law has arbitrarily (under the influence of politically and economically powerful factions) declared that commercial and industrial corporations are "persons" and entitled to the rights previously established for human beings who live, breathe, ingest nutrition, reproduce, have bodily organs whose function sustains organic beings, and eventually die. Human ideological constructs are not identical to human beings themselves, who are members of the biotic community; they are merely socioeconomic or sociopolitical inventions that were designed originally, in many cases, to benefit, not control, humans who live as social beings with social arrangements in human-organized and -oriented societies.

Freedom of "thought, conscience, and (spiritual practices)" (Art. 18), by contrast, should arouse no controversy; neither should the right for all intelligent species to participate in some form of governance "directly or through representatives freely chosen" whose authority depends on the will of regional or global politically cohesive intelligent species who are the constituents of those (to be) elected (Art. 21). Work-related rights should be mutually acceptable among intelligent species for their mutual benefit. These include the right to social security (Art. 22), and the right to work, to choose the job that will employ them and enable them to provide for the needs of themselves and others, rights to good working conditions and "protection against unemployment," to "equal pay for equal work," to "just and favorable remuneration," and to "form and join trade unions for protection of interests" (Art. 23); and the right to an adequate "standard of living" that would enable intelligent species to provide for themselves and for members of their extended forms of social organization (such as family, cooperatives, and communes) with essentials needed for "health and well-being"; these include food, clothing, housing, medical care, and necessary social services, and "the right to security in the event of unemployment, sickness, disability, widowhood, old age or other lack of livelihood" in circumstances over which they have no control (Art. 25). The promotion and preservation of these universal terrestrial and extraterrestrial rights require that each person from every intelligent species fulfill "duties to the community" (Art. 29).

The UN DHR, then, is readily modified from an Earth instrument to a cosmos document, a DIBR, in order to provide a "universal" statement of the rights of all intelligent beings and be incorporated into a Cosmic Charter.

The UN Antarctic Treaty

The Antarctic Treaty (hereafter: AT) integrates the territorial rights of humans organized into states with humans' rights to access and use Earth's natural goods and common places. It is a bridge between the DHR's advocacy of human-oriented rights and UN documents such as the World Charter for Nature and the UN-originated but independently developed Earth Charter, both of which argue that humans should equitably share in benefits derived from Earth's earth as place and shared space, and from the natural goods Earth provides. The Antarctic Treaty (1959), too, was a base for and a bridge toward the Outer Space Treaty (1967): the former document developed international relationships and policies for Antarctica that enabled agreement, in the latter document, on parallel provisions by member states in regard to exploration and colonization of Earth's solar system's planetary neighbors and their lunar satellites.

The Preamble to and Articles of the AT, as with the DHR, are amendable and adaptable for incorporation into a Cosmic Charter. The Preamble's modifications could include that it is "in the interest of all intelligent life that planets, moons, and other bodies of solar systems, and such celestial bodies as shall be found beyond, shall be used exclusively for peaceful purposes" (par. 2); that substantial scientific developments could result from cooperation of and contributions from all intelligent beings in collaborative research and development projects (par. 3); that ongoing cooperation in science is in accord with the "interests of universal science," and quests to understand and respond to the requirements of diverse biota and abiotic places in cosmic integral being will promote common progress (par. 4); and that the Cosmic Charter will facilitate use of space for "peaceful purposes only" and thereby advance Cosmic Charter "purposes and principles" (par. 5).

The Articles that follow would state, in a Cosmic Charter, that exploration in and use of the natural goods of space should be for "peaceful purposes" only. An integrated defense and peacekeeping force will be collaboratively developed or expanded, as needed, to preserve peace based on justice among signatories and with such other intelligent beings as might be found; except for collaborative operations between intelligent beings designed to defend Signatories jointly against possible aggressive invaders from outside their own places and their shared space, there shall be no military bases or maneuvers or weapons testing; however, because of the increasing danger of ever more sophisticated weapons that have a longer range, and the consequent enhanced danger to nations and whatever political entities ETI has, *no* members of any Signatory's military should be present on any base on a planet or planet's satellite that would threaten the well-being of those who live on nearby planets or celestial bodies: no accommodation should be made to militarists of whatever Signatory who insist on having such a presence on the moon or other places in range of Earth, for example (Art. I); "scientific investigation" and "cooperation" shall continue (Art. II); to sustain scientific cooperation, the Signatories should exchange "scientific programs plans," "scientific personnel," and "scientific findings" and should establish "cooperative working relations" with each other including through cooperative organizations, or shared agencies, that intelligent beings conjointly establish (Art. III); that no interpretation of the Cosmic Charter can mean that the Charter's associated beings have renounced rights of or claims to their respective places of origin in the universe, with which they are related biologically, historically, and culturally, or to its natural goods (Art. IV); that disposal of nuclear and other toxic and hazardous waste (should any be generated: perhaps TI and ETI would have developed technologically beyond their production) shall be in mutually agreed upon sites, and shall not endanger biota on those sites, if any, at the present or future stages of their natural evolution; and that intelligent beings should strive collaboratively to provide other natural goods or produce alternative manufactured goods manufactured alternatives that would enable elimination or at least substantial reduction of nuclear, hazardous, and toxic waste generation (Art. V); that Cosmic Charter provisions apply throughout space, and are not confined to places, or proximity to places, in areas that are distant from the places of origin of the respective intelligent species (Art. VI); that the respective Signatories noted in Article IX may designate their own observers who will have access to any Signatory's areas or facilities to inspect stations, installations, equipment, and transport spacecraft bringing in or taking out personnel and equipment, that every Signatory will have the right to aerial or satellite reconnaissance over all areas being explored or colonized at any time, and that the Parties will provide notice in advance of their exploratory and other expeditions on a celestial body's surface (as allowed by Art. I) that are established in space (Art. VI); that Signatories alone have authority over their observers, and any jurisdictional disputes should be resolved immediately into a "mutually acceptable solution" by Signatories (Art. VIII); that Signatories shall meet regularly on a commonly accessible place in space, or by reliable real-time audio-visual communications media, to exchange data, consult on "matters of common interest pertaining to places and space," and propose measures to further Cosmic Charter "principles and objectives," including that space be used solely for "peaceful purposes," that scientific research and international scientific cooperation be facilitated among Parties, and that "preservation and conservation of biota and their habitats in space" should be ensured (the phrase "living resources" in the original has been removed, since it reduced biota to instrumental value for humans without regard for biota's own intrinsic value) (Art. IX); should conflict emerge, Signatories should resolve disputes among themselves; any irresolvable dispute should be referred to a mediation body, a *Cosmic Charter Council* established by the Signatories as part of the overall governance structure of confederated worlds, whose decisions shall be final (Art. XI); the Cosmic Charter may be amended or modified by unanimous consent of the Signatories); and any intelligent beings encountered in the future should be invited to join a Cosmic Confederation should one be developed, and may accede to the Cosmic Charter if all of the Signatories consent (Art. XIII).

The AT's promulgation and international acceptance decades ago (and its ongoing operation and influence today) provides another segment of the foundation of a Cosmic Charter that might be acceptable and implemented on Earth and in the heavens. It would promote positive relations among Signatories from diverse intelligent species who have originated in disparate places in space and come into Contact in a specific place or places.

Declaration on the Rights of Indigenous Peoples

The UN Declaration on the Rights of Indigenous Peoples (DRIP) has implications well beyond its current application to Earth's native peoples and their traditional places. It could be readily amended, resulting in a cosmic Declaration on the Rights of Intelligent Beings (DRIB).

During the billions of years since the cosmos burst forth, its ongoing and unending dynamics have meant that new stars are born constantly, new planets come into existence and revolve around those stars, and new and diverse life emerges and evolves—likely including, over time and in some places, intelligent life. In this cosmic context, as life evolves its intergenerational evolution means that at any "moment" of cosmic time there might well be intelligent beings in diverse planetary places who are less technologically advanced due to historical circumstances, unavailability of particular types of natural goods, lack of mechanically oriented community perspectives and practices, impeding economic structures, deliberate cultural choices by entire communities or by a dominant segment of the population, or other factors. These indigenous peoples would be at the mercy (or lack thereof) of spacefaring species who are alien to their world, "extra" terrestrial to their native terrestrial place. If incoming species discovered upon landing or afterward arrive with an immature consciousness, their aggressive and acquisitive conduct toward indigenous peoples would endanger natives' well-being, the well-being of their place, and even their survival.

In the event that humankind as it voyages into space has not made Contact with more mature extraterrestrials prior to arriving on exoEarth planets, and has not learned how to engage in cosmically congenial Contact and interspecies community respect and collaboration, some humans on space voyages might well be prone not to respect but to dominate indigenous intelligent beings. This possibility suggests that, as mentioned so often before, a change of consciousness and conduct is a pressing need prior to space voyages, exploration, and colonization; in the absence of this conversion to a sense of cosmic community whose members are responsible for themselves and act responsibly toward others, an operative Cosmic Charter, accepted by common agreement among Earth's nations (as happened in Earth space treaties developed *a priori*), might prevent the most egregious behavior of humankind being redux in space as humanity comes into Contact with native inhabitants elsewhere in the cosmos.

Analysis of the DRIP Articles brings to the fore those that are most pertinent to advocate respect for indigenous peoples wherever encountered in a *cosmic commons*, and consequently redeveloped for a DRIB. This would stimulate activation of a consciousness and conduct that would promote the well-being of the cosmos and positive Contact with intelligent extraterrestrial beings. A DRIB would be readily adaptable to ubiquitous cosmic contexts.

In all cosmic milieus, indigenous intelligent beings should have a "right of self-determination," and be empowered to "freely determine their political status and freely pursue their economic, social, and cultural development" (Art. 3), and a "collective right to live in freedom, peace and security as distinct intelligent cultural beings" (Art. 7). Humans-as-aliens should establish among themselves "effective mechanisms" to prevent implementation of policies or practices which would facilitate actions (whether or not by design) that allow "dispossessing [indigenous intelligent beings] of their lands, territories," and natural goods, and ensuring that such policies and practices would not come to be in the future (Art. 8). Indigenous intelligent beings in exoEarth contexts should have the right to "maintain and develop their political, economic and social systems or institutions," and to securely exercise traditional subsistence and development practices and economic activities (Art. 20); the right to ongoing spiritual relationships with their traditional areas and to use securely, without interference, the natural goods there, through generations to come (Art. 25); the "right to the lands, territories and natural goods which they have traditionally owned, occupied, or otherwise used or acquired," and should have from Earth states recognition of this right; the recognition should respect indigenous peoples' "customs, traditions and land tenure systems" (Art. 26); arriving humans should follow a mutually agreeable, open, and transparent process, in which indigenous peoples are major (not token) participants in order to guarantee continuation of their rights to territory and natural goods (Art. 27); the right to redress violations of rights cited "by means that can include restitution" or at least a "just, fair and equitable compensation" for traditional lands and natural goods that are unjustly and illegally, and in violation of Cosmic Charter requirements, "confiscated, taken, occupied, used or damaged without natives' free, prior, and informed consent," and if indigenous intelligent beings agree, they might, to redress their grievances, receive territory or goods instead of monetary compensation, in an equivalent amount of land or in equitable quantities of needed natural goods (Art. 28); the right to decide and put into place their own priorities for use of their land and natural goods (Art. 32); the right to receive from Signatories and their successors recognition and enforcement of, and honor and respect for, treaties and other agreements and arrangements that have been made (Art. 37); the Cosmic Charter Council and its agencies shall assist in the realization of the DRIB, including by "financial cooperation and technical assistance," and establish means to ensure that indigenous peoples participate in resolution of issues (Art. 41); the Cosmic Charter Council will promote "full application" of the DRIB and monitor the extent to which its provisions are being effective (Art. 42); the rights recognized in the DRIB are only "minimum standards" needed to ensure that indigenous peoples not only survive, but have "dignity and well-being" (Art. 43); and all rights recognized in the DRIB are "equally guaranteed" to individuals and groups of all genders and sexual orientations, as culturally accepted, of indigenous communities (Art. 44).

The inherent and intrinsic rights of indigenous peoples should be recognized by and enforced in human space exploration, on new worlds discovered, and on exoEarth colonies, and overseen by representatives of the Cosmic Charter Council, whose particular responsibility is to ensure that space continues to be the "common heritage" of all intelligent beings who have become part of the *Cosmic Charter Confederation*, from whatever place(s) of origin. Discovery Doctrine should not be implemented or even under any consideration when humankind or ETI voyages into the cosmos; the rejection of Discovery would retard, if not prevent, abuse and oppression of intelligent life on or far from Earth or ETI origin planets.

International Human Ecological Rights and Responsibilities:
Cosmic Inter-Terrestrial Adaptation

As was the case with human social rights and responsibilities, the insights and guiding principles discussed in UN documents provide a solid, consensus-based foundation for development of human ecological rights and responsibilities in cosmos contexts. Here, too, the formulation of a responsible human consciousness expressed in concrete conduct will enable humankind to integrate well with new places and, should they be Contacted, newly found extraterrestrial intelligent beings.

In the pages that follow: first, ecology-directed UN documents that focus specifically on human actions on Earth will be related exoEarth to the habitable planets and celestial bodies found in the cosmos; and second, ecology-directed UN documents that focus on human actions in outer space will be analyzed to determine the extent of their ongoing application in exoEarth milieus.

The UN World Charter for Nature: "World→Worlds"

The World Charter for Nature (WCN), as should have been apparent in its earlier elaboration (chapter 3), should be applied in extraterrestrial settings. If "Nature" were to be extended beyond Earth to include what is "natural" on other celestial bodies, then the document's principles and themes are very relevant for promoting respectful extraterrestrial explorations, and the careful acquisition, development, and use of natural goods on other worlds. In that case, extraterrestrial socioecological ethics might assume a "united planets" or "united cosmos" perspective and propose a Worlds Charter for Nature. This would promote protection of biota and abiotic settings on newly discovered worlds during exploration and colonization. Other intelligent species might be expected to have a similar regard for the biota and abiotic places on planets beyond their places of origin. Intelligent beings' self-interest would also provide a rationale for recognition of the necessity that natural systems on all habitable celestial bodies should function properly to supply needed energy and nutrients for newcomers, beyond what is needed already *in situ* to sustain planetary ecologies and inhabitants.

On other planets, then, "natural processes" and a "diversity of life forms" should not be jeopardized by intelligent beings' actions. Nature's own rhythms and quality should be safeguarded, and natural goods ("resources") protected, through interstellar cooperation that ensures nature's intergenerational sustainability, and the ongoing availability of natural goods for intelligent beings and other biota.

All intelligent beings must recognize that they are part of nature and depend on natural systems, and that civilizations have originated in and are sustained by nature. As in the original WCN, this assertion breaks away from anthropocentric, androcentric, and hierarchic understandings of humans' relationship with nature that have been expressed in philosophical, political, religious, and economic ideologies for millennia. Instead, humans and ETI should see themselves as integrated with and related to other biota in coherent ecosystems; they are responsible to nature as a whole, and for nature, in regions they occupy or otherwise use, in its rhythms and processes. Humans and other intelligent beings are not situated over and superior to an externalized, inferior nature; they are immersed in and mutually engaged with nature—or should be—and are to respect natural biological processes and biotic diversity. Where these ideas are operative, humans and ETI would live respectfully and interdependently in "nature" in whatever planetary setting in which they reside and responsibly earn their livelihood.

The WCN as applied exoEarth would declare that there, too, biota are unique in themselves and as they relate interdependently to each other. It would advocate respect for organisms, as expressed in and guided by a "moral code of action," to be sustained whether or not specific biota are considered to have instrumental value for humans or ETI. Intelligent beings would deplore, and refrain from having, adverse impacts on nature caused by "excessive consumption and misuse of natural goods" and the lack of "an appropriate economic order" in local and interplanetary intelligent beings' communities—thereby displaying concern about issues surrounding both the use and the distribution of land and natural goods. New, equitable economic structures would have to be developed, intended to provide for the basic necessities of each and every member of intelligent beings' communities, from their inception and intergenerationally.

World Charter for Nature principles intended to ensure conservation of nature in the present and for the future should be operative in exoEarth contexts: respect for nature and protection of its "essential processes"; maintenance of genetic viability, including through safeguarding population levels sufficient to enable species to survive, and ensuring sufficient habitats to provide for their needs, with special attention paid to unique areas and ecosystems; security for nature against impacts caused by war; activation of social and economic efforts that include conservation of nature because of its intrinsic value in itself, and its instrumental value for humans and ETI as provider of natural goods to meet their needs and promote community well-being; planned development of planetary areas in such a way that biota and aesthetic beauty are protected; careful use of all natural goods; maintenance of soil productivity and water quality and supply; and efforts to ensure that non-renewal goods are "extracted with restraint" in a manner that is compatible with ecosystem functions. Land, natural goods, and natural living goods should be cared for and conserved for what they are in themselves in their niche as part of the natural world. Humans and ETI should responsibly use natural goods when meeting intelligent beings' needs. In the process of caring for and in discovered planets (and in their own places of origin), humans and ETI should develop a collaborative assessment of processes proposed to develop natural goods, including by deciding and acting responsibly, in context, to have minimal impacts on ecosystem dynamics.

Implementation measures would include incorporating principles collaboratively developed by TI and ETI into the laws and practices of humankind and ETI, educating the general public about ecology, and a Cosmic Charter Council governance structure that provides planning and financial support for continued scientific research, needed economic and administrative structures, monitoring of natural systems, and public participation in decision-making processes and project implementation, with communities' right to redress if their local environment is harmed.

The UN-Originated Earth Charter

The Earth Charter's principles' headings, as well as the principles and sub-principles themselves, might be restated to be incorporated into a Cosmic Charter. One helpful amendment would be to advocate for social consensus decision-making in political processes rather than democracy, since democracy can be subverted in social conditions of unequal wealth distribution and political influence, and a majority might not vote and act in the interests of society as a whole, including its ethnic minority and less financially secure populations, but primarily to benefit themselves and whatever biases they might have (including racism, classism, sexism, ethnocentrism). A proposed Earth Charter revision for incorporation of EC principles into a Cosmic Charter would be:

> I. Respect and Care for the Cosmic Community of Life
>
> 1. Respect the biotic diversity of origin planets and new habitable places; 2. Relate to the cosmic biotic community with understanding, compassion, and love; 3. Develop social consensus societies that respect the insights and practices of all intelligent beings, and are permeated by justice, governed in a participatory manner, dynamically sustainable over time, and embody and promote peace; and 4. Secure origin planets' and other worlds' bounty and beauty for present and future generations.
>
> II. Ecological Integrity
>
> 5. Protect and restore the integrity of cosmos communities' ecological systems, with special concern for biological diversity and the natural processes that sustain life; 6. Prevent harm as the best method of environmental protection and, when knowledge

is limited, apply a science-based and socially concerned precautionary principle; 7. Adopt patterns of production, consumption, and reproduction that safeguard all worlds' regenerative capacities, intelligent beings' rights, and biotic communities; 8. Advance the study of intra-world and inter-world ecological sustainability, and promote the open exchange and wide application of data and knowledge acquired that would effect cosmosocioecological sustainability.

III. Social and Economic Justice

9. Eradicate poverty as an ethical, social, and ecological imperative; 10. Ensure that economic activities and institutions at all levels promote intelligent beings' ongoing development in an equitable manner, including by equitable distribution of land, natural goods derived from the land, and wealth acquired through human productivity; 11. Affirm gender equality and equity as prerequisites to responsible development and ensure universal access to education, health care, and economic opportunity; and 12. Uphold the right of all, without discrimination, to a natural and social environment supportive of intelligent beings' dignity, bodily health, and spiritual well-being, with special attention to indigenous and ethnic minority populations' requirements for social, cultural, and ecological integrity.

IV. Consensus Governance, Nonviolence, and Peace

13. Strengthen community-oriented consensus decision-making institutions at all levels, and provide transparency and accountability in governance, inclusive participation in decision-making, and social, political, and economic justice; 14. Integrate into formal education and life-long learning the knowledge, values, and skills needed for a responsible way of life; 15. Treat all living beings with respect and consideration; and 16. Promote intelligent beings' cultures of acceptance, nonviolence, and peace—in the cosmos as a whole and in all inhabited worlds.

Since the EC (similarly to UN documents) was developed through the input of peoples from throughout the Earth, and agreed to by representative people from distinct continents and countries and diverse constituencies, it provides a process and a product that might be fruitful for the development of a Cosmic Charter. In the CC, interested parties might well include intelligent beings from throughout the cosmos with diverse cultures, political and economic structures, and philosophies. As with nations and peoples represented in the UN, they might have competing interests—but they might also be able to agree to core principles and guiding ideas for the development and successful operation of a Charter that would provide an acceptable consensus-based instrument, a foundation for a stable form of cosmic governance, and a helpful guide for inter-intelligent life interaction and inter-terrestrial locale integration, interdependence, and interrelation.

As discussed several times in *Cosmic Commons*, a dialogic relationship could be developed between Earth peoples who bear the principles and practices they use to safeguard humans' origin planet and home, and humankind in space, in the distant worlds to which they emigrate; similarly, with ETI. As pertinent provisions of UN ecology, human rights, and space documents become utilized on Earth, they could be similarly used on other intelligent beings' places of habitation and areas co-inhabited by humankind and ETI. Inter-terrestrial (in the sense of Earth⟵⟶exoEarth worlds) cosmosocioecological rights and responsibilities would assist and guide humans and ETI in all contexts, and might become influential, too, in the independently held theory and practice of ETI (to the extent that these ideas and practices are not already part of ETI culture[s]). If a Cosmic Charter comes to represent and guide intelligent beings' thought and action this would be facilitated, and be more expeditiously and extensively disseminated universally.

The Earth-oriented UN documents discussed to this point are complemented by the insights and principles developed by UN states related to humans' activities in space.

Treaty on Principles Governing the Activities of States in the Exploration and Use of Outer Space, Including the Moon and Other Celestial Bodies

This first UN document developed to elaborate principles for nations' extra-terrestrial explorations and other space activities is commonly referred to as the Outer Space Treaty (OST). Its provisions have served as general guidelines for forty-five years, having particular application to the United States, which has done the majority of space exploration through the Voyagers (now beyond the outer boundary of the solar system, and still broadcasting messages to Earth), Mars Rovers, and a lunar landing, among other national achievements, and its initiation of the collaborative construction, operation, and supply and ongoing resupply of the International Space Station, using labor and expertise provided by astronauts, scientists, and technicians from several countries: it is a shared international achievement. Russia is second only to the United States in space development. Other nations such as China and India are in the wings as their scientists and technologists (including in government and the private sector) develop space technologies and spacecraft, and as they train the scientists and technicians who will staff them. The OST, although intended specifically to elaborate policies and principles focused on spacefaring nations' Earth-based activities and the results thereof in their space ventures, provides wording that would be useful as a prelude to, and applicable for activities in, humankind's conscientious expansion. The latter premise implies an expectation that this would be accompanied by a transformed biotic and abiotic consciousness whose inherent benevolence would effect beneficial conduct in all contexts, beyond the moon and the solar system, into other solar systems and galaxies, and among distant stars.

The hopeful tone from the introductory section of the OST, when adapted to a Cosmic Charter, should be retained—it envisions "great prospects" for humankind in space. This should be expanded in the new textual context to include the understanding that those prospects are also "great" in the view of extraterrestrial intelligent beings as yet unknown; "all intelligent species," not solely "all peoples," would anticipate benefits from their space missions, would acknowledge a possibility of TI-ETI encounters, and would agree to collaborative and cooperative, mutually acceptable accommodation of each to the other—if their fundamental orientation and operations are essentially benevolent. Should that be the case, there is a high likelihood that space will be used for "peaceful purposes" rather than guided by strife-ridden competitive efforts to gain some advantage in acquisition of territory or natural goods to benefit exclusively one's own kind. "Friendly relations" might well result should this be the intent and practice of humankind when it meets likeminded members of other spacefaring species.

The OST principles that follow the opening section provide the UN member states' vision for humankind as a whole (not in terms of distinct nations), and describe and elaborate what should be states' intentions in space. The OST Articles discussed previously would be amended as follows to accommodate TI-ETI needs, and promote the cooperative conduct needed to meet them and assure equitable distribution of simultaneously discovered and mutually desired natural goods.

The exploration and use of outer space, in a rephrasing of Article I, "shall be carried out for the benefit and in the interests of all intelligent beings, regardless of origin and any disparities in economic structures, scientific development, and technological accomplishment" (intelligent beings might well want to evaluate the economic systems characteristic of their respective communities, in order to ascertain whether or not they want to amend their own or adapt aspects of others into their own worldview, if this were able to be socially acceptable and culturally adaptable). So, too, should outer space be "free for exploration and use by all intelligent beings without discrimination, on a basis of equality and in accordance with conjointly and consocially developed space laws." All intelligent life shall have free access to all areas of space not otherwise inhabited or used by resident intelligent beings, unless mutually agreed upon divisions are made for specific sites or regions. All intelligent species would benefit from space exploration, if they were to pursue their "common interest" while directly and transparently stating the expectations of results from their voyages. Since diverse space explorers (who might come from different contexts on many occasions when they have diverse Contact moments and experiences in the vast cosmos), their governments, and the inhabitants of their places of origin expect to receive benefits from space exploration, they must realize that this would best be achieved—and perhaps only achieved—when TI and ETI work together. On the "common ground" of space, their respective (or simultaneous) discoveries will result in some economic benefits, and conflicts over their fair and just distribution would need to be resolved amicably. Provisions such as the prohibition against TI or ETI military bases in space and related stipulations should be adapted too, but amended to state that no species might construct military bases in a locale near the place of origin or of colonization of another species, such that a military presence could pose a threat to the security or stability of planets or celestial bodies occupied by another species. Policies on and proposals for collaborative scientific research in shared scientific centers should be continued. Transparency of each species to the other vis-à-vis mutual rights to visit and learn from others' work should be continued.

Important provisions in succeeding Articles should be incorporated also. No intelligent beings may claim sovereignty over any part of space save its own places of origin and of colonization—especially places yet to be explored, or where resident species' habits and habitats are yet to be scientifically examined—prior to Contact. If parties in Contact agree to be signatories to a Cosmic Charter, this interstellar law will govern all participants when they have competing claims to territory or natural goods. People on Earth unhappy with any type of international agreement that conflicts with their particular perspective, but especially with UN documents such as the United Nations Charter and others cited in *Cosmic Commons*, will, of course, oppose the concept and practice of a legal and legally binding document oriented not toward their narrow nationalistic and economic interests or even their planet's integrated interests, but rather toward these as included in common TI-ETI interests, common interspecies well-being, and the common good of all.

Just as Earth nations' astronauts are instructed to be hospitable toward each other and to assist anyone needing support away from their nation's base, so, too, should intergalactic and interstellar astronauts succor members of other worlds in times of need. Each intelligent species, too, should take responsibility for the conduct of their respective representatives, and for impacts made by their vehicles, bases, or any equipment that cause harm to others. Astronauts in all their activities should avoid contamination of places they investigate and pathogenic transfers to or from biota on those places.[3] The latter stipulation is very important not only to avoid harmful impacts in space but also on the biota of Earth and other origin planets, since this would have extensive ecosystemic and even global biotic consequences. Similarly, in scientific research: protocols must be established, accepted, and observed. TI and ETI must take care that they do not pass on to others any communicable diseases. Organisms absent from one party's evolution and from their experience might inadvertently be introduced into the other's body and thereby unleash a pathogen that would have immediate and adverse impacts because there are no natural, biological or genetic barriers to its spread, and cause devastating harm to individuals and catastrophic harm to species, even more extensively so if they are transferred subsequently through space to other planets, including home planets. Representatives of individual nations or groups of nations coming from Earth, and representatives of intelligent extraterrestrial species coming from elsewhere in space are all bound by agreements multilaterally reached, and should respect and collaborate with each other accordingly.

The TI-ETI dialogue should include incorporating into a Cosmic Charter respect for the intrinsic value and interdependent autonomy, ongoing evolution, and material well-being of less complex forms of life, as well as conservation of their habitat. So, too, must abiotic being be treated respectfully, and altered only to meet species' life-sustaining needs and well-being, which can be accomplished with equitable distribution of commons goods for the common good, and not their exploitation to satisfy a species' or segments of species' insatiable wants. Biologically more simple organisms might need to be carefully conserved as they are, in place, for current and future ecosystemic and even planetary stability and well-being—as, too, might be some abiotic areas or beings. The cosmos as a whole must be regarded holistically, and its interdependent entities respected and conserved.

In cosmic space as on planet Earth there are enough natural goods to satisfy everyone's need, but not anyone's greed. It should be evident by now that those with a very nationalistic, planetaristic, jingoistic, xenophobic

---

3. The Curiosity rover on Mars has already created the possibility of a pathogenic transfer of potentially harmful biota, in the form of microbes, from Earth to Mars. Prior to landing, a sterilized container of drill bits was opened on board in a non-sterile environment, and a bit inserted into Curiosity's drill. Once the rover was exploring the Mars surface, the drill was used on rocks and crater walls.

ideology and a corresponding acquisitive perspective would strongly object to values such as sharing with other species in an equitable distribution of the natural goods and territories of the universe. Some object to sharing space goods even with members of their own species from a different nation, especially if they have less political, economic, or social power. Nationalistic and acquisitive individuals and groups in the United States with significant financial resources and access to government agencies and officials—elected and appointed—are pressing NASA and politicians even now for "private property" to be established in the distribution of territory and the natural goods therein on other solar system planets, the moon, and celestial bodies elsewhere in space—despite provisions to the contrary in UN treaties and statements to which the United States is a party.

Agreement Governing the Activities of States on the Moon and Other Celestial Bodies

The Agreement had a particular focus on the moon. Nations recognized more clearly that a lunar base would be desirable as a base for launching exploratory, including scientific, spacecraft further out into the solar system and beyond, but also as a military base from which a nation or nations might conduct military operations against other nations on Earth. There was also speculation that the moon itself might contain valuable goods, and suspicion that a powerful nation's astronauts and accompanying civilians and military personnel might extract them secretly for export back to Earth. People hoped, too, that valuable scientific information might be garnered from lunar exploration. Many of the Agreement's provisions, as was the case with OST principles and guidelines, might be reformulated and restated for TI-ETI cooperative use. Those previously discussed will not be included here, to avoid redundancy.

As discussed previously, too, Article 4 is complex and significant. Amendments for a Cosmic Charter would *not* include revising phrasing from "the Moon shall be the province of all mankind" to "the province of all intelligent species" because, as suggested earlier relative to the OST, intelligent species should have, as a precautionary security measure, sovereignty over space bodies near to their places of origin. Even if the ETI first Contacted is pacific and benevolent, this might or might not be the case with subsequent species encountered. However, phrasing should be retained concerning regard for intergenerational well-being, including in terms of improving living standards and promoting "economic and social progress and development."

Article 7 provides important cautions against causing environmental imbalance on the moon through making adverse environmental changes or by contaminating it through "extra-environmental matter," or on Earth by causing similar environmental damage through contamination by lunar-originating harmful matter. This Article is especially important because of the moon's relative proximity to Earth, and the likelihood that over time there would be more frequent journeys from and to Earth: to transport needed natural goods, for tourism, for resupply of the moon with goods unavailable in the lunar environment, to transport workers, and so on. The provision of "protected international preserves" might be amended to "protected inter-intelligent species preserves," with the understanding that the areas might be visited but not occupied by ETI or humans.

Article 11's statements regarding rejection of national sovereignty would not apply to establishment and maintenance of a unified Earth sovereignty, under an international legal system (such as the United Nations), to benefit all humankind. The sharing of natural goods among peoples of all Earth nations remains a worthy goal. It would provide benefits, too, to ETI communities and their places of residence.

Declaration on International Cooperation in the Exploration and Use of Outer Space for the Benefit and in the Interest of All States, Taking into Particular Account the Needs of Developing Countries

The UN General Assembly Declaration on International Cooperation (hereafter: DIC) stipulates that material benefits acquired through outer space exploration and celestial bodies' development must be equitably shared. In a Cosmic Charter this phrasing would be retained, and the equitable sharing would provide for the well-being of intelligent beings. Benefits would be distributed to the human community as a whole on its Earth home, and to ETI on its origin home and other places where it might reside at the time of Contact. Development would be for common "peaceful purposes" a stipulation to be accepted and observed by both TI and ETI. Space exploration and the places and goods found must promote the interests of all species' communities: places and goods discovered should be the "'province' of all intelligent beings." A primary principle and intent for their allocation should be to meet pressing material needs—including territory and goods—of the least developed terrestrial peoples and extraterrestrial beings. Importantly, Cosmic Charter Council representatives should develop "mutually acceptable" interspecies cooperation: in planning space explorations in areas about which they have a common interest or curiosity, and when they form interspecies collaborative teams to realize their plans. Teams should include representatives of less technologically advanced (which are probably less economically developed and financially stable) cultures, and use the best available organizational entities (e.g., government agencies or nongovernmental organizations [NGOs], commercial and profit-oriented or noncommercial and people-oriented enterprises [the Mondragón Cooperative in Spain's Basque country would be an exceptional example of the latter]). Should TI states and ETI communities be eager and willing to collaborate on these efforts, they should share a common goal of joint development of space technology and educating less developed entities in their respective worlds to develop their space capabilities; or, better yet, to work as cosmic cooperatives in which their distinctions become interactive as compatible characteristics which benefit collaborative technology development, natural goods acquisition, and distributions of places and goods. A Cosmic Charter would provide a firm foundation to initiate and continue these efforts in a way acceptable to all participants.

UN Space Principles and Guidelines in a Cosmic Charter

The principal proposals of the member states of the UN, as presented in UN documents developed as treaties or agreements of principles and approved by UN representatives in dialogue, would provide initial bases for principles of an Earth-developed Cosmic Charter. The existing documents, as just discussed, would be adapted to terrestrial-extraterrestrial considerations and contexts, and alter "international" to "interstellar" as a working term for humankind as it looks toward its future in space. The Charter should avoid all reference, direct or indirect, to elements of the globally lingering Discovery Doctrine, which could be invoked to seize indigenous peoples' territory on newly discovered worlds.[4]

Permission to employ military personnel on space sites, as granted in the Antarctic Treaty, is a weak link in the present chain of Earth-developed principles and requirements. Military scientists might make invaluable contributions to space exploration and research. However, it is also possible that the reason for their presence and for the instruments they employ is to coerce in some manner other nations' personnel. A compromise might be to allow a minimal presence of military personnel from intelligent species to be present on site in locales distant from origin planets and colonized worlds, except for sites they own or occupy exclusively, and participate in scientific research on

---

4. As discussed at length in chapter 1, Discovery claims have been used on Earth by Europeans and their Euroamerican descendants, and have been adopted by leaders of other powerful nations, to unilaterally authorize one culture to take other peoples' territory and all natural goods within its boundaries and to have exclusive trade rights with native inhabitants. Elements of Discovery included the practice of first arrival, first settlement, placement of symbols such as flags or other objects as monarchical territorial markers to claim that place and ownership of or exclusive control over its goods, etc.

celestial bodies that are not satellites of an inhabited or habitable planet, or within striking range of one. Their numbers should be limited, as should be their on-site time, and the amount of cargo they are allowed for their specific use. "Peaceful uses" would be better served if military personnel were prohibited altogether, or if they were required to be from diverse intelligent beings' communities. These stipulations might eliminate or at least partially diminish their potential to be a divisive and even aggressive force in space, especially on site on the moon and other celestial bodies.

Toward Initial, International Development of a Cosmic Charter

The documents discussed, when adapted to new times and places, provide a foundation for development of a new, initially international agreement (intended to become part of an interstellar agreement) focused specifically on terrestrial-extraterrestrial intelligent life Contact and consequent cosmic community initiatives to promote mutual well-being and collaborative projects. It would be beneficial, too, for human communities as a whole when nations' representatives explore the planets of the solar system, and venture forth for interstellar voyages, even before TI-ETI Contact is made.

What follows is not proposed as the draft for a Cosmic Charter. The writer does not egotistically think that such a document could be formulated by one person or one nation or one interested social organization. Rather, it is suggested to provide some "talking points" to be developed further by international organizations, perhaps particularly by NGOs. Government agencies, non-governmental organizations, individual scholars, and other citizens of distinct nations and cultures might well have developed already their own initial formulation of a document to govern human conduct vis-à-vis other space places and intelligent species, wherever encountered. If so, the time is ripe for an international exchange of such documents and for ongoing, internationally collaborative work to integrate, strengthen, and develop them further.

I enthusiastically relinquish the proposal presented and its initial principles to such other people and organizations as have far more expertise than I, for adaptation and incorporation, in whole or in part, in such documents in process, as might be useful for already ongoing efforts. (Some US government agencies' staffs are prohibited from even voicing the need to develop some basis for peaceful interaction with ETI on and after Contact because the US government declares "officially" now and again that UFOs and ETI do not exist; needed voices are thereby excluded from participating in such efforts, except anonymously.) The auspices of the United Nations, perhaps building on initial efforts by an NGO[5] or NGOs able and willing to host an international exchange of ideas online, might be the best vehicle for assembling international input on a developing Cosmic Charter.

Eventually, the Cosmic Charter could provide a basis for interplanetary dialogue among human beings located on Earth and on distant cosmic bodies, enable collaborative development of a shared vision, and catalyze projects oriented toward realizing it. This would promote planetary well-being and social well-being in all human-occupied and ETI-occupied contexts. It would effect further a continually, mutually formed and informed positive development of consciousness and conduct that would benefit both the human communities in dialogue, and ETI cultures with whom humankind might become consociated in cosmic contexts.

Cosmic Charter

Preamble

The cosmos is a vast and complex expanse with multiple dimensions in time and space. All comprise a single reality, and are related to each other in the integral being of all that exists. The existents, energies, elements, entities, events, and entropy of the universe are intertwined, integrated, interrelated, and interdependent. In incompletely known and complex ways, they influence and impact each other in time and through distinct dimensions of reality.

Intelligent beings are conscious, reflective, creative, curious, and contextually adaptable inhabitants of the cosmos, in whole or in part. They have developed modes of cosmic travel that enable them to voyage expeditiously and safely through what would be, in observable space-time, immense spatial distances and distinct zones of time: past, present, and future. In their voyages, they come into Contact with other intelligent beings who comprise diverse species with distinct technological capabilities and possibilities; disparate environmental contexts on sometimes distant planets or celestial bodies; and new (to them) and innovative native cultures and the institutional—political and economic—and transcending—philosophical and spiritual—social structures that comprise and organize their relationships to

---

5. The Center for UFO Studies (CUFOS), founded by astrophysicist J. Allen Hynek, would be an appropriate place for assembling insights and suggestions from around the world and integrating them preliminarily into a more comprehensive Cosmic Charter.

each other as social beings in social contexts. Intelligent beings consciously integrate with each other for mutual benefit, and to envision, develop, and realize common well-being. In newly discovered environmental and cultural contexts, too, they become aware of diverse types of ecological integration: among members of intelligent species, between communities of intelligent species, between intelligent species and less complex species, and between all species and the places that they inhabit.

In their initial encounters and engagement intelligent beings will seek to develop such relationships and social institutions as will enhance their consequent interaction. In order to proceed pacifically and to provide mutual benefits, intelligent species strive to effect equitable allocation of mutually sought territory and natural goods, as provided in various contexts by the dynamic cosmos, each or all of which might be necessary or useful for particular beings' organized community entities during some or many of their stages of social development: to meet the shared needs of their own communities or of other societies with which they come into Contact.

Signatories pledge to develop, establish, and support a Cosmic Charter Confederation, with associated governance, for mutual well-being and interactive support as an interrelated cosmic community. A Cosmic Charter Council will provide such governance, and oversee and facilitate implementation of Cosmic Charter provisions, such as would benefit each and every Signatory to the greatest extent possible in an interdependent and interrelated association. The membership of the Cosmic Charter Council shall include a representative selected by each Signatory or groupings thereof that might be developed.

Signatories of the Cosmic Charter agree that all planets, celestial bodies, and other habitable or natural goods-rich places that are discovered by any of them independently or collaboratively shall, if not inhabited at the time discovered, become the common territory of all Signatories, and its natural goods, if not currently under the authority otherwise of a governing entity, shall be the common goods of all Signatories. The common territories shall be administered by the Cosmic Charter Council.

Signatories agree that cosmic space, planets (including origin planets and discovered planets) and celestial bodies (including origin celestial bodies and planets' satellites) shall be used only for peaceful purposes.

The Cosmic Charter is a work-in-progress, subject to internal evolution—as approved by its Signatories after the Cosmic Charter Council has reviewed and considered changes and presented them to Signatories—stimulated by ideas and insights suggested by diverse intelligent beings in distinct social settings and disparate cosmic contexts.

The Cosmic Charter is presented for consideration by all intelligent beings in diverse planetary and other cosmic settings as a foundation for ongoing congenial and collaborative efforts to provide for the needs of all intelligent life; to safeguard the well-being of simple and complex biota evolving in distinct places; and to ensure the well-being of the commons ground and commons goods on which all life depends for its survival and communal well-being. To these ends, the principles that follow are proposed as a dynamic foundation for intra-species, inter-species, interplanetary, interstellar, and interdimensional collaboration and community.

I. Common Ground: Origin Planets

I. A. Each Signatory is responsible for securing the social, cultural, and ecological well-being of planets and other places on which they evolved or on which they resided when they signed the Cosmic Charter and became members of the Cosmic Charter Council and the Cosmic Charter Confederation, and which they inhabit or will inhabit.

Article 1. Respect and conserve planetary common ground: Planets and other inhabited places shall be carefully conserved, to maintain their abiotic contexts and biotic habitats.

Article 2. Provide territorial space and natural goods access needed to ensure species diversity, well-being, and survival in natural evolutionary processes, and to provide security against being extincted by intelligent beings.

Article 3. Conserve the integrity of cosmos communities' ecological systems, with special concern for biological diversity and the natural processes that sustain life.

Article 4. Conserve contiguous ecosystems necessary for species to survive and to thrive.

Article 5. Intelligent beings from more technologically advanced and economically stable societies should assist intelligent beings who lack the capability for space exploration but otherwise are sufficiently mature to do so responsibly and well should be assisted, without violating their culture or place, to develop this capability through enhanced scientific education and technological assistance.

Article 6. Transparency should characterize specific space missions' purposes and activities, and the dissemination of the results of exploration and research on planets and celestial bodies.

Article 7. Intelligent beings' politically organized bodies should exercise control over private enterprises under contract to them or under their sponsorship; and, intelligent beings' politically organized bodies should provide universal access for all intelligent beings' mutual education and potential collaboration—including in areas developed by private enterprises within

Signatories' jurisdiction—to their space exploration-related sites on origin planets or other places where they have been constructed.

II. Commons Good and Commons Goods: Origin Planets

II. A. Each Signatory shall care for the well-being of its places of residence and ensure equitable access, distribution, and use of biotic and abiotic planetary natural goods.

Article 8. Respect the biotic diversity and evolutionary integrity of all biota present as native inhabitants of planets and celestial bodies, whether found on first exploration, during colonization or natural goods acquisition, or at any other time during the course of development and use of the land, water, minerals, and other natural goods of planets or celestial bodies.

Article 9. Ensure that mining processes and mineral production and refinement are not destructive to natural environments and ecologies, and that mining sites, after responsible use, are restored to a viable state suitable for biotic introduction and reintroduction.

Article 10. Develop social consensus societies that respect the insights and practices of all intelligent beings, and are permeated by justice, governed in a participatory manner, dynamically sustainable over time, and embody and promote peace among all intelligent beings.

Article 11. Acknowledge and ensure the intrinsic value of all biota and all abiotic places, and only alter that status to instrumental value for limited members of a species and limited ecosystem areas during periods when, and in places where, they are needed to provide necessary sustenance or goods for intelligent life.

Article 12. Maintain sufficient habitat to safeguard the well-being of biota when they are displaced from their customary habitat, or when abiotic places are impacted by intelligent beings' acts.

Article 13. Strive to ensure that all intelligent beings' communities have access to and the ability to acquire such land and natural goods as are needed for their sustenance.

Article 14. Ensure that the least benefited intelligent beings' communities have priority rights to share common ground and common goods in order to improve their economic base, promote social well-being, and sustain their cultures.

III. Common Ground: Discovered Planets and Celestial Bodies

III. A. Each Signatory shall respect natural processes in place: geological and meteorological integrity, and biotic evolution.

Article 15. Respect all intelligent species' integrity, and all species' evolutionary development, in native niches.

Article 16. Equitably allocate planets' and celestial bodies' territory. Provide territorial space and natural goods access needed to ensure species' survival and well-being within natural evolutionary processes, to provide security against being extincted by intelligent life.

Article 17. Ensure that intelligent species have equitable access to, allocation of, and distributions from, territory and natural goods on newly discovered planets and celestial bodies, or parts thereof when they are sufficiently extensive to allow for non-competing claims to and development on or under an entire planetary or celestial body surface.

Article 18. Promote collaborative development and exchange of natural goods for mutual benefit and well-being, with particular regard for the least economically benefited intelligent beings' communities.

Article 19. Provide equitable opportunities for diverse species' exploration, colonization, and natural goods development on newly discovered planets and celestial bodies, or allocated parts thereof.

Article 20. Formulate and promulgate inter-intelligent beings laws, policies, guidelines, and other agreements that will provide for the Cosmic Charter Council, as a mediating third party, objective criteria by which to resolve competing claims on territory or natural goods; and, agree to accept the Cosmic Charter Council's mediation as an arbiter, and the Council's subsequent impartial decisions and determinations.

Article 21. Natural goods, when needed in places other than the setting in which they were found, should be removed carefully and conscientiously; areas from which they have been removed should be restored to their original state, to the greatest extent possible, with provision for native biota reintroduction or introduction into restored places or regions.

IV. Commons Good and Commons Goods: Discovered Planets and Celestial Bodies

IV. A. Each Signatory shall conserve natural characteristics and natural goods of worlds they discover, and equitably distribute or allocate all newly discovered territories and their natural goods.

Article 22. All places discovered by one or several Signatories immediately become common Cosmic Charter territory *unless they are occupied and used by indigenous intelligent beings*, in which case the original occupants shall remain the owners and users of the territory. Unoccupied newly discovered places enter under the jurisdiction of the Cosmic Charter Council, which will allocate or distribute them to benefit Signatories according to their need, prioritizing the survival or well-being of Signatories to favor those most in need of territory or particular natural goods.

Article 23. No intelligent beings' political entities shall have a claim of sovereignty over or ownership of discovered planets or celestial bodies, whether by first detection, first robotic research vehicle, first arrival, first placement of a particular flag or other cultural artifact, first scientific expedition, or first preliminary settlement with a base for research, mapping, or other initial assessment and demarcation of site geophysical characteristics, biotic habitation, or geographic boundaries.

Article 24. No private organization, corporation, association, agency, or individual shall have ownership of discovered places, but may have the use thereof as stipulated by statutes in or governing parties under the Cosmic Charter.

Article 25. Exploration of space generally, and of proximate celestial bodies particularly, is to be done collaboratively, carefully, and conscientiously.

Article 26. Where the survival or well-being of intelligent beings is threatened by bellicose invaders, the interstellar defense body of the Signatories, whose members shall be representatives of diverse intelligent beings chosen by the Cosmic Charter Council, should intervene with such force as is necessary to maintain Signatories' security and integrity, whether or not the team's members are of the same intelligent species as those threatened.

Article 27. Origin planets and their satellites, other planets and their satellites, and other celestial bodies should not be endangered or harmed as a consequence of intelligent beings' exploration or use.

Article 28. Special, protected science preserves should be established on pristine places at the earliest opportunity; and, data derived from scientific research and experiments by any intelligent life community entity should be shared among all, disseminated either directly or under the auspices of the Cosmic Charter Council.

V. Common Good on Common Ground

V. A. Each Signatory will make equitable provision for the habitable space and natural goods needs of all Cosmic Charter Signatories in all Cosmic Charter Confederation territories.

Article 29. The occupation and use of Confederation territories shall be available for all interested intelligent beings' entities to promote collaboratively the mutual common good, for themselves and other intelligent beings.

Article 30. Material goods and other material benefits derived from exploration and use are to provide particularly to meet intelligent beings' needs, including especially those in the least developed regions of the cosmos and least prosperous intelligent beings on all planets and celestial bodies; and, priority shall be given to intelligent beings most in need of a particular territory or of natural cosmos common goods found there; where there is equal need for places or goods, Signatories shall develop conjointly their equivalent and equitable distribution or prioritize, when necessary because of limited availability of territory or goods, the order in which recipients are to benefit from them.

Article 31. Species-open sites should be designated to provide areas in which intelligent species might collaboratively or independently explore for natural goods to be responsibly extracted or altered on site, or territory that would be habitable for intelligent species to be integrated with the setting and rhythms of their new context.

Article 32. Intelligent beings' politically organized bodies have ultimate ownership of and responsibility for objects they or those under their auspices have launched into space.

Article 33. Intelligent beings' politically organized bodies should avoid contamination and other adverse environmental impacts on others' inhabited or occupied places, and the biota and abiotic goods present on them.

Article 34. Signatories should provide access to all collaboratively developed areas of planets of origin and celestial bodies, and to intelligent beings' distinct places, following advance notice and without disturbing local ecologies, cultures, and inhabitants.

Article 35. Intergenerational present and future needs should be projected for intelligent beings, and responsible practices utilized to meet them, especially through the equitable conservation and distribution of material goods and beneficial places found in space.

VI. Interdependence, Integration, and Interrelationship

VI. A. Signatories shall consensually formulate and collaboratively implement such forms of governance and guidance as will provide for the equitable needs, safety, and cultural conservation of all Cosmic Charter Signatories, and responsible care of common places and of historical artifacts of intelligent life and cultures discovered in cosmic voyages.

Article 36. Cosmic Charter Signatories shall promote dialogic engagement in cosmic contexts as constitutive of cosmic community, and a way to promote a peaceful and mutually beneficial relationship among all Signatories.

Article 37. Interstellar law developed by the Cosmic Charter Council should be operative in all space explorations and all interactions among representatives of distinct intelligent beings.

Article 38. Cultural artifacts should be respected and left *in situ*, because they pertain to and contain the cultural memories of previously existing intelligent inhabitants of and civilizations in a place, and are part of the common heritage of all intelligent life in diverse stages of material and sociocultural evolution; and to respect and honor cultures of intelligent species that evolved in diverse places through eons of integral being.

Article 39. Collaborative space projects, and development plans that would foster collaboration on a mutually acceptable basis, should be undertaken by intelligent beings with diverse financial resources or technological capabilities.

Article 40. Sites proximate to places inhabited or used by diverse particular intelligent beings should be available for shared use as launch sites for ongoing and outgoing space exploration for other intelligent beings, and a place to which they or others might return, with mutually agreed upon arrangements for responsible use.

Article 41. Dangers discovered in space or on a space body that pose a threat to any intelligent beings on site or in their places of origin, or in colonies or bases elsewhere, should be immediately disclosed to all astronauts and their respective governing authorities.

Article 42. Astronauts, as interstellar representatives of all intelligent beings and not solely of their origin planet or other place, should be given hospitality by any and all intelligent beings' stations in times of need.

VII. Cosmic Community in Cosmic Commons

VII. A. A Cosmic Charter Council shall be established, comprised of a representative from each Cosmic Charter Signatory. The Council shall promote cosmic community in places where its Signatories reside in the present and future in the cosmic commons. Cosmic community integral well-being will be enhanced through establishment of a Cosmic Charter Confederation.

Article 43. The Cosmic Charter Council shall be established as an intergalactic and interstellar body charged with promoting peace and providing just allocations of territories and goods for which there are seemingly equal claims of discovery of planets and celestial bodies, or of goods on species-open sites. A Cosmic Charter Confederation shall be developed as the interstellar organization that includes all origin planets and subsequent worlds discovered and developed by Signatories. Its individual Signatories govern territories under their specific jurisdiction, while the Cosmic Charter Confederation integrates all Signatories into a cohesive, interrelated, mutually supportive entity.

Article 44. The Cosmic Charter Council shall make all decisions regarding allocations to enable Signatories' occupation and use of newly discovered places.

Article 45. The Cosmic Charter Council shall make all decisions regarding the distribution of natural goods on Charter territories. Natural goods are to be used in context or removed from context without disrupting planetary abiotic dynamics or biotic evolution.

Article 46. The Cosmic Charter Council shall ensure that if any intelligent indigenous beings exist on newly discovered territory, the areas in which they live and the natural goods in those or contiguous territories shall remain under their control and for their use. They shall be invited to participate as incorporated Signatories in the Cosmic Charter, the Cosmic Charter Council, and the Cosmic Charter Confederation.

Article 47. The Cosmic Charter Council shall ensure that planets that are intelligent beings' places of origin, or other celestial bodies and their abiotic context and biotic inhabitants, would not be endangered or harmed as a consequence of intelligent beings' exploration or use.

Article 48. The Cosmic Charter Council shall ensure that material goods acquired during space exploration and use are to be equitably distributed so as to benefit all intelligent beings according to their needs, without consideration of their economic system or financial well-being, or the stage of their scientific development.

Article 49. The Cosmic Charter Council shall ensure that material benefits derived from exploration and use are to provide particularly for intelligent beings' needs, including especially the needs of those in the least developed regions of the cosmos, and to benefit first the least prosperous intelligent beings on all planets and celestial bodies, to be ensured by concerted collaborative efforts to utilize space-generated benefits and goods to meet intelligent beings' responsible development goals for themselves (when they are in need) and for others (when politically and economically stable species seek to provide for the needs of less-developed species).

Conclusion

The Cosmic Charter offers a vision and a hope that the space and places of the cosmos shall be regarded as a common, shared, intergenerational benefit for the well-being of all intelligent beings. The Cosmic Charter Signatories pledge that when newly discovered worlds are not already the habitation of intelligent species, at whatever their evolutionary stage, they shall be available to and for all intelligent beings, allocated equitably to each as a whole, or to discrete members of their communities, according to their needs. Intelligent beings not already Signatories will be invited, upon or subsequent to Contact, to become Signatories of the Cosmic Charter in order to promote the Cosmic Commons Good, characterized by peace, justice, and ecological well-being, for mutual intergenerational benefit, throughout the cosmos.

Signatories will be especially concerned to do no harm to members of their own or another species, and to do no harm to the communities and metropolises into which intelligent beings have been organized or within which they have coalesced as social beings seeking and stimulating the social, biotic, and planetary commonweal.

The Cosmic Charter is a dynamic document that will evolve over time and in space as events—including encounters with newly evolved intelligent beings in diverse cultures—and continuing consideration of socially and ecologically progressive ideas, proposals, and projects, based on experience in context, are evaluated and then incorporated when determined to be beneficial to present, future, and potential Signatories.

A shared vision and hope prompt the Signatories to interrelate peacefully and to integrate diverse ways of thinking and acting into a coherent whole that acknowledges interdependence. Signatories pledge to strive collaboratively to realize cosmic unity and mutuality in pursuit of the cosmos commons good, the common good of the biotic community of which all are members, and the common good of Signatories' communities, for present and future generations.

Cosmic Confederation and Complementary Charters

It might well be that spacefaring intelligent species, where they exist, already have developed and implemented both a Cosmic Confederation and a Cosmic Charter that serves as its guiding document. Such species, it would seem, would seek other intelligent beings in the cosmos whom they might invite to be participants in the Confederation—when they are sufficiently mature. Characteristics of biologically and culturally evolved maturity would probably include, for mutual support and well-being, intelligent species' intraspecies peace in their planet of origin and wherever else they might have explored and colonized; acceptance of intraspecies diversity of whatever kind; conscientious consideration of and compassion for the "least members" of their species; openness to Contact and collaboration with diverse existing intergalactic and interstellar species, known and yet to be encountered; evident equitable sharing of territory, natural goods, and manufactured goods in a commonweal of intelligent beings open to all members of their culture, characterized by a focus on community well-being, including especially of those who are the least advantaged in culturally-spatially organized entities.

Needless to say, species that would qualify for such a Confederation would evidence a lack of aggressiveness and acquisitiveness and actively demonstrate that their global (and beyond) society has a consciousness and conduct suitable for Contact and cooperation for mutual benefit. This would be revealed particularly by just, collaborative, and compassionate species practices as embodied in its social structures, laws, and policies, its equitable economic arrangements based on common needs rather than on individual wants, and internal interaction that witnesses to their holistic community's acceptance of its interdependence, integration, and interrelationship—including with other biota and their shared abiotic context.

It is obvious that humankind has a long way to go to reach that maturity and acceptance status. If intelligent beings already have been observing human conduct and interaction, as credible witnesses claim, they would not be likely to openly and directly engage representative humans.

However, among some members of humankind seeds have been planted for conversion from competitiveness to cooperation, from individualism, selfishness, and self-absorption to common progress, sharing, and neighbor love—for every neighbor. Insights from religious, spiritual, and secular humanist thinkers, community organizers and activists, innovators who contribute their intelligence to promote the common good, and nonprofit organizations are evident in the ways they are embodied in communities around the Earth that are solicitous of the well-being not only of their own members, but of near and distant neighbors as well.

12

Cosmic Commons

Celestial Cohabitation, Conservation, and Compassion

Science has only begun to explore, in a very superficial way, the complexity of the cosmos. Optical and radio telescopes, among other instruments, have discovered that the universe in which humankind dwells is vast, deep, and very old. Unlike our ancestors in the not-so-distant past, humanity in the twenty-first century realizes that the sun, moon, and stars are not mere bright lights traveling in fixed tracks in a dome affixed atop Earth's surface, and that Earth and the universe have not existed for only thousands of years. We know that we have evolved, along with all that lives, over the course of approximately 3.5 billion years on a planet that is 4.5 billion years old and in a universe that is some 14.82 billion Earth years old. Earth has existed for slightly less than one-third of the time the cosmos as a whole has existed.

In 2009–10, using Ultra Deep Space infrared imaging, NASA's Hubble telescope found the oldest galaxy encountered to date: 13.2 billion years old (the data were examined, and the date confirmed, on the NASA Web site in September 2012). The galaxy was born, therefore, just 1.62 billion years after the Big Bang with which our universe began. The James Webb Space Telescope, under construction and due to be launched in 2018, is anticipated to be able to search back earlier in time. It will seek the first galaxies born after the Big Bang as it extends the human technological reach and scientific research further out into the inflating universe.

Humankind is beginning to understand, too, that we are part of the integral being of the cosmos. Just as all life on Earth is directly connected via the DNA that is the very physical core of its existence, so, too, all life and its home planet are connected, directly or indirectly, to the intricate flow of time in space, in a space-time continuum. The human sense of integration into the relational community of all that is enables people, even before they expand materially-physically into the universe, to expand mentally and psychically. We can sense that our personal being is extending outward; we might note this particularly on a clear, starlit night when we experience more fully and embrace the immensity before us. We can recall, too, that we are all related: we are kin to what we see (and beyond) in that we and they are stardust, we share the same primordial moment in cosmic birth. In such a consciousness-extending and -enhancing moment, we might ponder our role, our niche in the vast all-that-is, the being-becoming of the universe: as individuals, as diverse but interrelated, interdependent, and integrated communities, and as one species sharing the same Earth home. We are looking to dwell, in some future time, in other places within the sky that we see now, and in the more distant sky that we will see in the future with our advanced optics. We have reached a moment when our consciousness might transform our conduct, at home and abroad, such that it radiates the best of who we are as ourselves and as part of the integrally interrelated cosmic community.

Sacral Relational Community

We are conscious, thinking, self-reflective stardust. We recognize the truth in this when we reflect, however briefly, on what Georges LeMaître called "the day without yesterday," that singular explosion and expansion in which cosmic birth occurred, and the cosmos began to form and unfold dynamically. We humans are the result and the progeny of that extraordinary event. We were born long after the original and originating birth—billions of Earth years afterward—but we retain its residues in our very being. We contain within our materiality the elements and energies, or their intertwined descendants, which emerged from our cosmic birth.

We share with all other stardust and stars the sacrality of the universe, the inherent sacredness that permeates the cosmos because it is a Spirit creation and because it is a bearer of Spirit—what some religious teachers call a "temple" of the Spirit—the immanent Spirit who as transcendent Spirit imagined and created, and still continues to create, through the extensive and extending and evolving universe set into dynamic motion billions of years in the past. Creation continues more indirectly now, via the initial parameters, or their permutations, which were the origin seeds of all that is, which have been granted creative freedom by Spirit to be autopoeitic, to organize, diversify, and complexify over cosmic time, and to evolve over historical time. We share also with all stars and stardust an intimate interaction and interrelation with Spirit. This was conceptualized by Maximus millennia ago when he described the relationship between the creating Word and the offspring of the Word: *Logos-logoi*; we are all, biota and abiota, sparks of the eternal

One-Who-Is, One-Who-Will-Be, the Being-Becoming transcendent-immanent creative Spirit. We are all related, each being in the universe one to one and one to all others, in a cosmic extended family. The words of Lakota and other Indian elders express this well; they pray, teach, and greet each other and all creation thus: *Mitakuye oyasin*—We are all related.

We are stardust, then, not solely in that we originated in a primeval or primordial (although it might have had "ordo" in terms of its Creator-provided parameters) explosive birth. We share with all stardust the original, intense, unified existence in divine Being-Becoming, in the pre-form of the sacred cosmos, and continue to be in community in and with each other in the ever-sacred cosmos Presenced by divine immanence-transcendence. We as "microcosm" and the universe as "macrocosm" are bearers of Spirit, in the sense that Spirit's immanence is ever present within us because we exist in Spirit. We are not really "temples" of the Spirit: temples are human structures, and can be or perceived to be confining—physically to humans as material beings, and spiritually to humans who come to see them as the only places or the principal places to pray to Spirit. The perception that temples confine Spirit is acknowledged in initial biblical warnings regarding sacred structures, and in Israelites' fears that people will see Yahweh as present in a special way only in the Temple built in Jerusalem by Solomon. They might come to forget, or to regard as secondary, their belief that God accompanied their escaping Hebrew slave ancestors in their wilderness sojourn during their journey away from Egypt. Solomon's prayer at the dedication of the first temple, recorded in 1 Kings 8:27, states this well: "But will God indeed dwell on the earth? Even heaven and the highest heaven cannot contain you, much less this house that I have built!"

In all times and places, fragile and transient human beings cannot confine the sacred: we participate in the sacred, we are permeated by the sacred, we walk in the Presence of the immanent Spirit. We are, then, more than "temples": we are not the bearers of a divinity "confined" almost to materiality in human beings and human structures. Rather, we are presences of the divine Presence that immanents but transcends us. In the spirit of Maximus' teaching, we are called to recognize the *Logos* present in *logoi* in every part of creation, biotic and abiotic; in every member of the human community, among whom some experience and witness to diverse spiritual understandings, and express them in sacred teachings and texts which developed in diverse cultures through time and throughout Earth. A planet-wide ecumenicity among all people of religious faith is achievable if particular parts of the faith community come to recognize and respect that they and all others are *logoi* who bear within them sparks of the creating Spirit.

We are shared stardust with and kin to all being, in integral being. While on Earth and as we voyage into space, we live in a sacral social cosmic commons. We will encounter and Contact in new places and planets, and in new ways in new experiences, diverse (and even, perhaps, surprising) members of our extended cosmic family, as life and as intelligent life.

Earth Relational Community

A sense of relation within the community of all life, initiated on Earth, our terrestrial home, the planet and place of our origin, is enabled and experienced when people free their consciousness from an exclusive Earth community particularity and individualistic species self-deception. At that time, we are enabled to participate in an Earth-extended relationship integrated in community-in-commons. We come to understand more fully that we are related, each species and each individual within every species, not only as individual members to each other, but to all other species in the aggregate and as individuals—all the living beings on Earth's common ground, our common commons. If people are not able to relate each to the other accepting differences of race, ethnicity, gender, social class, religious or humanist perspective, sexual orientation, nationality, and so on, they will not realize peace and justice on Earth such that the common good is sought and common needs are met; and, they will not be open to accepting and respecting extraterrestrial intelligent life, and sharing equitably with their cosmic counterparts the natural goods that abound in cosmic common ground. On the cosmic level, as on Earth, there should be enough to meet everyone's needs but not to satisfy everyone's greed. We are citizens of Earth, yes; but we are citizens of the cosmos, too.

A complementary consideration inserts itself here, pressing for acknowledgment and acceptance: Spirit Presence is not confined to Earth and to humankind. Since Spirit cannot be confined to a single place of worship in one religion on one planet in one solar system, one galaxy, and one universe, how much less (to paraphrase Solomon) can Spirit be confined to a single planet, if there are diverse intelligent beings on multiple planets and other celestial bodies throughout the cosmos, who are receptive to immanent the Spirit's Presence and guidance? What other intelligent *logoi* (to permute Maximus minimally), for example, share in the conscious experience of *Logos*, perhaps most evidently those who have received a Spirit-focused calling, and are relationally in communion and community with Spirit?

It is fascinating to ponder the possibilities of transcending perceptions of ETI that exist in diverse cosmic contexts. Atheists in different Earth cultures have expressed their belief that theists are an "endangered species" that will be threatened by ETI: they assert that extraterrestrials are scientifically and technologically much more advanced than humankind, and will have evolved beyond the "necessity" of believing in Spirit. There is, of course, an alternate possibility: extraterrestrial intelligent life will have evolved to a deeper perception, experience, and relationship with Spirit. Should this prove to be the case (and it quite likely has the same probability as that of ETI having evolved "beyond" the "necessity" to believe in divine being and the sacrality of the cosmos), it might prove, ironically, to be "threatening" intellectually to those who believe that there is solely material existence and no spiritual transcendence, and project this belief onto unseen and unmet cosmic Others.

A similar displacement from being absorbed by a previously perceived "truth" would be the realization that extraterrestrial intelligent life had evolved to understand that community took precedence over individuality, without absorbing all people into a "melting pot" or "cyborg" civilization. Such an ETI understanding, with corresponding forms of social, political, and economic organization, would directly challenge—and even condemn—laissez faire economic systems that benefit a handful of the citizenry of a particular planet or of a politically distinct region of a planet. The latter unexpected occurrence would mean that to progress to be more advanced than they are currently, humankind would have to "go ahead to the past," proceeding to a future in which ancient communal forms of organization are present and operative such as in early Christian thought and practice and current Christian and other biblical ideals, and in still-present communal Earth cultures. The compassionate and communal Acts community shared all things in common. While this was discarded over time in favor of individual spiritual salvation through faith alone, and individualistic economic benefits through economic systems and practices that enshrined greed as a virtue and ensured the benefit of the few to the detriment of the many, the sharing community would return as an ideal using updated practices, but still characterized by collaborative cooperation for community well-being—on common ground for the common good. In such an event, terrestrials would be called back to their communal roots by extraterrestrials: God's will *will* be done "on Earth as in the heavens."

Cosmic Relational Community

In cosmic contexts—on the discrete celestial bodies where people explore, settle, and acquire and adapt available natural goods—as in their Earth context of evolution and development, humankind must (for its well-being and even survival) appreciate, advocate for, and apply understandings and practices of cohabitation, conservation, and compassion, all of which express, at their core, cosmosocioecological ethics.

Celestial Cohabitation

Cohabitation is the ability to consociate, in a shared habitat (environmental setting) and habitation (dwelling place), with others who have similar and dissimilar characteristics. Humans share much of their materiality in common with other members of their species, e.g., similar anatomical structure and identical DNA. Humans are dissimilar in relation to each other in terms of such secondary traits as color, facial arrangement, language, ethnic origin and culture, levels of intelligence, physical ability, and distinct institutional structures that have emerged from diverse cultures in disparate (but ecologically linked) areas of the globe. These include *political systems* (including democracy; monarchy; autocracy, whether based on political or economic power, or as theocracy, based on religious authority with substantial power); *economic systems* (including communal, cooperative, socialist, communist, capitalist; focused primarily on community well-being or on individual and individualistic "rights"); and *religions' institutions* (including Ba'hai, Buddhism, Christianity, Hinduism, Islam, Jainism, Judaism, Shintoism, Sikhism, and indigenous peoples' diverse and less institutionalized spiritual understandings). On Earth, these secondary characteristics in all their diversity and complexity have caused conflicts over occupation of place, possession of property, and expressions of ideology (as doctrines or dogmas in religions), many of which have resulted in war, assassination, torture, massacre, and even genocide against the perceived "Other."

In order to cohabitate peacefully and promote mutually beneficial prosperity, humans in all their diversity must do more than "tolerate" those they perceive as Other and dislike or despise despite their shared, primary, material characteristics: they must openly acknowledge, appreciate, accept, and accommodate the creativity and cultural perspectives in which these distinctions have emerged and evolved, and the people who embody them.

Similarly, in space. Potentially, there exists a far greater difference among species characteristics and forms of social organization than those operative on Earth. It is theoretically possible that some have a different DNA than what humans share in common among themselves, and to varying degrees with all other biota. As reiterated throughout *Cosmic Commons*, it is essential that prior to Contact humans would have undergone a profound conversion, a substantial transformation of consciousness and conduct. Indeed, some scholars, and those who describe encounters of various kinds with UFOs/UAP or ETI, speculate that ETI has not made direct, universally acceptable (including to the most skeptical scientists), and highly visible Contact on the ground with humanity because we are too primitive, as expressed in our unequal social arrangements, biases against each other, ecological irresponsibility (including overpopulation and climate change), and continuous preparations for war, including for international nuclear conflict which would destroy our Earth home and all life—all of which provide evidence of a far-from-mature species. In order to cohabitate—on Earth and in the heavens, on planets and on celestial bodies other than planets—humans must become open to the initial physical, cultural, or institutional "strangeness" of the "Other"—their inherent "otherness," which distinguishes them from us and from every other Other.

If we cannot cohabitate with other humans, we will not be able to cohabitate with intelligent beings who are far more different from all of us on Earth than we are from each Earth 'other' prior to Contact. ETI might share with us some or many of our material similarities, such as DNA or a complementary anatomy, but not be materially congruent with our form of being, let alone our diverse secondary physical traits and cultural expressions. Consequently, without the type of conversion described earlier, we will be unable to formulate, together with ETI, common perspectives on conservation of shared habitat, or compassion for the "least ones" among either or both species. Some individualistic humans in the United States already are advocating for private property on celestial bodies, despite UN documents (signed by the United States) that state specifically that space is the "common province" of humankind, without consideration of the implications of the existence of multiple intelligent interstellar species that the UN did not consider. UN instruments such as the Outer Space Treaty and its documentary descendants describe a fundamentally communal vision, a vision that expresses what has been historically the social organization, territorial understanding, and property arrangements of Earth's indigenous peoples. That UN communal vision and an indigenous peoples'-like tradition might well be already the reality of ETI, perhaps a result of their own material or cultural evolution as intelligent species. It might also be a principal reason why some diverse ETI cultures have been able to collaborate culturally and technologically on their home planet in order to produce and share space aboard common spacecraft venturing into space for intergalactic and interstellar voyages, and to be able to collaborate interspecially with other ETI with whom they have made Contact.

Celestial Conservation

Conservation, in celestial contexts, is living in responsible relationships through which species integrity (at whatever evolutionary stage a species might be) and species' habitats (including abiotic constituents of place, and resident biota complementary to and integral for species survival in place) are unaltered, or altered only to the extent necessary for species' survival and well-being.

Conservation is not advocated to become operative as preservation when the latter means striving to maintain a species at a particular, usually present stage of its evolution (which might alter or impede its evolution, or accelerate the extinction of other biota dependent on that species as a source of food). Conservation can be operative as preservation when this means to maintain a place in a way that enables it to continue to exist with the integrated and interrelated interactions through which it provides habitat for its myriad biota, in the present and into the future. Preservation of place is ordinarily beneficial (unless in- or out-migration of species in the course of evolution would be impeded or prevented), especially when humankind has been causing harm in a place; preservation of biota because humans value them particularly for what they provide for humans—for example, aesthetic beauty that humans appreciate—is a biota-adverse practice when it impedes or eliminates natural evolution. It is beneficial when biota would have been extincted because of adverse, irresponsible human conduct (done consciously or unconsciously) but would not have gone extinct in the natural course of biotic co-evolution.

Celestial Compassion

Compassion, in celestial contexts, is conscientiously striving to ensure the well-being of the community of life as a whole, as particular species parts of the whole, or as individual members of species. Humankind and other intelligent beings, no less than other self-consciously aware beings or those who have primarily a basic survival instinct, seek the well-being of their own species. Compassion means going beyond one's own species or one's self's survival and well-being, acting with a holistic consciousness of the relational community of which all are part. Cohabitation of intelligent beings each with the other requires that all be solicitously and mutually concerned about the well-being of intelligent life and of other life in the areas in which they cohabitate.

When intelligent beings are open to work toward mutual benefit in terms of cosmic natural goods and common accommodation on celestial bodies' 'terra' firma, they will have the capacity to determine how they might cohabitate particular areas of space and time, conserve the natural goods therein while using them to meet their respective needs, and act compassionately toward not only their kind but toward others, such that those less able to provide for themselves, who might historically have been oppressed by more powerful members of their own or other species, will be able to meet their needs in dignity and justice.

Relational Cosmic Community in Process

Intelligent species' mutual recognition of and respect for the rights and responsibilities of all intelligent life, and solicitous regard for less complex life forms, is essential not only for their own well-being, but for the well-being of their cosmic relations. As described earlier, socioecological praxis ethics will enable humankind to care for Earth and all terrestrial life in practical ways in terrestrial contexts. It will enable humans, too, to enter into dialogic relationships with other intelligent species, in order to strive collaboratively to engage in common projects for the universal common good—or at least for those parts of the universe in which their lives intersect and interact—with shared cosmosocioecological ethics, lived in diverse contexts as cosmosocioecological praxis ethics and, as respective perspectives are integrated, as cosmoethics.

Toward this end, it is important for humanity, as humankind begins space exploration and aims toward colonizing or at least toward mining or farming celestial bodies to meet human needs, to have that conversion of consciousness and conduct that is requisite for fruitful Contact when it takes place in a serious and sustained manner.

Our "thought experiment" has led us to the point of considering more in depth how we might acknowledge the reality of a cosmic commons woven together in integral being, and seek to be good citizens in our universe context. Should we have begun to achieve such a consciousness and conduct pre-Contact, and have the objective of acquiring it fully before, during, and after Contact, we would be doing and would have progressed well to have done that which is necessary for peace and justice to be realized in outer space. As this process continues, the dialogic relationship between our present time and place, and future times in distant places first as envisioned and finally as realized, will be mutually beneficial.

The visions, hopes, and ideals regarding each place will dialogically stimulate socioecological progress in all places. What humans hope they will do better elsewhere than they have done on Earth, that is, what their vision is for much improved socioecological conduct, will draw them forward to effect similar benefits on Earth. And, as humankind progresses socioecologically on Earth, its vision for and actions in the future will be constantly improved for the better.

It would be helpful, at this point, to project and reflect upon the rationale which might have impelled ETI, now roaming the cosmos, to begin their own journey; and, the social stability and planetary protection which they might necessarily have had to achieve in order to work well together to develop the technology and vehicles that have enabled them to travel for apparently vast distances at what has to be described currently as faster than the speed of light, for want of knowledge of cosmic complexities. This might help us, too, as we develop our tentative, dynamic socioecological ethics on our own not only now, but also later in cooperative engagement with ETI, in interrelated terrestrial-extraterrestrial commons contexts within cosmic integral being. Our quest for cosmic peace, justice, and biotic stability should prompt us, before Contact in theory and after Contact in theory and practice, to study more closely relationships in our Earth environment, on celestial bodies on which we disembark and as we develop the capability to do so, and onward and outward into space, among intelligent life's particular communities (including human communities): on their own home planet, on others' home place, on other celestial bodies in which they become common inhabitants, and in space itself when they encounter each other; between intelligent life and the rest of the biotic community; and between all life and its abiotic home and habitat, on any cosmic body. In this engagement there should occur a dynamic formulation, reformulation, or development of principles and practices that would help to sustain or stimulate ecosystemic integrity, ecological interrelationship, and biotic interdependence on Earth and celestial bodies.

But for now, in this time and place, humankind should diminish ever increasingly and finally eliminate entirely, its errors of anthropocentrism and anthropomorphism (as displayed, however inadvertently, by Hawking and others) that, if they had been at some past time similarly present extraterrestrially, might or might not have stimulated or even necessitated ETI space voyages. A hopeful note might be suggested at the outset: that other intelligent beings, unlike what humankind has done especially since the first sailing ships set forth to Discover lands, goods, and peoples in the fifteenth century, will not have formulated and acted in accord with its own Doctrine of Discovery.

Science Fiction Anticipation and Angst: Antagonism or Accord?

Science fiction narratives ordinarily might be placed within one of two broad categories: the aliens will help humankind with medical benefits, new technologies, and new ideologies to assist people to improve beyond current actualities and possibilities; or, the aliens will use their advanced technology to harm humanity, and thereby acquire Earth's territory and natural goods, and subjugate or extinct people who would pose a potential threat to meeting this objective. Stephen Hawking's comments cited previously demonstrate well the latter fear: he suggests eliminating SETI probes of space, and abandoning all efforts to contact ETI who might well have destroyed their own planetary home and deployed into space to seek another world to conquer: for resettlement, or to exploit in order to provide for themselves desired abiotic (or perhaps even biotic) goods.

As discussed earlier, ETI motives for intergalactic or interstellar travels are not known as yet; they might if fact be multiple and disparate. ETI might not all be of one accord in ideology or action, nor of solely one civilization or species. Competition or collaboration—or both—might characterize distinct attitudes and behaviors in space. Divergent ETI possibilities are presented in popular movies such as *War of the Worlds* and *Predator* (ETI as malevolent alien invaders attacking humans); *District 9* (ETI as initially benevolent aliens malevolently treated by humans); *Avatar* (ETI as benevolent residents and humans as malevolent alien invaders); and *Close Encounters of the Third Kind* (ETI as potentially benevolent explorers arriving on Earth, and TI humans as cautious and, in terms of scientists, curious about apparently benevolent ETI).

Which ETI or ETIs will come into Contact with us cannot be known with certainty in the present time in human history. We should be prepared to discover either or both as they discover us and our Earth home.

Considerations on Reasons for ETI Space Voyages

Intelligent species that voyage into space might do so for a variety of reasons. They might be on a quest for knowledge in general, or traveling intentionally to engage other intelligent beings and seek a positive encounter that would include a fruitful cultural exchange and equitable trading: for natural or manufactured goods or technologies, and medical knowledge and procedures. Or, they might intend to dominate a new world's indigenous life and exploit the new world's natural goods to satisfy their own needs and wants. They might have destroyed their home planet (or escaped natural catastrophes that destroyed it) and are living in an artificial environment on their space vehicle on a projected permanent basis, or in search of a new world in which to take up residence. They might be an exploratory party in search of territory and natural goods to benefit themselves or their extended families in communities on their home planet. They might be on an intergalactic or interdimensional journey as bellicose beings seeking new worlds and species to subdue, or as benign beings seeking new neighbors with whom to exchange goods, ideas, technologies, and other mutually enhancing benefits. Their trips might be funded by their home-grown military-industrial complex to enhance their weaponry or to appropriate the natural goods of other species, by scientific organizations seeking greater understanding of the complexity and diversity of the cosmos, by commercial enterprises seeking new trading partners or tourism destinations, or by compassionate ("humanitarian," on Earth) organizations whose own biological and cultural evolution has stabilized, to a great extent, and who seek to enable other species to progress to a new level of relational consciousness and relational community.

The diverse types of space vehicles that credible witnesses claim to have seen are of different sizes and shapes. There are several possible reasons: some might be cosmos-traversing, self-sustaining cities-in-space; some might be entirely robotic (including in their on-board personnel, who might have an appearance and types of behavior that make them seem living), others staffed by minimal scouting crews of a particular ETI; some might be exploratory, others potential colonizers, others on natural goods quests for commercial or industrial purposes, and yet others for tourism; some might be civilian craft, others military craft.

Several apparently distinct types of species, with diverse characteristics, have been reported: they might have originated from a single planet or multiple planets; some might have mutated over time in space, or after periods of time on diverse celestial bodies, as they adapted to altered circumstances and contexts. Some might be benevolent and benign, some malevolent and bellicose: perhaps both types have learned to coexist to some extent, in a balance of power and mutual respect. The possibilities are seemingly limitless. Until such time as more extended Contact and communication occurs with diverse forms of ETI, human conjecture remains just that.

TI and ETI, then, have the contradictory possibilities of either (potentially combative) competition for available "terrigenous" goods and places, or cooperative (potentially congenial) arrangements that enable ETI and TI to negotiate a mutually agreed upon equitable share in and development of such goods. The latter could lead to development and implementation of collaborative commercial practices and complementary industrial processes that produce manufactured goods because TI and ETI are able, and choose, to share independently developed technologies.

Expectations Regarding the Conduct of an "Other"

Humans (as perhaps is the case with other evolving and developing intelligent beings) tend to project onto potential ETI beings virtues and vices characteristic of themselves as they soar into space. When considering Contact possibilities they might hope to meet benevolent intelligent life, but fear meeting malevolent intelligent life. Realistically then, they would strive to be pragmatically prepared for either or both possibilities. A self-serving national government's propaganda, and fictional portrayals of encounters, might emphasize the dangers of ETI Contact: the former to control popular perception, and the latter to play on the fears and commercial success associated with "horror" genre films.

As humanity ponders the possible outcomes of ETI Contact, reflective people try to consider seriously and objectively what human conduct might be when people voyage through space: will explorers and settlers act according to their projected best hope for ETI conduct at initial Contact on Earth or other places, or will humans act in a manner consonant with humans' worst fears about ETI actions? Consideration of distinct possibilities for diverse consequences of ETI and TI consciousness and conduct produces a plethora of possibilities about what might characterize initial and continuing interaction between very distinct forms of evolved intelligence: will there ultimately be collaboration and even consociality and cohabitation to everyone's benefit, or competition and conflict to everyone's detriment?

Human speculation in this area suggests that there are multiple philosophical, religious, economic, political, ecological, and ethical implications of possible past and potential future encounters with ETI. Thoughtful speculation will be limited here to four types of consequences: ecological, economic, ethical, and ecclesial.

Contact: Ecological Considerations

Scenes of the devastation that human conduct—intentionally or unintentionally—has wrought on Earth should catalyze broad and deep human reflection on human impacts on ETI and ETI's homeland when Contact occurs in places far from Earth. It is evident that humans are not solicitous of the well-being of their home planet. With enthusiasm increasing for "discovering" other worlds—both in the original sense of the word, to come upon something previously unknown, and in the Discovery Doctrine sense of domination over peoples and places newly encountered, no matter what the extent of development which their traditional, indigenous civilization has achieved—humankind, even before landfall elsewhere in space, poses a danger to biota and to abiotic contexts throughout the integral being of the cosmos. If human consciousness remains "as is," human conduct will do likewise.

People must, by contrast, respect the biota and the places they explore. To do otherwise would be to ensure that humankind's will is done "in the heavens as on Earth"—and ecological suffering and harm will follow.

Humankind's change of consciousness would be facilitated if people were to think about their worst nightmares about an ETI arrival on Earth, perhaps along the lines suggested by science fiction films of late, which, in order to generate the greatest profits imaginable, use dramatic computer-generated battle scenes and alien monsters to heighten fears (and expectations?) that ETI likely will be malevolent, not benevolent. People should take their worries about what ETI would do to "us" and consider what intelligent beings' fears are in other cosmic places, regarding "extraterrestrials" arriving on their worlds, which is who and what we would be. Might ETI's fears about alien arrivals on their world be the same as humans' fears about alien arrivals on Earth, if inhabitants elsewhere have similarly devastated their home with destructive wars and ecological catastrophe? What might transpire if human explorers chance upon a benevolent civilization which has not come close to ecological destruction, and might, in fact, have no concept about such a thing because they have no history of similar conduct on their own or their ancestors' part, and cannot conceive that intelligent life would destroy what provides for its own and other life? How would humans act in such a setting? Would they take advantage of ETI *naiveté* and seek out the natural goods on ETI's world that would benefit humankind, no matter the impacts this exploitation of places and extraction of natural goods would have on ETI? Would humans take imperiously without asking, or take by force if their request is denied, because the indigenous ET peoples regard a particular natural good as beneficial in place, or because in order to exploit or remove it the human aliens would have to destroy something else of great value, such as the habitat of biota on the surface atop the desired natural goods? Will humankind learn to think natural goods rather than natural resources, intrinsic value prior to instrumental value, ecosystemic well-being instead of human wants, and shared places to serve the common good instead of private property to satisfy individual wants, purchased or seized for selfish exploitation?

Stephen Hawking's instruction that humankind must use its *deus ex machina* to escape from what humans have wrought upon and to Earth will prove fruitless and in vain, ultimately, if humans proceed to do in space as they have done on Earth. Can people stop their self-destructive and Earth-destructive ways here and now? If not, why would they not replicate in space what they have done and are doing on Earth?

Contact: Economic Considerations

In the Outer Space Treaty and subsequent documents, the United Nations declared that space is the "common province of [hu]mankind." The international instruments state further that the needs of less prosperous nations, and less prosperous people and peoples within nations, should be the first to benefit from the natural goods found in space in order that their economic well-being would be promoted and their economic situation bettered.

Considering, for the moment, solely *human* responsibilities that are required or suggested by these statements, it is evident once again that a new consciousness and conduct must come to characterize humankind's attitude toward territory in land and toward natural goods in place. There should be neither nationalistic nor trans*stellar* (extension of current trans*national*) corporate appropriation of the "common province" of humankind (and of other intelligent beings) in whole or in part. If celestial bodies, such as planets, are "common" to all peoples and intelligent species, then mutual, multicultural, multinational, multiplanetary, and multispecies beneficial policies and laws must be firmly established; they would be altered or amended only by the unanimous or substantial majority consent of Earth-based states.

The arrangement on celestial bodies, when only humankind is involved, might be similar to what was enacted in the Antarctic Treaty, with one important, notable exception: Antarctica was divided only among those nations who could establish a prior connection to the continent, such as by undersea contiguity with their national territory, or by prior arrival and at least minimal settlement. In the case of celestial bodies, *all* states and peoples, and ETI would have a claim, not just those who at this moment in time have superior telescopes, technology, or transport vehicles which enable them to be first to see potential habitable places, first to assess these sites and prepare structures and equipment for occupying and developing them, and first to travel to them to acquire their natural and manufactured goods because they have faster or otherwise more advanced modes of space travel. By contrast, the envisioned TI-ETI mutual concern would prompt each to be both on the lookout for natural goods or habitable places needed by the other, or to actually journey throughout the universe specifically to seek them out when such travel becomes possible.

Once settlement occurs, the new territory and its natural goods should be allocated more precisely, in context, after surveys and assessment determine who might benefit most from a place and occupy it at first or subsequently, and who might have the first share of newly available natural goods.

The common good of all, as a fixed objective, would prompt collaboration among the explorers, colonizers, and industrial developers (ideally, enterprises—commercial and industrial—should be cooperatives: each owned by those who work in them, each comprised of intelligent beings from distinct cultural and national backgrounds). All told, such enterprises might form a unified, integrated "cooperative of cooperatives" as modeled since 1956 by the Mondragón Cooperative Movement in Spain's Basque Country.[1] Cooperatives embody the best of capitalist and socialist theory and practices: the ownership is private but the owners are all owner-workers and owner-operators.

In the United States and elsewhere, agricultural cooperatives are paradigmatic for such a structure. In Wisconsin, for example, a dairy cooperative's owner-operators are family farmers whose individual production is pooled and transported to common collection places and plants. They might be sold to large corporate enterprises; or, the farmers might operate a joint dairy cooperative and produce and process their own milk, cream, cheese, and, if they have chickens, their own eggs. The profits are distributed among all owners equitably and proportionately according to what each has contributed to the common collection.

Ideas from John Wesley and John Paul II would be significant not only for socioecological ethics in terrestrial contexts, but also for cosmosocioecological ethics in extraterrestrial contexts and on Contact. When Wesley's three principles regarding the "use of money," and John Paul II's statement that care for the environment is an "essential part" of Christian faith are integrated, for example, they advocate care for the community and care for creation—and this consciousness and conduct might well be extended beyond Earth into the heavens. This recalls another point made earlier: when Christians pray the "Lord's Prayer" thoughtfully they will want to make congruent how they are working to effect God's will "on Earth" and how they are acting similarly "in the heavens."

Another important aspect of celestial bodies' territorial distribution and natural goods development would be to accept and embrace the understanding that the goal of each and every exoEarth settlement should be to "commonize the commons." That is, the celestial body is the commons and its land and goods are *commonized*, held in common, to benefit each and all. To consider this ideal is simultaneously to recognize its reality, to accept that we do indeed live in a creation commons whose existence Presenced by its

---

1. Begun in the Basque country during General Francisco Franco's Fascist dictatorship, gradually expanding, and enduring into the present, Mondragón was founded by a Catholic priest, José María Arizmendiarrieta, who sought to put Catholic social doctrine into practice. His efforts began in 1943 when he arrived in Mondragón. The historical development and current operations of Mondragón are explained on their Web site: http://www.mondragon-corporation.com/language/en-US/ENG.aspx. See also Wolff, "Yes, There Is an Alternative"; Burridge, "Mondragon Defies Spain Slump."

Creator awaits our acknowledgment and acceptance. For secular humanists, particularly those from the United States, Thomas Paine's statements should be relevant. He declared that the Creator did not open a land office and distribute land as private property; and he advocated that the needs of the most disadvantaged citizens should be met. For Christians, the teaching of Jesus in Matthew 25 and as embodied in the Acts Christian community provide together both a religious theory and a Christian embodiment of that theory. For Jews, the Torah has relevant passages from the prophets that Yahweh is concerned about the poor, and that people should be compassionate toward the poor. For Jews and Christians, the teachings of prophets such as Isaiah, Jeremiah, and Amos are relevant. For Muslims, comparable verses in the holy Qu'ran about Allah's solicitous concern for humanity and care for all creatures state that people should do likewise.

Peoples throughout the world, for whom communal orientation and community benefit have been embodied in their current culture or were embodied in past culture, recognize the commonization of the commons for the common good as a familiar idea and practice. Its loss or diminution has been lamented and remains still an ideal, not a reality, for regional or national practice.

Obviously, advocates of laissez-faire capitalism and so-called "free markets" (which are not really "free" for the vast majority of people, or even a majority of businesses) would find the commonize ideas, ideals, and practices unacceptable; they might be worshippers of the God of the Market described so clearly and eloquently by Harvey Cox, and the God of Discovery cited indirectly throughout this book. But idolatry and individual greed must be superseded in the interests of community, justice, and peace. It is absurd to believe in and advocate, and force other beings and other cultures to believe in and follow this capitalist ideology: they might have (or have memories of) more advanced forms of social structures in which compassion, consociality, community, and the common good supersede conflictive, competitive, combative, and individualistic forms of social organization advocated by the Earth minority that benefits from them at the expense of the Earth majority. The lives, livelihoods, and well-being of the common people, as evidenced for millennia in the West and in cultures it has influenced, have historically been coercively subordinated to the interests and desires of those with greater financial and political power. On Earth, in the current class war perpetrated by the "haves" against the "have nots," the gap continues to widen not only between the wealthy and the poor, but between the wealthy and the middle class. People and planet suffer as a consequence.

The economic arrangements and social organization of ETI, if more advanced (and yet, concurrently, more traditional) than what has been developed and established firmly on Earth to date, might well be a threat to those who benefit from the status quo and seek to expand their hegemony even into space—by proposing, for example, that the lands and goods on celestial bodies should not be common property, the "common province of humankind," but private property, the privileged domain of individuals and the corporate enterprises with which they are associated and from which they benefit.

Consider again our worst fears about ET invaders, and turn them around such that we are the invaders, the "extraterrestrials" or "aliens" arriving in other worlds. How might we not, at Contact, be the "world" or the intelligent beings that we project others to be? What if those Contacted have a communal society? Will we respect their communalism or will we, like the Europeans and Euroamericans who arrived on Turtle Island, claim that private property concepts make human civilization "superior" to any other civilization (and, in a parallel way, that religious beliefs on Earth are "superior" to those elsewhere in the cosmos, because of the Christian claim that God became incarnate solely on Earth, or that atheist beliefs are "superior" to or "more advanced" than whatever spiritual or religious beliefs are present among intelligent species in the cosmos)?

We should be concerned about being the "worst nightmare" for the Other on other worlds. If we are not, then we are implicitly acknowledging that Others have the right, if they have superior technology and a Discovery Doctrine, to be our worst nightmare. "Do not do to others what you would not want others to do to you" can be at least a caution here, if not a principle, regarding what destruction humans might wreak in space if they carry forward in their consciousness a human Discovery Doctrine to guide their extraterrestrial conduct in the cosmos.

Contact: Ethical Considerations

In the same vein as the previous section and topic explored, the community-oriented social arrangements developed by ETI over time, founded on an ethics based in and benefitting communities as a whole rather than individuals within communities, would confront the individualism and greed-as-virtue and greed-embodied-in-social class that is the self-determined entitlement characteristic of elements of humankind, particularly in Western nations and other nations under their ideological, economic, and political sway. Might this confrontation of diverse and conflictive ethical perspectives, and humans' refusal to acknowledge the oldest traditions, which advocated or mandated communal life and property, cause ethical conflict and thereby contextual conduct conflict?

Once again, the recurring point: humans must change their consciousness and conduct on Earth, where it has caused such ecological and economic catastrophe. They must act conscientiously and compassionately toward other humans, and conscientiously and with care toward Earth, and do this in equivalent places and with intelligent and simple species throughout the cosmos.

Contact: Ecclesial Considerations

The most profound impacts in TI-ETI Contact, as discussed in earlier chapters, might well be in Christian churches. A majority of Christians, according to polls, are amenable to and do not feel religiously or doctrinally challenged by the idea of an ETI entrance into their lives. (Some atheist scientists, however, foresee and foretell threats to religious—especially Christian—faith.) There is room in the thought of many, including clergy, for there to be other intelligent beings in the vast cosmos. However, it might prove to be the case that clergy would be especially impacted by alternative spiritual perspectives that seem to call into question the doctrines, rituals, and structures of particular religious systems, and that laity, while untroubled in the abstract when reflecting on how ETI would affect their thinking and practice, would be traumatized if Contact were to take place. Christian clergy might be especially threatened by Contact with ETI because the theological rationale for their vocation as clergy might be disrupted if God has in some way become present in another species' likeness or other species' likenesses on another world or multiple worlds eons ago: the "sacrifice" of Jesus and "salvation" in Jesus could not be extended universally by Christian clergy, nor could particular Christian rituals and dogmas. This trauma could be tempered by awe in light of divine creativity, and an appreciation for the Spirit who is and has been present in a special manner throughout the cosmos, not solely on Earth.

Ironically, atheists might be troubled at Contact if ETI has developed a more profound spiritual life, based on their own experiences and ideas and not confined to institutional constraints and control, than that lived on Earth by humans.

As discussed earlier, for Christian clergy and laity alike new understandings of divine creativity and Incarnation, and the role and teachings of Jesus, might have to be developed and disseminated in order to reconcile Christian beliefs and doctrines with the spiritual insights of ETI. (Similarly, Orthodox Jews might have to rethink aspects of Moses' teachings, and Orthodox Muslims might have to reinterpret Mohammad's life and teachings in a cosmic context.) With some soul-searching, adherents of Earth faith traditions might be able to find common ground in a common core of human religious beliefs and ethical insights that would be reconcilable with what ETI brings to Contact.

Cosmic Commons

Ethical, ecological, economic, and ecclesial implications and impacts of Contact emerge, and engage humankind's attention and reflection, even if people only consider, as a thought experiment (as has been done in *Cosmic Commons*), an "as if" scenario. Actual Contact between terrestrial intelligent life whose origin is on Earth, and extraterrestrial intelligent life whose origin is elsewhere in the vast and complex cosmos, would highlight even more the distinctions between terrestrial and extraterrestrial intelligent life and perspectives. Some of these implications and impacts have been elaborated, however summarily, in these pages.

There are likely to be no "spiritual" impacts or implications of Contact. Spirituality is the most personal and yet the most universal engagement of being with Being. Each being has its own experience with the sacred, however culturally understood; all beings have that sacral experience with the singular cosmic Presence of immanent-transcendent Spirit, aspects of whose Being are understood incompletely and to varying degrees by adherents of particular faith traditions—whatever their context and wherever they are located in the cosmos. What is likely is that diverse intelligent biota would see the complementarity of their respective spiritual ideas and practices and be in awe of the diversity and extent of *Logos-logoi* relationships.

Just as the center of the universe is the Spirit (despite erroneous concepts of a local center projected as the cosmic center in one or another historical heliocentric ideologies, astronomies, or metaphysical constructs), there is no cosmogeographical center, some specific site's sacred star: separate stars hold their own planets in orbit, and are sacred as the cosmically evolved suns that are part of ongoing divine creativity—they do not hold the Spirit exclusive of other solar systems. The Spirit (and spiritual) center is simultaneously and eternally present to all beings in integral being: Being presences and permeates being in all its diverse and distinct cosmic contexts. Spirit's reality and material manifestations are not localized, or, perhaps better stated, are localized in every locality and universality.

Terrestrial and extraterrestrial lives are originally centered in their own limited individual and cultural understandings of the Creating Spirit. In their respective courses of species evolution they have retained the cosmic sacrality that has been present in their ancestors, their primordial biological species antecedents. In the imagery of Maximus, each original and consequently emergent being is one of the *logoi* in which the divine *Logos* has self-extended. When they have materially and culturally reached a stage of consciousness open to embracing Spirit beyond their originating contexts, every intelligent being is enabled and invited to transcend their particular context and become consciously part of the cosmic context in which they exist.

Sacred Cosmos, Sacred Commons

All life, in integral being, lives in a *sacred cosmos*, a universe made sacred by divine Presence, with whom and in whom all are intertwined as a relational community. As a consequence of divine creativity, all has emanated from the Creator Being-Becoming. The universe is their sacred common ground, their shared space for encountering others and the Other, with all of whom they are in relation. The sacred cosmos is therefore also a sacred *social cosmos*. It is the locus of interaction between terrestrial and extraterrestrial intelligent life, to benefit each and all equitably through conscientious sharing of available places and natural goods.

The sacred cosmos is a *sacred commons*.

On our Earth common ground, our local commons, we are more than temples, the designated sacred places, particularly human-built structures, to which we have been linked by analogy. We represent all beings, not just humans from our own culture and with our own religious sentiments, beliefs, and rituals. With the Spirit present in us we are all places, constructed by humans or by nature, that have been used to worship divine being as particularly perceived by specific people, and that have originated from and to serve belief systems and religious institutions in specific cultures as separated "sacred space."

We are bearers of Spirit. We have the capacity to be theophanies of Spirit, in whom and to whom we are present. We enable such theophanies to be manifest by the way in which we live, the manner in which we interact with others (in whom Spirit is present and Presence), by what we do and say, and to the extent to which we are open to Spirit's Presence. We can become at such times mediations to humanity, other Earth biota, Earth, and intelligent life, wherever encountered. We must be conscious that what we communicate by word and action might be intended by Spirit for only a moment in time (however short or long that might be), and for only one or several places, cultures, worldviews, and contexts; and, that it is culturally conditioned and expressed due to such factors as our personal and social limitations of historical understandings, language, geographical-geopolitical milieu, and limiting (even if only unintentionally) spiritual or religious consciousness.

As bearers of Spirit we enable, through our theophanic moments and to the degree in which we and others are open to their meaning, Spirit-to-spirit interaction and interrelation within us; Spirit-to-spirit and spirit-to-spirit interaction outward to others; and spirit-to-Spirit and spirit-to-spirit interaction inward from others. We express in these moments Spirit bonded to our spirit, Spirit-uality mediated through materiality: not in a type of Incarnation as in the Christian doctrine pertaining to Christ Jesus as both divine and human in one person, but more in the way of Abraham, Muhammad, Krishna, and Buddha, each of whose materiality mediated Spirit Presence, each of whose words were messages mediated by the era, culture, area, and language from which and through which they were expressed, with all the limitations of such contextual communication.

Interpersonal and intersocial interactions always presenced by Spirit (whether or not that Presence is understood or acknowledged by those who experience it or by others, at that or a later historical moment and place) and sometimes enabled or enhanced by mutual openness, lead to personal and community interrelation and integration in local settings within integral being. They are sacred moments in sacred places in the sacred commons of the sacred cosmos.

Sacred cosmos-sacred commons are integrated as a *sacred cosmic commons*. In the sacred cosmic commons people, whether conscious or unconscious of Spirit within them and in whom they are, as sacred beings (beings in whom Spirit is present, who have emerged from divine creativity, in form after form of *logoi* emanating from the *Logos*, originating in the primordial sacred context from which the cosmos burst forth and through stages of cosmic dynamism and biological evolution ever thereafter) are ever-present in sacred places and related to other sacred beings. In these cosmosocio-ecological sacred contexts, the human species and all intelligent species are charged with Spirit presence and charged to provide responsibly for their mutual common good (their personal and species well-being) and the common good of the sacred context of the sacred cosmos which they share and from which they hope to provide for the necessities of the culturally and materially distinct members of their respective species. Just as on planetary common ground, so too on cosmic common ground should intelligent beings be conscious of the needs of the economic, ethnic, and geopolitical/cosmopolitical "least ones," and conscientiously and compassionately share with them commonly encountered natural places (space on planets and on celestial bodies) and natural goods. Interspecies compassion, cooperation, and collaboration are necessary in order for this to be effected consocially and congenially in cosmic community.

The unseen energies and relationships present and experienced in the integral being of the cosmos provide parameters for intelligent life conduct, no matter the degree of theoretical expertise and technological advancement operative for distinct intelligent beings.

Spirit, science, and space become interactive, intertwined and inter-influential in the cosmic commons. Their integration might be unacknowledged, or its extent not understood. In space, some might reject out of hand the idea that a Spirit being is present; others might formulate a nonreligious or nonspiritual understanding of a unifying presence or energy in the cosmos. In space, too, all should accept that scientific understandings are essential, on so many levels, for responsible, interactive cosmic citizenship.

In any case religious fundamentalisms and claims to sole possession of "truths" about divine Being and the "way" divine Being expects us to be, and notions of cultural "superiority" expressed globally or universally (cosmically) must be overcome and rejected, and replaced with a sense of cosmic mutuality and openness to new cosmic insights about the Being-Becoming Presence in which we live, move, and have our being—whether such insights are expressed from within our terrestrial human family or from extraterrestrial intelligent life in our extended cosmic community.

Scientific fundamentalisms must be rejected too, when those whose focus is rightfully upon the materiality of being in all its complexity seek to impose their metaphysical perspectives (often not understood as such) upon those who affirm and even experience unseen dimensions of reality which they regard as spiritual.

Hope for a Common Cosmos

Humankind does not know what lies ahead. There have been indications of the presence of other intelligent beings in the cosmos and even on or near Earth, as described earlier, but there is no globally accepted evidence of this. We conclude, at least for this work, the thought experiment with which we began. As a consequence of our reflection on it, we might commit ourselves to continue to act "as if" Contact had occurred, the better to sharpen and deepen human consideration and contemplation of what this means or might mean, and who we are or might become in a cosmic community.

It appears that in the present moment of human history humankind might regard Contact with more anticipation than apprehension. No technologically more advanced civilization has aggressively attacked or even interfered with humanity to determine the course of human history—even when social or ecological issues prompted people to hope that some exterior power would guide or coerce them away from operative ideologies and catastrophic actions—including religious ones—that have been leading them inevitably to their and their planet's cataclysmic destruction.

The future realization of cosmosocioecological ethics in places where humans colonize, as well as on humans' origin planet, appears to a great extent to be up to humankind as an intelligent and spatially responsible species.

On Earth, people should find ways to accommodate human differences of whatever divisive type. People should affirm socioecological thought and organization that enable global well-being for human communities, the biotic community as a whole, and Earth, their shared home and common ground—and envision projects to realize what they affirm.

In cosmic community in the cosmic commons, spirit, science, and space are integrated. In integral being and the relationships it generates, intelligent beings could envision and collaborate to effect over time, in the heavens and on Earth, a commonweal simultaneously planetary and cosmic.

Conclusion

At Home on Earth, Reaching for the Stars

In the origin of our species, humans had simple beginnings prior to evolving to become the diverse and complex creatures we are today. Some tens of thousands of years ago, humans began to walk upright and then stride forward. They had a perceptive consciousness that separated them intellectually from the primates from which they emerged. Initially, the very first humans were probably a single family from a common biotic ancestor. They branched out: most varieties went extinct, but *homo sapiens* survived, continued to complexify, migrated globally, and eventually flourished as distinct ethnicities and diverse cultures dwelling in places distant from each other. As human history and cultures developed further and once-distant human communities came into contact, disputes arose when territory or natural goods were competitively desired—at first to meet survival needs and then, for some, to satisfy wants substantially beyond needs.

On the threshold of extensive space exploration, humankind needs to become a single family once again. We must be an integrated species among ourselves, and related respectfully to all other biota and to our sacred Earth common ground. We must work together or we will perish together—along with Earth and all life. We need a profound change of consciousness and conduct in order to relate each to the other in extended communities, to become a new family on a new Earth and in an ever-new cosmic reality. In so doing, we will be, at first only in our imaginative projections but then in reality, cosmic beings: in the spaces in space *in* which we reside and *from* which we acquire natural goods needed to provide for our human common good.

We must learn to be truly "at home" on Earth as one family. We should recognize that we are interrelated with the biotic community as a whole, of which we are a part, and not placed atop a pyramid from which we survey our Earth home, over which we have dominion, in a cosmic context provided solely for humans' benefit. When this material, psychological, and spiritual foundation is in place, we can with greater confidence and a holistic perspective reach for the stars in the vast universe that extends far beyond us, and find our place, too, not only in other cosmic settings but as members of a sacred cosmos in integral being. We would embody and extend a cosmosocioecological ethics to continually guide our consciousness and our conduct as we settle in new places and wherever we might have Contact with other intelligent beings.

Dialogic Relational Communities in Context

A beneficial dialogic relationship between present and future, and Earth context and cosmos contexts, will take place only if humankind in each place provides insights and examples of how humans are fulfilling their terrestrial and extraterrestrial responsibilities. Social (ethical, economic, and ecological) and spiritual (ethical and ecclesial) shifts are required for that to happen. (*Ecclesial* here represents, by analogy, all religions' doctrines and practices.)

Ethically, we must learn to think, as a species, more in terms of "we" than "me." Among some cultures, community is still paramount. This does not require absorption into some mass, some amorphous cultural "melting pot" in which cultural diversity and personal individuality are lost. Rather, it means focusing in a holistic way on the *common good*, on the well-being of all people and peoples in the *commonweal*. Most cultures historically held this perspective; some still do, rather than solely or even primarily focus on individual advantage and benefit. When the community benefits, all of its members benefit; when solely distinct individuals benefit, or one social class alone benefits, the commonweal suffers.

Economically, we must accept that some form of common ownership, particularly in the form of independent cooperatives rather than collective enterprises, would benefit the commonweal. Capitalism's focus on individualism, and markets that are "free" for only a handful of entrepreneurs, have caused poverty, hunger, poor health, and environmental degradation, among other ills, intranationally and internationally. The commons contains commons goods for biotic communities to share; all would benefit when each is able to provide at least for their minimal needs and well-being. Private appropriation by avaricious individuals and monopolistic businesses has harmed creation and community. A new vision for a renewed Earth and renewed community must be formulated, developed, and implemented.

Ecologically, we must recognize—and act as if we recognize—our shared responsibility to promote Earth's well-being. Christians who claim to be "images of God" must see, think, and act as God does, as portrayed biblically: all creation is "very good" and our service to creation is a requirement for us, a service localized in every place and time. We are all part of cosmic integral being; all exists, in some way still primarily unfathomable, in interdependent interrelation; we are all stardust. We have to reappropriate our

"Garden" understanding and conduct, idealized in regard to the past, and yet to be realized in terms of the future, as our utopian vision.

Ecclesially, we must not intellectually or doctrinally limit the divine Spirit's concern for all that emerged from divine creative love solely to one species on one planet in one solar system in the entire universe, nor should Christians anthropomorphically claim that every intelligent being in the cosmos has "sinned" as has humankind, or anthropocentrically claim that to atone for such sin God has chosen to enter God's creation in a unique incarnation solely on Earth, for all cosmic beings for all time—including during the more than two-thirds of the life of the universe that preceded Earth's very existence. Divine being has communicated and communed on Earth with humans of diverse faiths (including secular humanists) in particular ways. Spirit has inspired spiritual leaders in distinct cultures through the ages, and likely in other places in space in such manner as is appropriate for other beings. The human spirit can soar, in fact, in contemplating some form of divine presence throughout the ever-inflating, expanding cosmos.

Logos, Logoi, *and Cosmic Community*

In the understanding of Maximus noted frequently in these pages—and well worth considering in regard to cosmic relationality—the *Logos* flows to and into and is part of all creation and creatures, all of whom emerge from *Logos* creativity as *logoi*. A *dialogic relationality* is established between Creator and creature; this *Logos-logoi relationship* is realized in diverse celestial bodies; the Logos-Transcendent becomes incarnate in unique ways in intelligent beings, in whatever form that intelligent life has evolved over billions of years, in whatever place in which this has occurred. The same Logos takes on or permeates in an unusual way a material form, and cosmically is incarnate in different material forms. Extending what Maximus taught then, the *Logos*-Immanent, in all creation as a whole, in all individual creatures, and specially in some beings, can become embodied in any intelligent species, anywhere and anytime—and in multiple places and times, even concurrent ones—in the cosmos.

The *Logos* is not, then, diminished by being present in diverse and multiple worlds—or even multiple universes. The understanding of "Incarnation" is extended on a cosmic scale. It is not necessary for a particular segment of Christianity, or even of a unified Earth Christianity, to baptize aliens as if, anthropocentrically, the Logos only became flesh in human form and only on one planet and at one local cosmic time, determined by Earth's stage of existence and humanity's level of evolution, rather than being "all things to all 'people'" in all planetary places where intelligent life evolves.

This understanding articulates the extent and extension of divine love in the universe.

People of religious faiths should be able to come to acknowledge that since God, the Creator-Spirit, created humans out of love (literally and figuratively), such a magnanimous Spirit would likely have created other life on other worlds, life which would in some times and places eventually evolve into intelligent material-spiritual being. The Spirit would relate lovingly to all life in the cosmos, solicitous of their well-being, and enjoy being their conversation partner, at whatever level and in whatever manner their capabilities would enable. The Spirit might choose too, in these places, to be present with biota materially and visibly as well as be a transcendent-immanent spiritual Presence, as each might need and to the extent to which they are capable of apprehension. A loving Spirit would not limit divine indwelling in material form solely to the physiology of one intelligent species on one planet in one galaxy in a plethora of galaxies in the cosmic expanse and dynamic expansion-inflation that is God's continuing creation.

Cosmic Visions and Community Prospects

Planetary, interplanetary, and universal well-being require that humans develop a new sense of their place in, and integrality within, Earth and cosmic contexts. When people have a cosmic consciousness and the theoretical and practical foundations essential to ongoing development of terrestrial-extraterrestrial ethics, their perspectives will be informed, enhanced, and stimulated as they explore cosmic realities, extend common responsibilities, and establish parameters for realistic concrete projects. The result will be responsible scientific research and outreach, respectful terrestrial-extraterrestrial engagement, and religious, philosophical, and ideological reconciliation with the implications and impacts of Contact. If this were to be the case, the likelihood for cosmic interrelationship, integration, and interdependence would be enhanced.

People might, then, consider in their present terrestrial context what socioecological vision they have as an ideal for colonies in extraterrestrial contexts. In dialogic relationships with both their present and projected settings, while people reflect on current harmful human impacts on Earth they might consider ethical principles and conduct that might prevent similar harm in space, and envision a better world on an extraterrestrial site. They might consider in turn why the present Earth reality is not congruent with what they envision for their ideal world, and what they envision for a renewed Earth. In light of the foregoing considerations, people might envision a dialogic relationship that could be established between future and present visions and contexts such that the envisioned extraterrestrial ideal and the envisioned Earth inform action on present Earth, and prompt humanity to alter human consciousness and conduct on Earth in order that the envisioned Earth might be realized (made real). The extraterrestrial ideal, then, becomes the present and future terrestrial ideal—and, what is currently a conceptualized and anticipated reality becomes the future concretized reality in cosmic planetary settings. In this dialogic process, the future extraterrestrial context—as envisioned and as gradually realized—influences and transforms present and future Earth, and informs future interplanetary exploration and development while promoting the well-being of communities in context on other worlds. In the cosmic commons contexts, *cosmosocioecological ethics* would provide an evolving basis for exploratory projects, partnerships, and productions.

Such a possibility might not seem very far-fetched when complementary claims regarding the foregoing, made by Robert Wright, are considered. In his *New York Times* opinion piece "Ethics for Extraterrestrials," Wright suggests two reasons why ETI might be benevolent rather than malevolent. First, he describes Peter Singer's view that humans have, historically, progressed greatly in social systems formulation and social conduct, as evidenced by changes made from the time of the Greek city-states in which citizens in opposing states were regarded as subhuman, to the twentieth century when people of diverse races, colors and creeds are regarded as human. Singer believes this indicates a natural process in the evolution of intelligent beings: they eventually use their reason and interaction to discern that moral concern is socially beneficial, and they will extend such concern to all sentient being. Second, Wright expresses his own perception that "pragmatic self-interest," not reason, promotes moral progress. Integrating both perspectives, Wright concludes that humankind will need such moral enlightenment—however evolved—to progress sufficiently—technologically and socially—to journey from its solar system farther into space. He observes that what he and Singer theorize to be necessary for human development can be extrapolated to ETI: for them to have the technology to venture vast distances, they must have evolved along these lines. In a hopeful contrast to Hawking, Wright anticipates that ETI would not be the menacing, conquering species that people on Earth would not want to encounter.

Current Earth concerns stimulate speculative consideration about beneficial dialogue in *time*, in which the present and a projected future (on Earth and in the cosmic heavens), and a future that eventually will be a new or renewed present, will mutually inform each other; in *context*, in which terrestrial and extraterrestrial environments will provide mutually helpful scientific and social data; in *intergenerational dialogue*, in which present cosmic citizens (in each succeeding present) will reflectively consider the needs of their progeny and of other biota and their shared contexts; and in *intercultural dialogue*, in which human consciousness and conduct, and the consciousness and conduct of extraterrestrial intelligent life, will each enrich the other and create bonds to promote community rather barriers than provoke conflict.

Common Ground in Earth and Cosmos Contexts

The proposal of Harvard biologist emeritus and Pulitzer Prize recipient Edward O. Wilson put forth in *The Creation: An Appeal to Save Life on Earth*[1] is apt here. It might be applied not only to people of different ideologies (Wilson, a secular humanist, writes the book as an open invitation to a representative Southern Baptist pastor; Wilson grew up in the Southern Baptist Church) who can find common ground on Earth by agreeing to work together, despite their differences on such issues as divine existence and biotic (especially human) evolution, to stop the ongoing ecological destruction of Earth and Earth's biota. Wilson's statement should be part of human consciousness in our present and future efforts to do better in space than we have done terrestrially in addressing and preventing socioecological injustices; and, equally it should be part of human consciousness as people make Contact with and strive to relate positively, openly, and justly to extraterrestrial intelligent beings. Said Wilson:

> Let us see, then, if we can, and you are willing, to meet on the near side of metaphysics in order to deal with the real world we share. . . . I suggest that we set aside our differences in order to save the Creation. The defense of living Nature is a universal value. It doesn't rise from, nor does it promote, any religious or ideological dogma. Rather, it serves without discrimination the interests of all humanity. . . .
>
> [R]eligion and science are the two most powerful forces in the world today, including especially the United States. If religion and science could be united on the common ground of biological conservation, the problem would soon be solved. If there is any moral precept shared by people of all beliefs, it is that we owe ourselves and future generations a beautiful, rich, and healthful environment. . . .

---

1. Wilson, *Creation*. Disclosure: Wilson and I wrote complementary complimentary endorsements for each other's books. I wrote a blurb for *The Creation*; Wilson wrote one for my book *Sacramental Commons*. Coincidentally, both books were published at about the same time in 2006.

> [T]o protect the beauty of Earth and of its prodigious variety of life forms should be a common goal, regardless of differences in our metaphysical beliefs.[2]

On Earth, not only do differences in religious (non)belief separate people, but also differences in race, culture, social class, gender, sexual orientation, and more. In order to continue space exploration and colonization successfully, humankind must "meet on the near side" of all that divides people, to encounter and embrace commonality and community. When we recognize that we have a common biotic origin and that we can revive the Edenic hope of living in harmony with each other and all creation, this becomes possible.

If we change our consciousness to eliminate biases toward those who are not "just like us," and unite around common themes of ecological responsibility (as Wilson proposes) and economic equitability we will establish peace and justice not only on Earth but wherever we explore and colonize. A new beginning need not be attempted only when in space or on another celestial body, as Stephen Hawking seemed to imply. Rather, it should be initiated on Earth as a prelude to a fruitful dialogue between space places, with the visions and work of humankind in each place helping enhance the well-being of peoples and planets in all places. So, too, will it be beneficial whenever and wherever Contact occurs.

In the heavens, too, ETI explorers and settlers, even if lacking an Edenic myth and vision toward which to aspire, likely have overcome prejudices and bellicosity (assuming, perhaps erroneously, that they once had these social evils on their home planet), and are open to "meet on the near side of metaphysics." We both can recognize that we share a common origin—from which we have evolved respectively in different ways, places, and cosmic time—in the "stardust" that burst forth in seminal form from the Singularity, from whose primordial explosion and expansion we have all emerged in space and time, in the eons-old Day in which the Singularity is our instantaneous yesterday. We would have in common intelligence, reflective consciousness, and creativity. All might have, too, a sense of transcendence, spiritual or otherwise. We would share scientific inquisitiveness and achievement as means by which we understand better our cosmic materiality. We might have DNA in common too, at least in part; while this is not certain, it would provide additional common ground for a new relationship as an interstellar community. We might equally appreciate in a time when electron microscopes examine the minutest details of simple organisms, and telescopes in space examine distant nebulae, the words of poet William Blake (1757–1827) in "Auguries of Innocence":

2. Ibid., 4–7.

> To see a World in a Grain of Sand
> And a Heaven in a Wild Flower
> Hold Infinity in the palm of your hand
> And Eternity in an hour.[3]

Our scientific theories, and technological research and development, might well lead us to "hold infinity in the palm" of our hands, and "eternity in an hour": not in actuality, but in what we come to grasp intellectually and experience spiritually, and in the ways in which we come to traverse space.

Thought Experiment to Incontrovertible Evidence of Material Existence?

In the preceding pages, we began with a profound "as if," a thought experiment in which we have speculated together regarding ethical, ecological, economic, and ecclesial implications of terrestrial-extraterrestrial intelligent life Contact. In order to catalyze substantial thought on what these implications might be, a strong case was presented that such Contact had in fact already occurred or, if not, might occur in the not-so-distant future. Such a case is not congruent with present, publicly available concrete evidence. Despite all that has been written here, some will continue to declare, "Show me the proof." Witnesses, however credible, unless known personally or professionally, remain just that: witnesses to what they state they have seen and heard. Unless artifacts that people declare they have seen but that the US government has appropriated and hidden are released by that government, skeptics will remain unconvinced. Most people today, including those in the United States, think that UFOs/UAP have indeed been viewed; most governments outside the United States (and perhaps officials in the United States) think so too, and have released much of the scientific data that they have assembled.

Beyond Belief

Presently, as noted several times in *Cosmic Commons*, scholars and researchers in universities and research institutions, as well as members of the general public, fear to mention experience with, or express curiosity about, the existence of UFOs and ETI. They are concerned that doing so might imperil their professional position or make them subject to ridicule—including, perhaps, from friends and even family members.

---

3. The complete text is available on the Web site of the William Blake Archive; see www.blakearchive.org.

One impediment to relating one's own experience, conjecturing about others' experiences or claimed experiences, or expressing curiosity about the possibility or reality of UFOs/ETI is "belief" language. Consider the question, do you believe in UFOs? Why must UFOs and ETI be put in the realm of "belief" rather than consideration or curiosity? There is resistance or at least reluctance in some circles, particularly academic and especially scientific, of expressing a "belief" in anything. The language here is very important and influential. A shift is necessary from continuing the conversation in the somewhat nebulous realm of "belief" to engaging in the more intellectually situated and acceptable (particularly in academia) language and realm of "thought" that might lead to knowledge. The response a very intelligent person might make could be different if the above question were posed differently: "Do you think that intelligently controlled UFOs exist?" or, a bit softer, "Are you curious about stories (or a particular news story) describing sightings of UFOs?"

The next step in progressively more direct language in intellectual discourse would be firmly in the realm of knowledge; in today's culture it would be answered positively by few: "Do you know that intelligently controlled UFOs exist, because of your own experience or credible witness reports?" Astrophysicists and astronomers, air traffic controllers and radar operators, among others—when they are retired professionally and are adventurous regarding reactions from family and friends—might respond affirmatively to this query.

It is more than "about time" to shift language, at least, from "belief" to "think" or "curiosity"; it might (or might not, depending on the person asked) continue further to the language of "know."

In any case, let's go beyond questions about religious-like belief and engage in an intellectual discourse, a conversation based on our shared materiality rather than on metaphysical speculation, when we wonder ourselves, and wonder if others wonder, about UFOs and ETI. I would hope that such a linguistic shift would stimulate increasing numbers of professionals to come forward, as several did regarding sightings in the Hudson River Valley area, at least to express their curiosity, if not to relate their personal experiences. Others might be catalyzed to do likewise, and serious discussion would be extended socially.

The case presentation is over and the witnesses continue to increase in number but the jury is out—still a hung jury at the moment. Some of its members want a stronger case, and continue to pore over the evidence that has been presented and that, in this particular and extraordinary case, is still being presented. The final decision might well be given a jump start by statements from highly credible, well-known people or by some dramatic event . . . or not.

Extraterrestrial Humankind

Once humanity ventures into space, away from *terra firma*, we become "extraterrestrials." That day is fast approaching—"fast" relative to how slowly we developed technologically prior to the twentieth century, and how rapidly our technology has improved during the last six decades of the twentieth century.

The farthest reaches of the complex cosmos draw us outward on a starry night. In awe before their beauty, and conscious of their extent in space and time, we can experience simultaneously both a sense of being inconsequential and insignificant within this vast integral being, much of which is visible solely through telescopes, and a sense of wonder that we are individually and as a species a reflective consciousness in and of the cosmos. We are interrelated participants in a cosmos that is capable of communicating with other forms of reflective consciousness, and we could become and be responsible members of an integrated and interdependent cosmic community in the cosmic commons. The experience of integral interrelationship can be enthralling and exhilarating, rather than frightening, as we consider its implications for ourselves and generations after us. We might even long, in hope of seeing it accomplished in our lifetime, that some of us will indeed be "extraterrestrials" as we journey from our Earth home to voyage among the stars we see.

It has been said that we humans are the Earth reflecting on itself, which first became possible when humankind evolved to the point of pondering its own existence and, indeed, the existence of Earth, too, and the starry skies above Earth. Or, that we are the cosmos reflecting on itself for similar reasons. However, deeper consideration leads to another idea or at least a deeper insight into the first idea.

We voice the consciousness of Earth and cosmos. We are able to express, through our faculty of speech but also in the thoughts emerging in our minds, Earth consciousness and cosmic consciousness—at least those parts of Earth with which we are familiar through our material existence and experience, and our consideration of what we have seen and heard. The human voice of just one part of Earth, in all its evolved complexity, embodies the myriad diverse voices of all Earth's animate and inanimate creation, blended in a harmonious way—an integration of individual solos and a collective chorus.

Several years ago, I first formulated phrasing expressing the dialogic relationship between present and future, topia and utopia, vision and its realization:

The present is the mother of the future;
the future is the mother of the present.

What we humans do cooperatively in the present world will help determine the future world, Earth's context to come: the extent of its abiotic well-being and the degree to which it is able to provide the natural goods to meet everyone's needs, including those of biota who gradually and continually come into their own niche. Our present consciousness, vision, and conduct will determine the future state of cosmic contexts in which humans settle, and in which they might interact with ETI.

What we humans envision for the future world will influence what we do in the present world. On the cosmic scene, this means that what we think and do on Earth, and envision for Earth, will influence our consciousness and conduct, and the extent of our conscientiousness and community, on Earth now and in the heavens beyond in worlds to come as they become our new present places and times. What humans envision communally for the future will determine Earth's current context, to the extent that we strive to make our vision of the best that we might become and our planet might come to be, emerge as a reality in the nearest future time imaginable. Our *relative utopias*, as envisioned by terrestrials on Earth and by terrestrials and extraterrestrials in the heavens, might then gradually be realized over time, and culminate in the *absolute utopia* to which we aspire in common.

Creative and Creating Imagination

Phillis Wheatley (1753–84) was a slave (her first name is taken from the slave ship upon which she was taken from Africa and brought to Massachusetts in 1761) and an insightful and original poet. She was a contemporary of Adam Smith, John Wesley, and Thomas Paine (who republished her poem on George Washington in the *Pennsylvania Gazette*). Wheatley's acclaimed poem "On Imagination" was published in 1773 while she was still a slave (she was freed in 1778 as stipulated by the will of her master, John Wheatley). In the second stanza Wheatley, describing imagination, wrote:

> We on thy pinions can surpass the wind,
> And leave the rolling universe behind;
> From star to star the mental optics rove,
> Measure the skies, and range the realms above.
> There in one view we grasp the mighty whole,
> Or with new worlds amaze th' unbounded soul.[4]

---

4. Wheatley, "On Imagination." The complete text is available on the Web site of Virginia Commonwealth University; see www.vcu.edu/engweb/webtexts/Wheatley/imagination.html.

The world is perceived—and able to be pondered—in far different ways today than in the eighteenth century when the poem was written. In the twenty-first century, humankind expects to roam the cosmos not only in imagination, through "mental optics." This we do still, but now our minds are capable of going beyond just imagining roaming the cosmos, since our eyes enable us to roam in actuality in what we see through the Hubble, Kepler, and forthcoming Weber telescopes. Their technologically engineered and enhanced optics take us where no one on Earth has ever gone before. We "range" in our imagination, stimulated today by scientific data and theory, and by literary works. As our technology has progressed so, too, have our abilities to see farther and to travel further. While currently we travel only to the moon as exploring humans and to Mars through human-manufactured machines, we expect to go beyond: to space outside the solar system's chaotic boundary bubble, where Voyager vehicles now traverse places billions of Earth miles from the sun. We are surmounting the pull of gravity and surpassing the winds of Earth, and envision truly leaving behind the "rolling universe" we once thought we knew. We hope to guide our spacecraft to rove "from star to star." In order to do so, we hope to develop, or learn from another, more technologically advanced cosmic civilization, new modes of interstellar travel not bound by the physics we now know or theorize imaginatively.

Wheatley writes reflectively that we might "grasp the mighty whole" in our imagination; current science provides ways to understand the integral being of the cosmos in its complexity and its incomprehensibility, in greater depth and with greater clarity than we can imaginatively conceive. Future science will carry such exploration further and deeper. We recognize as we travel in both our mental and material exploratory journeys that the "whole" might not ever be understood completely because of cosmic complexity, distance, and time. Our "unbounded soul" will continue to be "amazed" by the new worlds, once only imagined, that we find—conceptually and concretely, imaginatively and experientially, and conjecturally and collaboratively, speculatively and materially.

Despite—indeed, because of—our technology, we must continue to let our imagination soar. We must imagine not only what we might accomplish, but how we will act before, during, and after our efforts to achieve ever-advanced goals. We must not leave our humanity behind. We must continue to envision new social and spiritual utopias, and refine our Earth-imagined cosmosocioecological ethics over time and in new contexts; therein and thereafter we will guide effectively our efforts and adapt our principles in new planets and other places, and with new intelligent beings should they be Contacted. We might hope, throughout, that we are not alone in the spaces of space, that we will make congenial and collaborative Contact with other intelligent life, however different than we they might be in their materiality and culture.

We have reached a time when our human vision for our common cosmic future is still in a formative state. To see clearly while we develop this vision, we must reject and transcend archaic, discredited anthropocentric claims to species and cultural superiority, on Earth and in the cosmos. With Phillis Wheatley, we must let our imagination, our consciousness, and our spirit soar.

Humanity as a whole, when all its diverse members come together conscientiously and compassionately, and collaborate as one community acting in concert, can communally envision and "grasp the mighty whole" of integral being as humankind reaches for the stars seeking its place on cosmic common ground. This is most likely to happen if humankind "puts its own house in order"—develops the consciousness and conduct needed to conserve and care for its Earth home, the human community, and the broader biotic community, and strives to carry its new *praxis* vision and action into space. People with this enhanced cosmic consciousness will experience their "unbounded soul" in all its depth as they see wonders yet unimagined in their voyages in the solar system, the Milky Way Galaxy, and interstellar space beyond. If on those journeys—or, even before they begin—humanity comes into Contact with congenial and consocial extraterrestrial intelligent life, such benevolent beings will be disposed to welcome into the cosmic community a human species that has similarly evolved spiritually, scientifically, and socially, and carries into space what it has developed and implemented on Earth—a similarly advanced benevolent consciousness and conduct.

In striving for the stars, humankind should seek simultaneously the enlightenment it needs to be an integral member of the wondrous cosmos, and ably become more than a reflective consciousness only on Earth. In space, humankind will become to an even greater extent a part of the cosmos reflecting upon itself. That role, with all its socioecological responsibilities, is one to which humankind should aspire. Hope for and action to reach it will benefit Earth community now, and celestial communities in the worlds to come.

We should recognize that Spirit, science, and space are intimately integrated in a relational community in the cosmic creation commons that is immanented by a sacred and active Spirit Presence. The vision we develop consequently will stimulate collaborative human community efforts to go forward into the vast cosmos with confidence. We will take with us on our voyages our wonder and wondering, and our openness to the diversity, abilities, and achievements of extraterrestrial intelligent life. We will experience an ever-increasing awareness of our place in cosmic creation, permeated by our extended and deepened cosmic consciousness and spiritual insight.

It is ultimately in the cosmic creation commons that Spirit, science, and space are integrated in a relational community permeated by a sacred and active Presence. As we reach for the stars, the relational cosmic community in all its diversity and complexity awaits the arrival and presence of enlightened human beings in the cosmic commons.

Bibliography

Aquinas, Thomas. *Summa Theologica.* Part II-II (*Secunda Secundae*). Translated by Fathers of the English Dominican Province. Project Gutenberg Book, EBook #18755, July 4, 2006.

Armstrong, Neil, et al., with Gene Farmer and Dora Jane Harblin. *First on the Moon—A Voyage with Neil Armstrong, Michael Collins, and Edwin E. Aldrin, Jr.* New York: Barnes & Noble, 2002.

Associated Press. "New Estimate Suggests Billions of Planets the Size of Earth in Milky Way Galaxy." *Helena Independent Record*, January 8, 2013, 4A.

Ayala, Francisco J. *Darwin and Intelligent Design.* Minneapolis: Fortress, 2006.

———. *Darwin's Gift to Science and Religion.* Washington, DC: Joseph Henry, 2007.

BBC. "Brazil Air Force to Record UFO Sightings." August 11, 2010. Online: www.bbc.co.uk/news/world-latin-america-10947856.

———. "Churchill Ordered UFO Cover-Up, National Archives Show." August 5, 2010. Online: http://www.bbc.co.uk/news/uk-10853905.

———. "Ministry of Defence Files on UFO Sightings Released." August 11, 2011. Online: http://www.bbc.co.uk/news/uk-14486678.

———. "MOD Releases UFO Files." May 13, 2008. Online: http://news.bbc.co.uk/2/hi/programmes/newsnight/7399717.stm.

Beach, Waldo, and H. Richard Niebuhr, editors. *Christian Ethics: Sources of the Living Tradition.* 2nd ed. New York: Ronald Press Company, 1973.

Beazely, C. Raymond. "Prince Henry of Portugal and the African Crusade of the Fifteenth Century." *The American Historical Review* 16 (1910–11) 11–23.

Berry, Thomas. *The Christian Future and the Fate of Earth.* Edited by Mary Evelyn Tucker and John Grim. Maryknoll, NY: Orbis, 2009.

———. *The Dream of the Earth.* San Francisco: Sierra Club, 1988.

———. *The Great Work: Our Way Into the Future.* New York: Bell Tower, 1999.

———. *The Sacred Universe—Earth, Spirituality, and Religion in the Twenty-First Century.* Edited by Mary Evelyn Tucker. New York: Columbia University Press, 2009.

Berry, Wendell. *The Unsettling of America: Culture and Agriculture.* San Francisco: Sierra Club, 1977.

Bird, Joan. *Montana UFOs and Extraterrestrials: Amazing Stories of Documented Sightings and Encounters.* Helena, MT: Riverbend, 2012.

Boff, Leonardo. *Cry of the Earth, Cry of the Poor.* Translated by Phillip Berryman. Maryknoll, NY: Orbis, 1997.

———. *Ecology and Liberation—A New Paradigm.* Translated by John Cumming. Maryknoll, NY: Orbis, 1996.

Borenstein, Seth. "US Set Heat Record in 2012." *Helena Independent Record*, January 9, 2013, 2A.
Boyle, Alan. "Hawking Goes Zero-G: 'Space, Here I Come.'" *MSNBC.com*, April 26, 2007. Online: http://www.msnbc.msn.com/id/18334489.
Buber, Martin. *Paths in Utopia*. Translated by R. F. C. Hull. Boston: Beacon, 1958.
Bullard, Robert D. *Dumping in Dixie—Race, Class, and Environmental Quality*. 3rd ed. Boulder, CO: Westview, 2000.
Burridge, Tom. "Basque Co-operative Mondragón Defies Spain Slump." *BBC News Europe*, August 13, 2012. Online: www.bbc.co.uk/news/world-europe-19213425.
Butler, Smedley D. *War Is a Racket*. Port Townsend, WA: Feral House, 2003.
Carey, Thomas J., and Donald R. Schmitt. *Witness to Roswell—Unmasking the Government's Biggest Cover-Up*. Rev. ed. Franklin Lakes, NJ: New Page, 2009.
Casas, Bartolomé de las. *Bartolomé de las Casas: A Selection of His Writings*. Translated and edited by George Sanderlin. New York: Knopf, 1971.
Catholic Bishops of Appalachia. "This Land Is Home to Me: A Pastoral Letter on Powerlessness in Appalachia by the Catholic Bishops of the Region." Catholic Committee of Appalachia, 1975. Online: http://www.ccappal.org/CCAbook040307.pdf.
Chryssavgis, John, editor. *Cosmic Grace, Humble Prayer: The Ecological Vision of the Green Patriarch Bartholomew I*. Grand Rapids: Eerdmans, 2003.
Churchill, Ward. *A Little Matter of Genocide: Holocaust and Denial in the Americas, 1492 to the Present*. San Francisco: City Lights, 1997.
———. *Struggle for the Land—Native North American Resistance to Genocide, Ecocide and Colonization*. San Francisco: City Lights, 2002.
Climate Communication. "Heat Waves and Climate Change." June 28, 2012. Online: http://climatecommunication.org/new/articles/heat-waves-and-climate-change/overview/.
Catholic Bishops of the Columbia River Watershed Region. "The Columbia River Watershed: Caring for Creation and the Common Good." Washington State Catholic Conference, 2001. Online: www.columbiariver.org.
Corso, Philip J., with William J. Birnes. *The Day After Roswell*. New York: Pocket, 2008.
Cox, Harvey. *The Feast of Fools—A Theological Essay on Festivity and Fantasy*. New York: Harper & Row, 1969.
———. "The Market as God—Living in the New Dispensation." *The Atlantic*, March 1999, 18–23. Online: http://www.theatlantic.com/magazine/archive/1999/03/the-market-as-god/6397.
Cronon, William. *Changes in the Land: Indians, Colonists, and the Ecology of New England*. New York: Hill & Wang, 1983.
Crowe, Michael J., editor. *The Extraterrestrial Life Debate, Antiquity to 1915: A Source Book*. Notre Dame: University of Notre Dame Press, 2008.
Cuénot, Claude. *Teilhard de Chardin: A Biographical Study*. Translated by Vincent Colimore. Edited by René Hague. Baltimore: Helicon, 1965.
Dick, Steven J., editor. *Many Worlds: The New Universe, Extraterrestrial Life, and the Theological Implications*. Philadelphia: Templeton Foundation Press, 2000.
Dussel, Enrique D. *A History of the Church in Latin America: Colonialism to Liberation (1492-1979)*. Translated and revised by Alan Neely. Grand Rapids: Eerdmans, 1982.
Eagleson, John, and Philip Scharper, editors. *Puebla and Beyond: Documentation and Commentary*. Translated by John Drury. Maryknoll, NY: Orbis, 1979.

Earth Charter Initiative. "The Earth Charter." 2000. Online: http://www.earthcharterinaction.org/content/pages/Read-the-Charter.html.
Farrell, John. *The Day Without Yesterday—Lemaître, Einstein, and the Birth of Modern Cosmology*. New York: Thunder's Mouth, 2005.
Friedman, Stanton T. *Top Secret/MAJIC: Operation Majestic-12 and the United States Government's UFO Cover-Up*. Cambridge: Da Capo, 2008.
Friedman, Stanton T., and Don Berliner. *Crash at Corona: The U.S. Military Retrieval and Cover-Up of a UFO*. New York: Paragon, 2004.
Funes, José Gabriel. "L'extraterrestre è mio fratello." Interview by Francesco M. Valiante. *L'Osservatore Romano*, May 14, 2008. Online: http://www.vatican.va/news_services/or/or_quo/interviste/2008/112q08a1.html.
Gillis, Justin. "It's Official: 2012 Was Hottest Year Ever in U.S." *New York Times*, January 8, 2013. Online: http://www.nytimes.com/2013/01/09/science/earth/2012-was-hottest-year-ever-in-us.html?_r=0.
Glatz, Carol. "Vatican-Sponsored Meeting Discusses Chances of Extraterrestrial Life." *Catholic News Service*, November 10, 2009. Online: http://www.catholicnews.com/data/stories/cns/0905002.htm.
Glave, Dianne D., and Mark Stoll, editors. *"To Love the Wind and the Rain": African Americans and Environmental History*. Pittsburgh: University of Pittsburgh Press, 2006.
Gramsci, Antonio. *Selections from the Prison Notebooks of Antonio Gramsci*. Edited and translated by Quintin Hoare and Geoffrey Nowell Smith. New York: International Publishers, 1971.
Gutiérrez, Gustavo. *A Theology of Liberation: History, Politics and Salvation*. Translated and edited by Caridad Inda and John Eagleson. Maryknoll, NY: Orbis, 1973.
Hanke, Lewis. *Aristotle and the American Indians: A Study in Race Prejudice in the Modern World*. Chicago: Henry Regnery, 1959.
Hart, John. "Cosmic Commons: Contact and Community." *Theology and Science* 8 (2010) 371–92.
———. "Crisis on the Land: Agribusiness vs. Agriculture." *Christianity and Crisis*, April 19, 1985, 130–35.
———. *Encountering ET: Aliens in* Avatar *and the Americas*. Eugene, OR: Cascade, forthcoming.
———. *Sacramental Commons: Christian Ecological Ethics*. Lanham, MD: Rowman & Littlefield, 2006.
———. *What Are They Saying about Environment Theology?* Mahwah, NJ: Paulist, 2004.
Harvey, Van A. "Ludwig Andreas Feuerbach." In *The Stanford Encyclopedia of Philosophy*, edited by Edward N. Zalta. Online: http://plato.stanford.edu/archives/fall2011/entries/ludwig-feuerbach/.
Heartland Regional Catholic Bishops Conference. *Strangers and Guests: Toward Community in the Heartland*. Des Moines, IA: Heartland Project, 1980.
Henderson, Neil. "UFO Files Reveal 'Rendlesham Incident' Papers Missing." *BBC News*, March 2, 2011. Online: http://www.bbc.co.uk/news/uk-12613690.
Herd, George. "UFO Wales: New X-Files Shed Light on 'Alien' Sightings." *BBC News Wales*, July 12, 2012. Online: http://www.bbc.co.uk/news/uk-wales-18798862.
Hessel, Dieter, and Larry Rasmussen, editors. *Earth Habitat: Eco-Justice and the Church's Response*. Minneapolis: Fortress, 2001.
Hochschild, Adam. *King Leopold's Ghost: A Story of Greed, Terror, and Heroism in Colonial Africa*. New York: Houghton Mifflin, 1999.

Holder, Rodney D., and Simon Mitton, editors. *Georges Lemaître: Life, Science, and Legacy*. Astrophysics and Space Science Library 395. Berlin: Springer, 2012.

The Holy Qur'an: Text, Translation and Commentary. Edited by Abdullah Yusuf Ali. U.S. ed. Elmhurst, NY: Tahrike Tarsile Qur'an, 1987.

Hui, Sylvia. "Hawking Says Humans Must Go into Space to Survive." Associated Press, June 14, 2006. Online: http://usatoday30.usatoday.com/tech/science/space/2006-06-13-hawking-humans-space_x.htm.

Hynek, J. Allen. *The Hynek UFO Report*. New York: Barnes & Noble, 1997.

———. *The UFO Experience: A Scientific Inquiry*. Chicago: Henry Regnery, 1972.

Hynek, J. Allen, et al. *Night Siege: The Hudson Valley UFO Sightings*. 2nd ed. St. Paul, MN: Llewellyn, 1998.

"Intervention of the Special Guest, Prof. Werner Arber." Online: http://www.vatican.va/news_services/press/sinodo/documents/bollettino_25_xiii-ordinaria-2012/02_inglese/b13_02.html.

Jessup, Philip C., and Howard J. Taubenfeld. *Controls for Outer Space and the Antarctic Analogy*. New York: Columbia University Press, 1959.

John Paul II, Pope. "Address of Pope John Paul II to the Participants in the Vatican Conference on Cosmology." July 6, 1985. Online: http://www.vatican.va/holy_father/john_paul_ii/speeches/1985/july/documents/hf_jp-ii_spe_19850706_conferenza-cosmologia_en.html.

———. "The Ecological Crisis: A Common Responsibility." Message for the Celebration of the World Day of Peace. January 1, 1990. Washington, DC: United States Catholic Conference, 1990.

Johnson, Kirk. "A Start-Up Sees a Gold Rush Among the Stars." *New York Times*, December 24, 2012. Online: http://www.nytimes.com/2012/12/25/science/space/washington-company-is-working-to-mine-asteroids.html?_r=0.

Kaminski, Bob. *Lying Wonders: Evil Encounters of a Close Kind*. Mukilteo, WA: Wine Press, 2006.

Kean, Leslie. *UFOs: Generals, Pilots, and Government Officials Go on the Record*. New York: Three Rivers, 2010.

Kirk-Duggan, Cheryl A. *The Sky Is Crying: Race, Class, and Natural Disaster*. Nashville: Abingdon, 2006.

Leake, Jonathan. "Don't Talk to Aliens, Warns Stephen Hawking." *The Sunday Times*, April 25, 2010. Online: http://timesonline.co.uk/tol/news/science/space/article7107207.

Mannheim, Karl. *Ideology and Utopia: An Introduction to the Sociology of Knowledge*. Translated by Louis Wirth and Edward Shils. New York: Harcourt, Brace, 1954.

Marcel, Jesse, Jr., and Linda Marcel. *The Roswell Legacy: The Untold Story of the First Military Officer at the 1947 Crash Site*. Franklin Lakes, NJ: New Page, 2009.

Marx, Karl. "Contribution to the Critique of Hegel's Philosophy of Right." In *The Marx-Engels Reader*, edited by Robert C. Tucker, 11–23. New York: Norton, 1972.

———. "Critique of the Gotha Program." In *The Marx-Engels Reader*, edited by Robert C. Tucker, 383–98. New York: Norton, 1972.

———. "Inaugural Address of the Working Men's International Association." In *The Marx-Engels Reader*, edited by Robert C. Tucker, 374–81. New York: Norton, 1972.

McGrath, Alister E. *The Reenchantment of Nature: The Denial of Religion and the Ecological Crisis*. New York: Doubleday, 2002.

Michaud, Michael A. G. *Contact with Alien Civilizations: Our Hopes and Fears about Encountering Extraterrestrials*. New York: Springer, 2007.

Miller, Robert J. *Native America, Discovered and Conquered: Thomas Jefferson, Lewis and Clark, and Manifest Destiny.* Native America: Yesterday and Today. Westport, CT: Praeger, 2006.

Miller, Robert J., et al. *Discovering Indigenous Lands: The Doctrine of Discovery in the English Colonies.* Oxford: Oxford University Press, 2012.

More, Thomas. *Utopia.* Edited with revised translation by George M. Logan. 3rd ed. New York: Norton, 2011.

MSNBC. "Hawking: Aliens May Pose Risks to Earth." April 25, 2010. Online: http://www.msnbc.com/id/36769422/ns/technology_and_science-space/.

Nasr, Seyyed Hossein. *Man and Nature: The Spiritual Crisis of Modern Man.* New ed. Chicago: ABC International, 1997.

———. *Religion and the Order of Nature.* 1994 Cadbury Lectures at the University of Birmingham. New York: Oxford University Press, 1996.

Newcomb, Steven T. *Pagans in the Promised Land: Decoding the Doctrine of Christian Discovery.* Golden, CO: Fulcrum, 2008.

O'Connor, J. J., and E. F. Robertson. "Georges Henri-Joseph-Edouard Lemaître." Online: http://www-history.mcs.st-andrews.ac.uk/Biographies/Lemaitre.html.

Paine, Thomas. *Agrarian Justice.* In *The Life and Major Writings of Thomas Paine*, edited by Philip S. Foner, 605–23. Secaucus, NJ: Citadel, 1974.

———. *The Life and Major Writings of Thomas Paine.* Secaucus, NJ: Citadel, 1974.

———. *The Rights of Man, Part Second.* In *The Life and Major Writings of Thomas Paine*, edited by Philip S. Foner, 345–458. Secaucus, NJ: Citadel, 1974.

Peacocke, Arthur. *Creation and the World of Science.* Oxford: Clarendon, 1979.

Peruvian Bishops' Commission for Social Action. *Between Honesty and Hope: Documents from and about the Church in Latin America.* Translated by John Drury. Maryknoll, NY: Maryknoll, 1970.

Plato. *The Laws.* Translated by Trevor J. Saunders. New York: Penguin, 2004.

———. *The Republic.* Translated by Desmond Lee. 2nd ed. New York: Penguin, 2007.

Randolph, Richard, et al. "Reconsidering the Theological and Ethical Implications of Extraterrestrial Life." *CTNS Bulletin* 17 (1997) 1–8.

Salas, Robert L., and James Klotz. *Faded Giant.* BookSurge, 2005.

Smith, Adam. *Theory of Moral Sentiments.* 6th ed. Mineola, NY: Dover, 2006.

———. *The Wealth of Nations.* New York: Modern Library, 2000.

Stannard, David E. *American Holocaust: The Conquest of the New World.* New York: Oxford University Press, 1992.

Stoeger, William R., editor. *Theory and Observational Limits in Cosmology: Proceedings of the Vatican Observatory Conference Held in Castel Gandolfo, Italy, July 1–9, 1985.* Vatican City: Specola Vaticana, 1987.

Swimme, Brian, and Thomas Berry. *The Universe Story: From the Primordial Flaring Forth to the Ecozoic Era—A Celebration of the Unfolding of the Cosmos.* San Francisco: HarperSanFrancisco, 1992.

Tawney, R. H. *Religion and the Rise of Capitalism: A Historical Study.* Holland Memorial Lectures, 1922. New York: New American Library, 1954.

Teilhard de Chardin, Pierre. *Christianity and Evolution.* Translated by René Hague. New York: Harcourt, 1974.

———. *The Divine Milieu.* Rev. ed. New York: Harper & Row, 1968.

———. *The Human Phenomenon.* Translated by Sarah Appleton-Weber. Eastbourne, UK: Sussex Academic, 2003.

———. *Letters from a Traveller.* New York: Harper, 1962.

———. *The Phenomenon of Man*. Rev. ed. New York: Harper & Row, 1965.
Tinker, George E. *American Indian Liberation: A Theology of Sovereignty*. Maryknoll, NY: Orbis, 2008.
———. *Spirit and Resistance: Political Theology and American Indian Liberation*. Minneapolis: Fortress, 2004.
Toolan, David. *At Home in the Cosmos*. Maryknoll, NY: Orbis, 2001.
Union of Concerned Scientists. "Findings of the IPCC Fourth Assessment Report: Climate Change Impacts." Online: http://www.ucsusa.org/global_warming/science_and_impacts/science/findings-of-the-ipcc-fourth.html.
———. "Findings of the IPCC Fourth Assessment Report: Climate Change Mitigation." Online: www.ipcc.ch/publications_and_data/ar4/wg3/en/contents.html.
———. "Findings of the IPCC Fourth Assessment Report: Climate Change Science." Online: http://www.ucsusa.org/global_warming/science_and_impacts/science/findings-of-the-ipcc-fourth-2.html.
United Nations. "The Antarctic Treaty." October 15, 1959. Online: http://www.ats.aq/e/ats.htm.
———. *The Declaration on the Rights of Indigenous Peoples—With an Introduction for Indigenous Leaders in the United States*. Tucson, AZ: University of Arizona Indigenous Peoples Law and Policy Program, 2012.
———. "Millennium Development Goals." Online: http://www.un.org/millenniumgoals/ www.un.org/millenniumgoals/pdf/mdg report 2010 en r15-low res 20100615-.pdf.
———. "Universal Declaration of Human Rights." Online: http://www.un.org/en/documents/udhr/index.shtml.
———. *United Nations Treaties and Principles on Outer Space*. New York: United Nations, 2002.
———. "World Charter for Nature." Online: www.un.org/ga/search/view_doc.asp?symbol=A/RES/37/7.
U.S. Conference of Catholic Bishops. "Economic Justice for All: Pastoral Letter on Catholic Social Teaching and the U.S. Economy." November 13, 1986. Online: www.usccb.org/upload/economic_justice_for_all.pdf.
———. "Renewing the Earth: An Invitation to Reflection and Action on Environment in Light of Catholic Social Teaching." In *Renewing the Face of the Earth: A Resource for Parishes*, 1–10. Washington, DC: U.S. Catholic Conference, 1994.
Vallee, Jacques. *Dimensions: A Casebook of Alien Contact*. San Antonio, TX: Anomalist, 2008.
White, Lynn, Jr. "The Historical Roots of Our Ecological Crisis." *Science* 155 (1967) 1203–7.
Wilson, Edward O. *The Creation: An Appeal to Save Life on Earth*. New York: Norton, 2006.
Wolff, Richard. "Yes, There Is an Alternative to Capitalism: Mondragón Shows the Way." *The Guardian*, June 24, 2012. Online: www.guardian.co.uk/commentisfree/2012/jun/24/alternative-capitalism-mondragon.aspx.
World Commission on the Ethics of Scientific Knowledge and Technology (COMEST). *The Precautionary Principle*. Paris: UNESCO, 2005.
Wright, Robert. "Ethics for Extraterrestrials." *New York Times*, May 4, 2010. Online: http://opinionator.blogs.nytimes.com/2010/05/04/the-moral-alien/.

Index

Agreement Governing the Activities of States on the Moon and Other Celestial Bodies, 225, 230, 236–40, 354–55
Anaya, James, 62 n. 5, 111
Antarctic Treaty, UN, 103, 105–7, 121, 234, 240, 341–43, 356, 383
Aquinas, Thomas, 43, 44, 49, 50, 54, 170, 174, 254, 266
 On theft, 43–44
Arber, Werner, 265–66
Armstrong, Neil, 117, 118, 119, 120, 325
Asimov, Isaac, 285
Avatar, 13, 14, 25, 126–28, 137 n. 15, 194, 205, 379
Ayala, Francisco J., 96
Being, integral, 7, 15, 22, 25, 40 n. 20, 94, 97, 98, 99, 100, 142, 175, 178, 180–84, 187, 193, 223, 224, 229, 297, 299, 303, 331, 336, 341, 358, 365, 369, 372, 377, 378, 382, 388, 389, 390, 391, 392, 394, 402, 404, 405
Berry, Thomas, 22, 91 n. 1, 156, 165, 169–73, 175, 294
Berry, Wendell, 29, 31,
Blake, William, 399–400
Bruno, Giordano, 256–57, 260, 266, 273, 323
Buber, Martin, 210–12
Bullard, Robert D., 56–57
Butler, Smedley Darlington, 86–87
Calvin, John, 44

Capitalism, 44, 45, 48, 58, 86, 108, 125, 203, 212, 385, 394
Chalmers, Thomas, 163, 260–62, 274, 293, 294
Churchill, Ward, 61–62, 203
Columbus, Christopher, 13, 64, 72, 74, 75, 81, 138, 239
Community, relational, 186, 369, 370–74, 376, 380, 389, 405, 406
Cone, James, 56, 57
Copernicus, 3, 40, 143, 144, 247, 256, 323
Corso, Philip J., 247 n. 8, 274–75, 325, 326, 327
Cosmic Charter proposal, 358–67
Cosmoethics, 22, 222, 224, 225, 226, 227, 329, 333, 377,
Cox, Harvey, 33–34, 52, 208–9, 385
Coyne, George V., 293–94
Creatio ex Dei, 92, 93, 94
Creatio ex nihilo, 61, 92, 93, 94
Creationism, 94–95, 146,
Cronon, William, 69–70,
Darwin, Charles, 3, 145–46, 156, 157, 266, 320, 335
Davies, Paul C. W., 285–86, 321
De Duve, Christian, 286–87
De las Casas, Bartolomé, 75–76
De Victoria, Franciscus, 63 n. 10, 76
Declaration of Legal Principles Governing the Activities of States in the Exploration and Use of Outer Space, Including the Moon and Other Celestial Bodies, 230, 240–41

Declaration on International Cooperation in the Exploration and Use of Outer Space for the Benefit and in the Interest of All States, Taking into Particular Account the Needs of Developing Countries, 230, 242–43, 355–56
Declaration on the Rights of Indigenous Peoples, UN, 18, 63 n. 8, 85, 103, 110–12, 114, 133, 343–45
Deere, Phillip, 258, 323 n. 30
Deus ex machina, 11, 33, 34, 35, 93, 382
Dick, Steven J. Dick, 284, 290, 294
Discovery, Doctrine of, 17–18, 21, 22, 37, 44, 45, 56, 60–61, 62, 63, 64, 65, 66, 70, 74, 76, 77, 79, 80, 81, 84–88, 105, 110, 111, 112, 141, 216, 284, 289, 356 n. 4
    In space, 87–88, 117–28, 129–39, 142, 179, 186, 192, 194, 205, 218, 228, 229, 234, 239, 240, 241, 243, 247, 251, 252, 283, 288, 290, 322, 325, 339, 345, 356, 378, 381, 385, 386,
    Racism in, 81–84
Dystopia, 22, 142, 189, 200, 202–6, 208, 214, 217, 218
Earth Charter, The, 22, 102, 103, 112–13, 114, 171, 246, 329, 335, 336, 337, 338 n. 2, 341, 348–50
Enlightenment, Age of, 36–37, 42, 43, 94, 174
Ethics, praxis, 7 n. 7, 22, 186–90, 195, 221–22, 224, 225–27, 329, 332, 333, 334, 335, 377,
Feuerbach, Ludwig, 122, 258, 292
Funes, José Gabriel, 263–65, 266
Galileo, 3, 37, 143, 144–45, 146, 157, 173, 247, 255, 266, 293
Gramsci, Antonio, 209–10, 214
Gutiérrez, Gustavo, 55, 207 n. 5
Hanke, Lewis, 74 n. 32, 75, n. 34

Hawking, Stephen, 7–14, 24, 35, 93, 128, 138–39, 140, 163, 190, 249, 263, 283, 318, 326, 378, 379, 382, 397, 399
Hudson River Valley UFOs, 22, 253, 267, 275–80, 282, 310, 333, 401
Hynek, Josef Allen, 266, 268, 275–76, 277 n. 67, 278, 295 n. 132, 297, 304, 305–16, 320, 358 n. 5
    Close encounters categories, 308–10
    UFO definition, 307
Inclosure, 71–73, 339
Inspiration, biblical, 149–50, 176
Intelligent Design (ID), 92, 94–96, 146, 274
Inter caetera, 64
Intergovernmental Panel on Climate Change (IPCC), 32
International Indian Treaty Council (IITC), 14, 18, 62 n. 5 and 6, 63 n. 8, 77 n. 38, 110, 111 n. 11, 258 n. 10,
Interpretation, biblical, 92, 94, 129, 143, 144, 145, 148–51, 156, 168, 176
John Paul II, Pope, 35–36, 55, 262–63, 266, 384
Johnson v. M'Intosh, 79–80
Kean, Leslie, 18–19, 267–68, 269, 275, 280 n. 77 and 78, 281
Kepler, Johannes, 3, 13, 143, 144, 199, 247, 274, 404
Lemaître, Georges, 40 n. 20, 165–69,
Logos, 92, 97, 98, 99, 163, 171, 370, 371, 372, 388, 389, 390, 395–96
Lord's Prayer, 61, 122, 123–24, 321–22, 384
Malmstrom AFB UFOs, 206 n. 4, 253, 281–2, 295–96
Marcel, Jesse, Jr., 270, 272–74, 295–97
Marcel, Jesse, Sr., 270–72, 273
Marx, Karl, 53–54, 203,
McGrath, Alister, 37, 41–43,

INDEX 415

McKay, Christopher, 140, 290,
McMullin, Ernan, 292–94
Means, William, 62 n. 5, 77 n. 38, 111 n. 11
Miller, Robert, 63, 64 n.11, 65 n. 13, 66, 76, 77, 79 n. 40, 84 n. 49, 85, 129, 131, 137 n. 16,
Mondragón, 356, 384
More, Thomas, 71–74, 75, 81
Myths, biblical, 151–53
Nasr, Seyyed Hossein, 37–41, 42, 43, 172
Newcomb, Steven, 63, 77, 80, 82–83, 129–30
Nicholaus of Cusa, 255–56, 323
Outer Space Treaty (OST), 121, 136, 225–26, 230, 233–35, 341, 350–55, 375, 383
Paine, Thomas, 45, 48, 50–52, 55 n. 61, 91, 99, 258–60, 274, 385, 403
Panentheism, 97–8
Peacocke, Arthur, 291–92
Peking Man, 156, 159, 160
Plato, 72–73, 81, 254
"preferential option for the poor," 46 n. 40, 51 n. 51, 55–56
Project Blue Book, 131, 266, 295 n. 132, 305, 313–14
Project Grudge, 305
Project Sign, 305
Racism, 59, 81–84, 128, 348
    ecological, 56–57, 155
Rees, Martin J., 290–91
Requerimiento, 74–5
Renaissance, 38–9, 174, 323
Rendlesham Forest, 269
Roswell, 22, 136, 161, 247, 253, 267, 270–75, 282, 295–96, 297, 325, 333
Salas, Robert, 281–82, 296–97, 326
Singer, Peter, 397

Smith, Adam, 45–48, 49–53, 86, 192, 258 n. 12, 403
    "invisible hand," 45–52, 132
Socioecology, 30 n. 2, 184
Stannard, David E., 14, 61, 203
Swimme, Brian, 91 n. 1, 171–72
Tarter, Jill Cornell, 292–93
Teilhard de Chardin, Pierre, 22, 40, 156–64, 165, 168, 169, 172, 173, 175, 262 n. 27, 274, 286, 293, 294, 323
    Extraterrestrial worlds, 161–64
Terra nullius, 65, 78, 79, 133
Terraforming, 129, 247–51
Tinker, George "Tink," 67–69, 77 n. 38
Toolan, David, 299
Tordesillas, Treaty of, 64, 83
Townes, Emilie, 56, 57
Unidentified Aerial Phenomena (UAP), 5–6, 253, 268, 282, 306, 316
Unidentified Flying Object (UFO), 5 n. 5, 18–19, 20, 22, 131, 161, 253, 266, 267, 268, 269, 271, 274–82, 295–97, 304–27, 357–8, 374, 400–401
Universal Declaration of Human Rights, UN, 103, 104–5, 338–40
Vacuum domicilium, 65–6, 78
Vallee, Jacques, 100 n. 4, 295, 297
Vorilong, William, 255
Wesley, John, 45, 48–50, 51, 52, 53, 384, 403
Wheatley, Phillis, 403–5
White, Jr., Lynn, 21, 35–36, 42
Wilson, Edward O., 175, 336, 398–99
World Charter for Nature, UN, 103, 107–10, 114, 189–90, 341, 346–48
Wright, Robert, 10 n. 12, 397

www.ingramcontent.com/pod-product-compliance
Lightning Source LLC
Chambersburg PA
CBHW021928290426
44108CB00012B/760